X-rays

X-rays

B. E. STERN
BSc, CEng, FIEE, FInstP
Director of the Radiation Unit
University of Surrey

D. LEWIS
MSc, PhD, AInstP
Reader in Crystallography
University of Surrey

PITMAN PUBLISHING

First published 1971

Sir Isaac Pitman and Sons Ltd
Pitman House, Parker Street, Kingsway, London, W.C.2 B 5PB
P.O. Box 6038, Portal Street, Nairobi, Kenya

Sir Isaac Pitman (Aust.) Pty Ltd
Pitman House, Bouverie Street, Carlton, Victoria 3053, Australia

Pitman Publishing Company S.A. Ltd
P.O. Box 9898, Johannesburg, S. Africa

Pitman Publishing Corporation
6 East 43rd Street, New York, N.Y. 10017, U.S.A.

Sir Isaac Pitman (Canada) Ltd
Pitman House, 381–383 Church Street, Toronto, 3, Canada

The Copp Clark Publishing Company
517 Wellington Street, Toronto, 2B, Canada

ISBN 0 273 40097 5

Made in Great Britain at the Pitman Press, Bath

G1—(T.989)

Preface

This book is an attempt to fill a gap in the literature by presenting a balanced account of the field of X-rays. The authors believe that the subject has suffered from the lack of contact between different user groups, such as crystallographers and health physicists, and industrial and medical radiologists.

It follows that the authors have had to make each topic intelligible to the non-specialist in that topic. They have tried to hold also the specialist's interest, and hope that the rather extensive references will make it useful to him.

Although this is primarily a users' book, the authors have been conscious of the need for a book for students in which neither the crystallographic nor the medical field is over-emphasized. They have borne in mind the requirements of their own students of crystallography and radiation protection.

The authors thank those who have helped them during the preparation of the book, and in particular Mr. E. J. Wheeler (University of Surrey) for constant assistance, criticism and encouragement, and Professor J. F. Fowler (Hammersmith Hospital) who read some of the early drafts. None of the authors' colleagues is, of course, responsible for any opinions expressed in, or defects of, this book.

Thanks are also due to the many authors and publishers who have allowed the reproduction of data and illustrations.

B.E.S. & D.L.

Contents

Preface v
1 X-ray Optics 1
2 X-ray Emission Spectra 19
3 Absorption and Scattering 31
4 Measuring Methods and Instruments. 55
5 Calibration of X-ray Sets 84
6 Dosimetry 98
7 X-ray Photography 108
8 Radiography 125
9 Radiotherapy 164
10 X-ray Crystallography 185
11 X-ray Chemical Analysis 202
12 Uses of Large Doses 212
13 X-ray Equipment 219
14 Radiation Protection 248
Glossary 287
List of Manufacturers 291
(*References to this list are indicated by superscript numbers*)
Index 293

Explanatory note

The Literature References appearing at the ends of chapters are referred to in the text by numbers in square brackets, e.g. Mendeleev [13].

The List of Manufacturers given on page 291 is referred to in the text by superscript numbers, e.g. image intensifiers.[4,15,20,22]

Units other than SI are given when still in common use; for example Å in crystallography.

1 | *X-ray Optics*

The discovery of x-rays by Röntgen [1] on 8th November, 1895 caused great excitement in scientific circles, and the journals of 1896 contain many accounts of experiments to investigate the properties of the rays. There was an early reaction of disbelief in the existence of rays so extraordinary that they could penetrate through the body, but this was followed by unrestrained speculation as to their nature. Many of the great men of the day had their theories [2]. Michelson advocated ether vortices; Sutherland suggested internal vibrations of the electrons in the atom; Stokes favoured a theory of irregular ether pulses and Bragg supported a particle theory, the particles resembling cathode rays. Other workers such as Sagnac, believed that x-rays were short ether waves resembling ultra-violet light.

The Nature of X-rays

It soon became the object of many workers to resolve the rival particle and wave hypotheses, and experiments were made to see if any wave-like properties could be detected, in particular diffraction effects. Others attempted to refract and reflect the radiation. It was not until 1912, when Laue [3] and his co-workers succeeded in diffracting x-rays, that the wave-like nature of the rays was convincingly demonstrated. Many further resemblances between x-rays and light were also noticed, and in the 1920s the essential identity between x-rays and light was established.

One of the most far-reaching scientific discoveries in recent times has been the dual nature of electromagnetic radiation; this can be thought of either as a wave motion or as consisting of discrete and discontinuous amounts or "quanta" of energy.

The full intellectual revolution following these ideas is not yet complete. For practical purposes, the context determines which theory is relevant. In general, the quantum theory is most convenient when individual particles are being considered, as in atomic theory. Wave theory is used when considering radiation in bulk, when many millions of quanta are being examined, as in diffraction. We are somewhat in the position of a traveller with two maps: one a six inch to the mile survey and the other a six inch diameter globe of the world. Both are correct, but each has its own application.

WAVE MOTION

In 1864, James Clerk Maxwell [4] read a paper before the Royal Society in which he showed theoretically that varying electric and magnetic fields should form the basis of a form of radiation which would travel through free space, or through various media with a velocity depending upon the magnetic and electric properties of the media. These predictions were confirmed by Heinrich Hertz [5] about 1887 at Karlsruhe, when he showed that an oscillatory electrical discharge produced radiation which could travel considerable

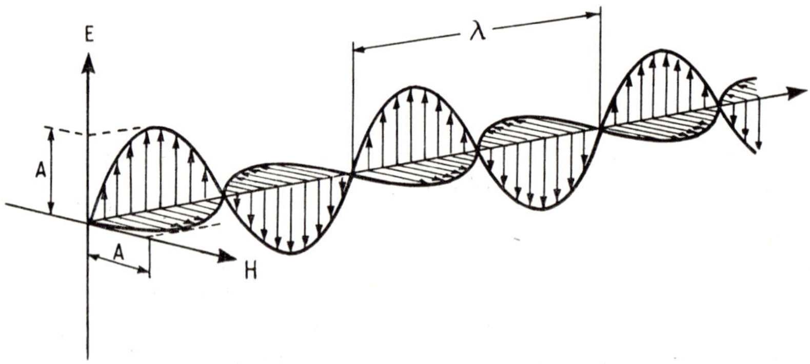

FIGURE 1.1 *Electromagnetic Wave* (Transverse)
A, amplitude; E, electric field; H, magnetic field; λ, wavelength.

distances. He showed further that this radiation had many properties in common with light, such as reflection and refraction.

Electromagnetic radiation was thus established as a form of wave-motion consisting of electric and magnetic fields oscillating in phase. When the wave is travelling through insulating media, the magnetic and electric fields are perpendicular to each other and also to the direction of wave propagation (Figure 1.1).

It is now known that there is a complete spectrum of such radiation from the long waves radiated by coils rotating in magnetic fields to shorter radio waves, infra-red radiation, visible light, ultra-violet light, and x-rays and gamma radiation. These radiations all have a similar nature and the differences in their properties follow from their respective wavelengths, which vary from many kilometres to small fractions of an Ångström unit, Å (10^{-10} metre).

X-rays occupy part of the electromagnetic radiation spectrum between the somewhat arbitrarily defined limits of about 0·02 to 10 Å. They merge into ultra-violet rays at the longer wavelengths and overlap with gamma-rays at the shorter wavelengths. The terms x-rays and gamma rays are used to distinguish between the modes of generation, gamma rays being produced by nuclear transformations and x-rays from extra-nuclear electron energy changes.

Wave motion may be regarded as the propagation of a disturbance through a medium without permanent displacement of that medium. The particles of the medium may oscillate in the direction of propagation of the wave as in sound waves, when the waves are described as *longitudinal* (Fig. 1.2).

In *transverse* waves, the particles oscillate perpendicularly to the propagation direction; this is the case for the wave motion along plucked strings, and in electromagnetic radiation.

A study of wave motion in deep water shows that water particles are not transferred with the wave, but that they move about an equilibrium position

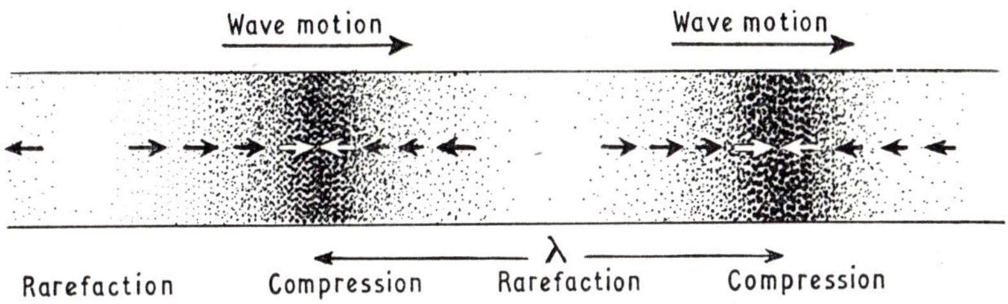

FIGURE 1.2 *Longitudinal Wave*
The individual particles each oscillate about a mean position to give regions of compression and rarefaction which move along with the wave velocity. The energy moves with the wave, but the medium only oscillates.

in elliptical orbits. Only the wave contour and the energy of propagation move with the wave (Figure 1.3).

The distance between corresponding points of the contour in adjacent waves is known as the wavelength, λ (Figure 1.1). The velocity v of propagation of the wave, is the distance covered by the contour in unit time. The

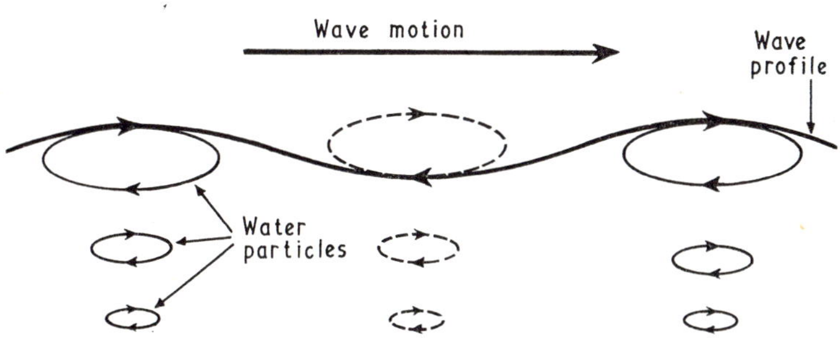

FIGURE 1.3 *Deep Water Waves*
The water particles move in ellipses, but the wave profile moves forward continuously.

time taken for a complete oscillation is known as the periodic time T. Hence

$$\lambda = vT = \frac{v}{\nu} \tag{1.1}$$

$$\nu = \text{frequency of oscillation} = \frac{1}{T}. \tag{1.2}$$

The maximum displacement from the equilibrium position is called the amplitude A of the wave. The amount of energy carried by the wave is proportional to A^2.

The energy flow per unit time across unit area perpendicular to the direction of wave propagation is known as the energy flux density or intensity, I. It is formally [6] defined in terms of the energy fluence F given by

$$F = \frac{\Delta E_f}{\Delta a}$$

where E_f is the sum of the energies, exclusive of rest energies, of all the particles which enter a sphere of cross-sectional area Δa.

The energy flux density or intensity I is given by

$$I = \frac{\Delta F}{\Delta t}$$

where ΔF is the energy fluence in the time Δt.

Energy will be absorbed from radiation traversing a medium. If I_0 is the intensity at a point in the absence of the medium and I is the intensity at the

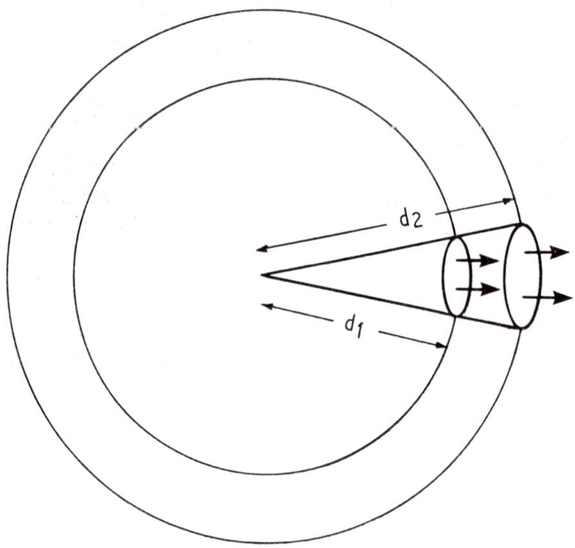

FIGURE 1.4 *Inverse Square Law*

same place, but with a thickness d of medium interposed between point and source then, provided that secondary or scattered radiation does not contribute appreciably to I,

$$I = I_0 \, e^{-\mu d} \tag{1.3}$$

where μ = linear absorption coefficient in the medium of the electromagnetic radiation of wavelength λ (*see* Chapter 3). This is known as Lambert's law [7].

If absorption is negligible and the radiation diverges from a point

$$\frac{I_1}{d_1^2} = \frac{I_2}{d_2^2} \tag{1.4}$$

where I_1 and I_2 are the intensities at distances d_1, d_2 from the point. This is known as the inverse square law and can readily be deduced from considerations of the constant total energy flux across spherical surfaces of radius d_1 and d_2 (Figure 1.4).

THE QUANTUM THEORY

The picture of electromagnetic radiation as wave motion was complicated by the advent of the quantum theory. It had already been shown that certain optical effects, such as the "anomalous" Zeeman [8] effect, could not be explained by the wave theory and further difficulties arose in explaining the distribution of energy in the spectrum from a perfectly black body. Many of these paradoxes were resolved by Max Planck [9] when in 1900, he made the revolutionary proposal that the energy of radiation is emitted in discrete packets, or quanta, of value

$$h\nu$$

where h = Planck's constant

ν = frequency of radiation.

Since $c = \lambda\nu$, where c = velocity of light *in vacuo*,

$$\text{Energy quantum} = \frac{hc}{\lambda} \tag{1.5}$$

The value of the energy quantum thus varies inversely with the wavelength λ of the radiation: the quantum of energy is greater for blue light than for red. The experimental evidence in favour of the quantum theory is indisputable, and much of this evidence is derived from X-rays.

A further development of these ideas led Louis de Broglie [10], in 1924, to postulate another hypothesis: that matter itself has a wave-like aspect. Consequently phenomena such as diffraction could be expected from small particles such as electrons, and this was in fact observed by Davisson and Germer [11] in the U.S.A.

RECTILINEAR PROPAGATION

The laws governing the propagation of light are summarized in Fermat's principle [7]. Fermat, who lived in the seventeenth century, suggested that light travels between two points by the path which takes the light the minimum time. In 150 B.C., Hero of Alexandria had shown that this law governed the reflection of light at plane surfaces, and Fermat demonstrated that it was true also for light refracted at plane surfaces (Figure 1.5). It does not hold necessarily for reflection or refraction by curved surfaces, where the time of travel between two points may be a maximum. However, all these cases are covered by the general principle that the time taken is at a stationary value. The classical optical laws of light propagation can be deduced from Fermat's principle, and the principle applies to the propagation of waves in all media. Thus it was important to supporters of the wave hypothesis of the nature of X-rays, to demonstrate that this radiation also conformed to the principle.

The early observations that X-rays cast shadows of objects in their path was an important indication that the rays were similar in nature to visible light, in that they travelled in straight lines. This phenomenon is, of course, the basis of radiography (Chapter 8).

Owing to the very short wavelengths of X-rays and their consequent high frequency compared with the natural frequency of oscillation of the electrons

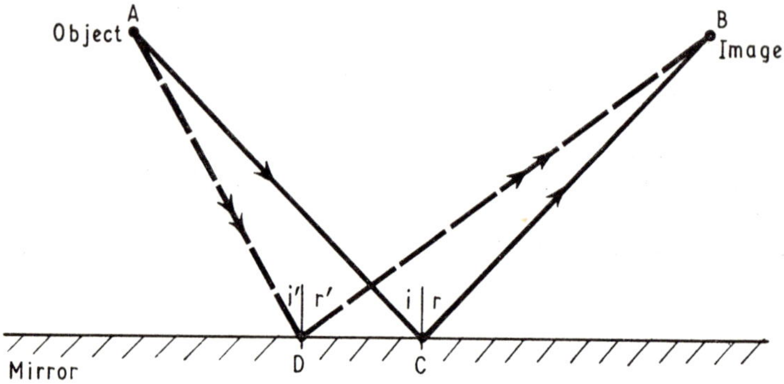

FIGURE 1.5 *Fermat's Principle*
Path $AC + CB$, where the angles of incidence (i) and reflection (r), are equal, is shorter than any other path, for example $AD + DB$ where the angles of incidence and reflection (i' and r' respectively) are unequal.

of the medium through which the rays are propagated, the values of refractive indices of media for X-rays differ very little from unity. Reflection and refraction are thus difficult to demonstrate and are not usually of practical importance.

REFLECTION AND REFRACTION

When radiation strikes a surface dividing two media, its energy is divided between reflected and refracted rays in a manner depending on the angle of incidence and the *refractive index*, μ. The refractive index is a constant depending only on the two media and the wavelength of the radiation; it is equal to $\sin i/\sin r$ (Figure 1.6). When μ is greater than unity, the refracted ray is bent towards the normal. When μ is less than unity, the refracted ray is bent away from the normal. For large angles of incidence the refracted ray may be absent; this is called *total internal reflection*.

Röntgen [2] tried, but failed, to detect any refraction on passing X-rays through slabs of various materials. He also passed X-rays through powdered salt and through powdered silver, but found no indication of reflection or refraction at the thousands of surfaces presented by the powders. Attempts by him to make X-ray lenses from glass and ebonite were equally unsuccessful. As reflection and refraction play a central role in optics, their apparent absence with X-rays was disconcerting to supporters of the wave theory.

The refractive index of a medium with respect to a vacuum is a measure of the ratio of the velocity of a wave *in vacuo* to its velocity in the medium. This ratio depends upon the frequency of electromagnetic waves and the natural frequencies of the electrons in the refracting medium.

In classical optical theory [7], the refractive index μ of a medium for light of frequency ν is given by

$$\mu^2 = 1 + \frac{e^2}{\pi m} \sum \frac{N_s}{(\nu_s{}^2 - \nu^2)} \tag{1.6}$$

where N_s = number of electrons per unit volume with a natural frequency ν_s and where the radiation is incident from a vacuum on to the medium.

For visible light $\nu < \nu_s$ so that $\mu > 1$.

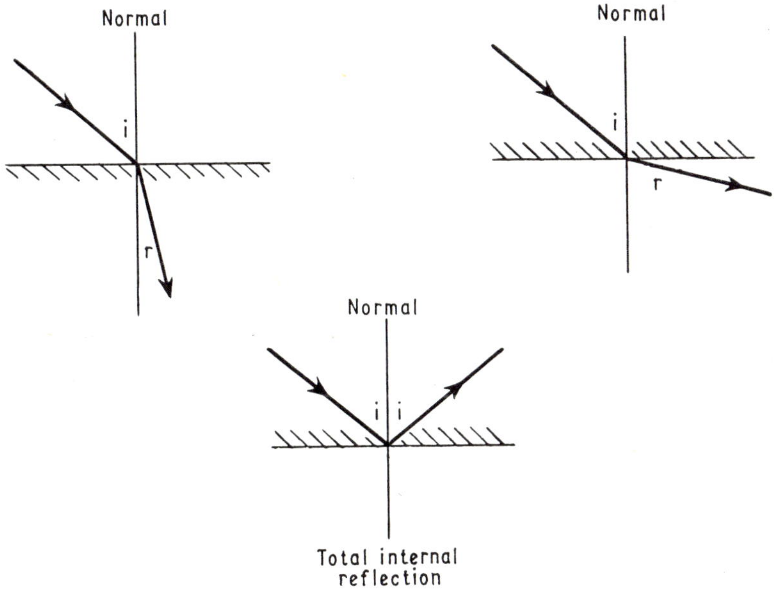

FIGURE 1.6 *Refraction*
(*top left*) $\mu > 1$; (*top right*) $\mu < 1$.

For X-rays $\nu \gg \nu_s$ so that, using the binomial expansion,

$$\mu = 1 - \frac{Ne^2}{2\pi m\nu^2} \tag{1.7}$$

where N = number of electrons per unit volume their natural frequency being unimportant, or

$$\mu = 1 - \delta \tag{1.8}$$

where $\delta = Ne^2/2\pi m\nu^2$.

For X-rays, δ is of the order of 10^{-6} so that μ is very slightly less than unity. Thus refractive indices of materials for X-rays are so near unity that it is impossible to detect refraction and total internal reflection by direct methods, such as those used by Röntgen.

The refractive index of glass to X-rays has been measured by Compton [12] from critical angle reflection. Compton obtained a good estimate of the wavelength of tungsten L radiation by reflecting a fine beam of X-rays from a piece of polished crown glass. The angle of grazing incidence was less than ten minutes of arc.

For total internal reflection the rise in intensity of the reflected ray is not very sharp, especially with softer radiations; this is predicted by the classical electromagnetic theory [7].

Demonstrations of refraction with prisms, which is the conventional optical method, in X-ray work require large-angled prisms and a glancing angle. The method has been used to estimate e/m [13].

The best method of measuring μ is by refraction in crystals. This was achieved by Stenström [14]. He found that the wavelength of molybdenum radiation apparently varied with the order n in crystal diffraction. He had, however, applied Bragg's law (Chapter 10) in the usual form

$$n\lambda = 2d \sin \theta \qquad (1.9)$$

and had not allowed for the refraction of the radiation in the crystal. To include this effect, by classical optical refraction theory [7]

$$\mu = \lambda/\lambda' = \cos \theta/\cos \theta' \qquad (1.10)$$

In the crystal

$$n\lambda' = 2d \sin \theta' \qquad (1.11)$$

Using these equations, and eliminating θ' and λ', we have, assuming $(1 - \mu)$ small,

$$n\lambda = 2d \sin \theta \left\{1 - \frac{1-\mu}{\sin^2 \theta}\right\} \qquad (1.12)$$

This equation (1.12) should be compared with equation (1.9). From equation (1.12), μ can be determined experimentally.

INTERFERENCE

Consider light arriving at P from a point A (Figure 1.7). Its amplitude varies sinusoidally with time. Similarly for light from the point B. The net amplitude

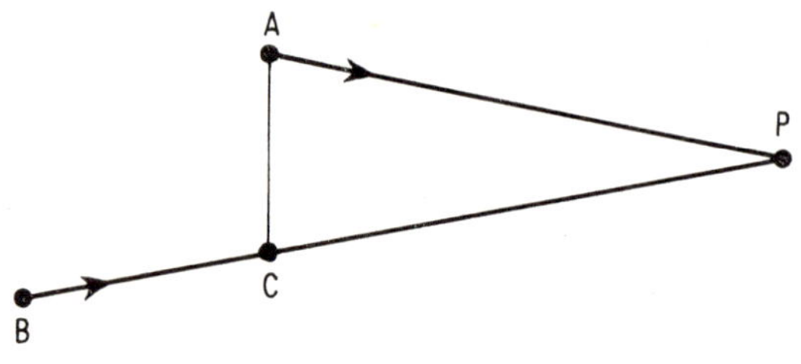

FIGURE 1.7 *Interference of Two Waves*
The two wave sources, A and B, are interacting at P. The wave from B travels an extra distance BC, where $AP = CP$.

at P is the vector sum of these two amplitudes. If the two wave trains are in phase at P, then they will reinforce each other. For a constant phase relationship, A and B must arise from the same original source. Thus, the resultant amplitude of the light at P depends on the path difference BC between PA and PB and the wavelength of the light. Consequently, if P is moved, it will pass through regions of maximum and minimum amplitude. This phenomenon is known as interference.

If $PB - PA = BC = n\lambda$ where n is an integer, the two waves will reinforce each other. For X-rays, λ is small and therefore interference is difficult to demonstrate. Linnik [15] obtained interference between the X-rays from a source and its image reflected from a mirror in an adaptation of the Lloyd's mirror effect in optics [7]; Kiessig [16] used multiple reflections from thin metal films. Both methods employed glancing angles to obtain total internal reflection.

DIFFRACTION

If a distant source of light is viewed through a fine silk handkerchief, the central image of the light will be surrounded by a regular pattern of weaker images. When the handkerchief is rotated, the images rotate with it. When

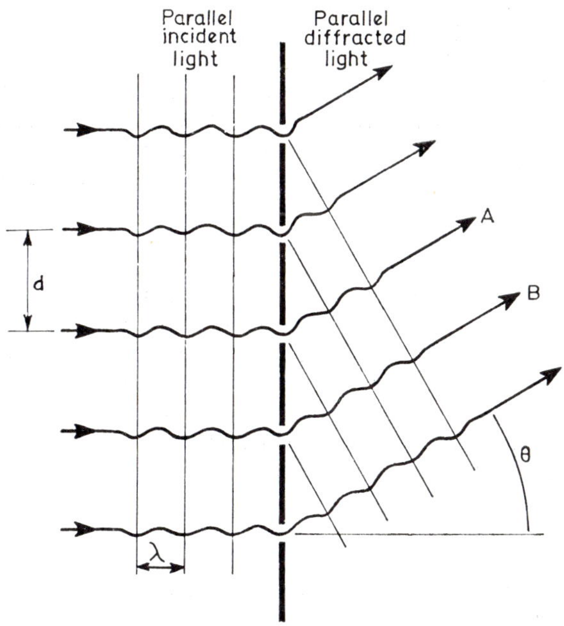

FIGURE 1.8 *Diffraction Grating*
λ, light wavelength; d, grating element; θ, deviation angle; n, order of diffraction, i.e. number of complete waves leading or lagging behind adjacent diffracted rays such as A and B; there is one wave in this case, so $n = 1$.

the handkerchief is inclined to the beam, the spacing of the images alters. The spacing of the images depends on the colour of the light. This experiment illustrates diffraction.

When there is an obstacle in the path of any kind of wave motion, the wave spreads into the region within the geometrical shadow. The angle of deviation into the shadow depends on the wavelength.

If there is a regular array of obstacles, waves diffracted from each member bear a uniform phase relationship with those from other members. Thus at certain angles of deviation, reinforcement may occur. When the spacing of the obstacles is of the same order of magnitude as the wavelength, an observable diffraction pattern, containing maxima and minima of intensity, will occur. The regular array is known as a diffraction grating (Figure 1.8).

Diffraction phenomena are important in the study of visible light, when the diffraction grating has spacings of a few thousand Ångström units. For reinforcement when the incident light is parallel

$$d \sin \theta = n\lambda \tag{1.13}$$

where n is an integer called the order of diffraction. It is equal to the number of waves which lead or lag between light diffracted from adjacent lines of the grating.

X-ray diffraction phenomena are of interest in crystallography because the atomic spacings of the crystal are a few Ånström units, and X-rays are easily obtainable with wavelengths in this range. The atomic arrangement of a large number of crystalline materials has been determined from a study of the deviation and intensity of diffracted X-ray beams.

An important application of X-ray diffraction is the accurate measurement of one of the fundamental constants of nature, Avogadro's constant (N). This is defined [17] as the number of atoms in 16 g of ^{16}O and, for any substance, is equal to the number of atoms per gramme-atom or the number of molecules per gramme-molecule.

In the method, X-ray diffraction is used to measure the volume of a unit cube of sodium chloride; its mass can then be calculated from the known density of sodium chloride. The number of sodium and chloride ions and hence the effective number of sodium chloride molecules in the unit cube is also known, enabling the weight of a molecule of sodium chloride to be calculated. Avogadro's constant, expressed as the molecular weight in grammes divided by the weight of one molecule, can then be found.

Now the charge on the electrons can be determined from N by electrolysis; the Faraday (F) is the charge required to liberate one gramme-atom and

$$Ne = F \tag{1.14}$$

if the ions carry only one charge each.

F can be measured accurately, and hence an accurate value of e calculated.

For a long time there was a small but significant discrepancy of 0·2 per cent between values of e obtained from X-ray measurements and those obtained from the Millikan oil-drop method [17]. The latter is based on balancing an electrostatic force on a charged oil drop against the weight of the drop. It involves a knowledge of the viscosity of air, because the radius of the drop is determined by allowing it to fall under conditions in which Stoke's law [17] applies. Many interesting speculations were advanced to explain this discrepancy, but it was resolved, about 1936, when the viscosity of air was redetermined with greater accuracy.

Absolute values of X-ray wavelengths are obtained from measurements using ruled gratings. As glancing angles are necessary, the accuracy is limited. However, when crystals are used as gratings, angles of deviation are not so restricted, and can be measured with accuracy. In this case, absolute values of wavelength require a knowledge of the spacing of the

Bragg planes, and this spacing cannot be directly measured. For these reasons, crystal gratings are used to determine the relative values of X-ray wavelengths to high precision.

Siegbahn [18] used calcite as his standard crystal and defined his standard of wavelength in terms of the basal plane. He called this the X-unit, its value being approximately 10^{-10} mm. This is inconveniently small, and the kX unit, a thousand times larger, is the practical unit. The Ångström unit (10^{-7} mm) is derived from the standard metre. Thus, the kX unit is almost equal to the Ångström unit, but it is slightly larger. The best value [19] for the relationship is

$$1 \text{ kX} = 1{\cdot}002\,063 \pm 0{\cdot}000\,007 \text{ Å}$$

As the conversion factor depends on experimental results, its value changes as more accurate determinations of Avogadro's constant and of X-ray wavelengths are made. The SI unit for wavelength is the nanometre (nm) which is 10^{-9} m; therefore 1 Å $= 10^{-1}$ nm.

Concave Gratings

More recent work in X-ray spectroscopy has been performed with concave gratings [20]. If a concave grating (Chapter 11) of *radius* R is mounted tangentially on a circle of *diameter* R and with the slit or source on the circumference of the circle, the spectra produced will be focused on that circumference, and therefore also on a detector lying along it. Concave gratings can be wide, to give high intensity and resolving power. In X-ray applications, grazing angles are necessary, and concave gratings are used in the long wavelength range. The X-rays are obtained from a spark discharge. The method is used to investigate energy levels in the solid state. Plasma physics and fluorescent analysis of light elements are other fields of study employing concave gratings.

Image Formation Using X-rays

As the refractive index of material for X-rays is very little less than unity, the focal length of a lens would be many orders of magnitude greater than its radii of curvature. Lens construction is therefore impracticable. Reflection systems are possible, however, and these have been attempted by some workers [21]. On the whole, the results are disappointing. Images have been formed, but such problems as coma (unequal magnification by different mirror areas), are proving difficult to solve, as the designer is always limited by critical angle considerations.

In an X-ray mirror a divergent beam of X-rays is allowed to graze two perpendicular concave cylindrical surfaces in succession (Figure 1.9). Very smooth surfaces to a tolerance of a few Ångströms are required. The width of the lens aperture is a limiting factor, the usual incompatibility of wide aperture and aberration freedom being more of a problem with X-ray lenses than with light lenses. It is possible to use bent crystals instead of mirrors, to obtain a focusing effect by Bragg reflection, and then grazing incidence is not required. The large and perfect crystals necessary are difficult to obtain.

Little attention seems to have been paid to the possibility of using a Fresnel zone plate for focusing. This is an arrangement of alternating opaque and translucent annuli, used in optics [7] to make an image of a point object by diffraction. As the focal length of the zone plate is inversely proportional to the wavelength, the method is suitable only for monochromatic radiation.

THE X-RAY MICROSCOPE

The resolving power, y, of the objective lens of a microscope is given by [7]

$$y = \frac{\lambda}{2\mu \sin \alpha} \tag{1.15}$$

where y = distance between two points which are just resolved
λ = wavelength of the light (in air)
μ = refractive index (relative to air) of the medium between the lens and the object.
α = half the angle subtended at the object by the microscope objective.

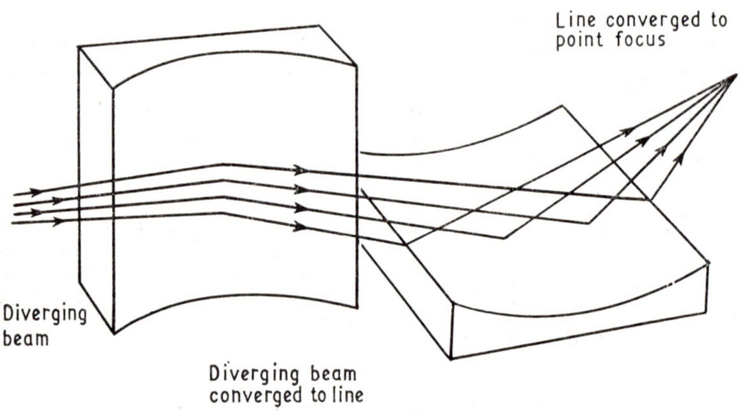

FIGURE 1.9 *X-ray Microscope*
Diverging X-ray beam converged to point focus after reflection at grazing incidence by two concave cylindrical mirrors.

The magnitude of α is limited by the difficulty of constructing a perfect lens of large diameter and short focal length, but in practice $\sin \alpha$ may approach unity. Thus, y is essentially limited by λ.

This fundamental limitation in resolving power has led to a search for microscopes using radiations other than light.

The wavelength λ in Ångströms associated with an electron is given by the working relationship

$$\lambda = \sqrt{\left(\frac{150}{V}\right)} \tag{1.16}$$

where V = electron-accelerating voltage. Therefore short wavelengths are obtainable using modest voltages, although the voltage must be high enough to produce electrons of useful penetrating power. An electron beam is easily focused using magnetic fields, and electron microscopes with high resolving powers, for example 8 Å, are available commercially. Disadvantages are that unless a surface replica is used the specimen to be examined must be thin; it has to be in a vacuum and it may be damaged by energy absorbed from the electron beam.

A reflecting x-ray microscope would not have these disadvantages, but it is not a practicable proposition because x-rays cannot be suitably focused. It is possible to use long x-ray wavelengths such as aluminium and carbon characteristic K-radiation and a system of image-forming x-ray mirrors, but resolutions [21] are still of the order of thousands of Ångström units.

X-RAY ASTRONOMY

Many sources of x-rays are thought to exist in the universe, but the technical difficulties of observing them are great.

Hoyle and Bates [22] have shown theoretically that a hot gas such as the sun's corona, should emit soft x-rays. In 1949, Burnight [23] detected these from the blackening of photographic plates carried in rockets. X-ray observations on the sun's corona have also been made [24] using pinhole cameras carried by rockets.

Hard x-rays should be produced in space by the "synchrotron" action of very high energy electrons spiralling in intense local magnetic fields. Such x-rays have been observed in the Crab nebulae [25], where a supernova appeared in A.D. 1054. They could also be produced by temperatures of several million degrees. Such conditions may explain the radiation from a source in Scorpius, which could be a neutron star of the type discussed by Cameron [26].

Telescopes for hard x-rays are difficult to design. Observations on the source in the Crab nebula were made by waiting for the precise moment that the moon occulted the nebula, and detecting the decrease in x-ray flux using Geiger counters carried in a rocket [25].

It may prove possible to build an x-ray telescope in an orbiting space laboratory, using crossed-mirror or zone-plate x-ray lenses [27].

Polarization of X-rays

J. J. Thompson [17] applied the principles of classical mechanics to the deceleration of an electron on striking a target. He showed that owing to the time taken for disturbances to travel along electrostatic lines of force, the decelerating electron should radiate electromagnetic waves. The deceleration is in the direction of the cathode stream, so that the electric field changes are in the same direction (Figure 1.10). This radiation is said to be plane-polarized.

Consider the electrons moving from the cathode to the anode. There will be slight divergence of the electron beam about *CA*. Moreover, on striking the anode, the electrons make multiple collisions, so that there is a range of directions of deceleration. For these two reasons, there is a spread in the directions of the electric field associated with the primary x-ray beam. The spread of electric field direction is reduced when thin transmission targets are used, and with high-velocity electrons.

When the primary x-ray beam strikes the scatterer *S*, the electric field E_1 excites the electrons of the scatterer into vibration. The direction of their vibration is parallel to E_1. Consequently all scattered x-rays have their associated electric field parallel to E_1. Consider the scattered x-rays in two directions *SP* and *SQ*. *SP* is perpendicular to the plane containing *CA* and *AS*. The associated electric field of *SP* must be perpendicular to the direction *SP*. As there is a direction E_2 which fulfils this condition and is also parallel

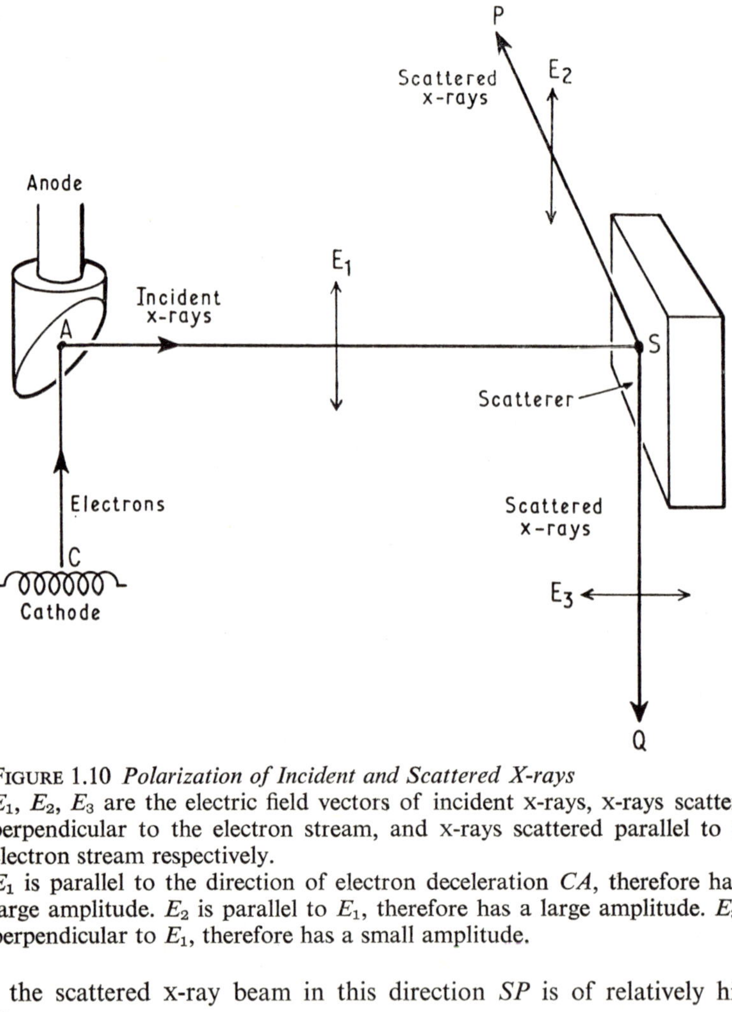

FIGURE 1.10 *Polarization of Incident and Scattered X-rays*
E_1, E_2, E_3 are the electric field vectors of incident x-rays, x-rays scattered perpendicular to the electron stream, and x-rays scattered parallel to the electron stream respectively.
E_1 is parallel to the direction of electron deceleration *CA*, therefore has a large amplitude. E_2 is parallel to E_1, therefore has a large amplitude. E_3 is perpendicular to E_1, therefore has a small amplitude.

to E_1, the scattered x-ray beam in this direction *SP* is of relatively high intensity.

Now consider the beam in the direction *SQ*. *SQ* is perpendicular to the incident beam *AS* and parallel to the electron beam *CA*. The associated electric field of *SQ* must be perpendicular to the direction *SQ*. There is no component of E_1, which fulfils this condition. But there is a small component E_3 of electric field resulting from the spread about the direction E_1. Hence, the x-ray beam scattered in the direction *SQ* is of relatively low intensity.

Hence, the intensity of scattered radiation and the direction of polarization are markedly dependent on the direction of scatter. For this reason, a scatterer may be used to analyse a polarized beam of X-rays in a manner analogous to the use of an analyser in the study of polarized light.

The Thompson radiation theories, and also the electromagnetic nature of X-rays, were confirmed by Barkla [28]. In his experiments Barkla scattered the X-rays from a carbon block at a scattering angle of 90°, to analyse the polarized radiation using an ionization chamber. Compton and Hagenow [29] confirmed Barkla's early results using superior experimental conditions.

The electric field E_ϕ of the radiation at a distance r from the scattering medium is given by [17]

$$E_\phi = \frac{ea}{rc^2} \sin \phi \tag{1.17}$$

where ϕ = the angle between the direction of the incident electric field and the direction r, and a is defined below.

If E is the strength of the incident electric field, the radiating electron will have an acceleration a given by

$$a = \frac{Ee}{m} \tag{1.18}$$

so that

$$E_\phi = \frac{Ee^2 \sin \phi}{rmc^2} \tag{1.19}$$

It is readily shown [8] that the intensity of the wave, which is a measure of the radiating energy, is proportional to the square of the electric field strength. If I is the incident wave intensity and I_ϕ the scattered wave intensity.

$$\frac{I_\phi}{I} = \frac{E_\phi{}^2}{E^2} = \left(\frac{e^2}{mc^2}\right)^2 \frac{\sin^2 \phi}{r^2} \tag{1.20}$$

Assuming that the incident wave is propagated along the x axis, then, if it is unpolarized, i.e. the orientation of the electric vector varies rapidly and randomly, the average electric vectors along the y and z directions will be equal. Let the amplitude of these vectors be E_y and E_z respectively, then

$$E^2 = E_y{}^2 + E_z{}^2 \tag{1.21}$$

$$I = I_y + I_z \tag{1.22}$$

where I_y and I_z are the respective intensities.

As $I_y = I_z$

$$I_y = I_z = \tfrac{1}{2} I \tag{1.23}$$

For the radiation scattered by I_y and I_z respectively,

$$I_{\phi_1} = I_y \left(\frac{e^2}{mc^2}\right)^2 \frac{\sin^2 \phi_1}{r^2} \tag{1.24}$$

$$I_{\phi_2} = I_z \left(\frac{e^2}{mc^2}\right)^2 \frac{\sin^2 \phi_2}{r^2} \tag{1.25}$$

If δ is the angle of deviation of the scattered radiation with respect to the incident radiation, and the axes are oriented so that the scattered ray and x and z are in the same plane, $\phi_1 = 90°$, $\phi_2 = 90 - \delta$ (Figure 1.11).

$$I_{\phi_2} = \tfrac{1}{2}I\left(\frac{e}{mc^2}\right)^2 \frac{\cos^2 \delta}{r^2} \tag{1.26}$$

$$I_{\phi_1} = \tfrac{1}{2}I\left(\frac{e}{mc^2}\right)^2 \frac{1}{r^2} \tag{1.27}$$

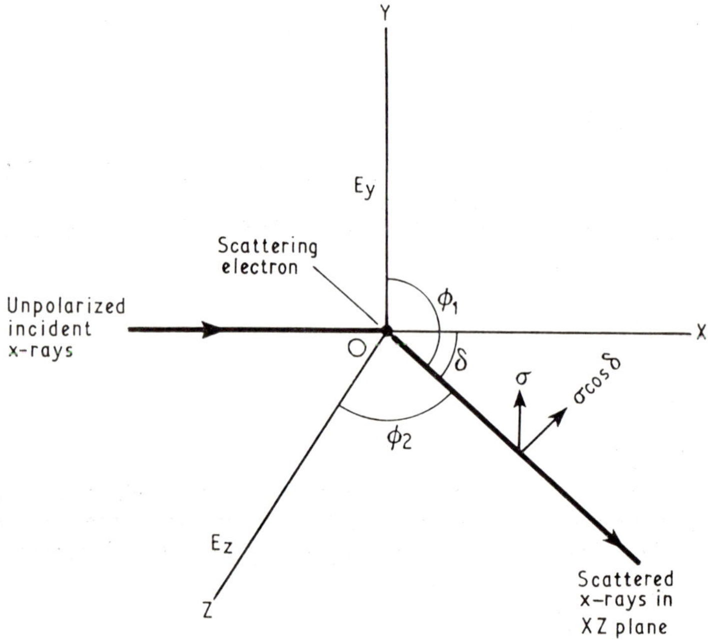

FIGURE 1.11 *Polarization of X-rays Produced by a Scattering Electron*
The electron is scattering unpolarized x-rays in the XZ plane, at an angle of deviation δ to the incident beam. The electric field of both the incident and scattered radiation is always perpendicular to the beams. The component of the electric field (σ) parallel to OY in the scattered wave is perpendicular to both incident and scattered wave, and is therefore unaffected by the angle of deviation δ. The electric component in the XZ plane perpendicular to the scattered wave is less than the OY component by a factor $\cos \delta$. Thus, the amplitude of the electric field of the scattered wave is not uniform in all directions perpendicular to the wave direction, and is said to be polarized. Total scattered intensity is reduced by a factor $\tfrac{1}{2}(1 + \cos^2 \delta)/2$ (*see* eq. 1.28).

The scattered radiation intensity therefore equals

$$I_{\phi_1} + I_{\phi_2} = I\left(\frac{e^2}{mc^2}\right)^2 \frac{1 + \cos^2 \delta}{2r^2} \tag{1.28}$$

Hewlett, Compton [30] and others have investigated this formula, and shown that for medium-wavelength x-rays, it is accurate except for small values of δ, when phase difference effects produce excess scattering.

This formula is used in crystal structure analysis for the determination of electron densities from the intensity of scattered x-rays (Chapter 10). The integration of equation (1.28) over a sphere of radius r gives the rate of emission of energy from one electron (see also Chapter 3, equation (3.34) *et. seq.*).

Literature References

1 RÖNTGEN, W. C., "Uber eine neue Art von Strahlen", *Ann. Phys. u. Chem.*, **64,** pp. 1, 12 (1898). Reprinted from *Sitzber. Würzburger Physik-Medic. Ges.*, 1895.

2 KAYE, G. W. C., *X-Rays* (London, Longmans, 1929).

3 FRIEDRICH, W., KNIPPING, P., LAUE, M., "Interferenz-Erscheinungen bei Röntgen-Strahlen", *Sitzber. Bayer. Akad. München*, p. 303 (1912).

4 MAXWELL, J. C., "A dynamical theory of the electromagnetic field", *Proc. Roy. Soc.*, **13,** p. 531 (1864).

5 HERTZ, H., "Uber einen Einfluss des ultravioletten Lichtes auf die electrische Entladung", *Ann. Phys. u. Chem.*, **31,** p. 983 (1887).

6 INTERNATIONAL COMMISSION ON RADIOLOGICAL UNITS AND MEASUREMENTS (ICRU), Report 10*a* (1962), *NBS Handbook* 84 (Washington, D.C., Nat. Bur. Stand., 1962).

7 LIPSON, S. G. and LIPSON, H., *Optical Physics* (Cambridge, University Press, 1969).

8 JENKINS, F. A. and WHITE, H. E., *Fundamentals of Physical Optics* (New York, McGraw-Hill, 1937).

9 PLANK, M., "Uber das Gesetz der Energieverteilung in Normalspectrum", *Ann. d. Phys.*, **4,** p. 553 (1901).

10 DE BROGLIE, L., "A tentative theory of light quanta", *Phil. Mag.*, **47,** p. 446 (1924).

11 DAVISSON, C. and GERMER, L. H., "Diffraction of electrons by a crystal of nickel", *Phys. Rev.*, **30,** p. 705 (1927).

12 COMPTON, A. H., "The total reflexion of x-rays", *Phil. Mag.* **45,** p. 1121 (1923).

13 BEARDEN, J. A., "A determination of e/m from refraction of x-rays in a diamond prism", *Phys. Rev.*, **54,** p. 698 (1938).

14 STENSTRÖM, W., "Experimentell Untersuchungen der Röntgenopektra. 1. Prüfung der Beziehung $n\lambda = 2d \sin \phi$. 2. M-Reihe", Dissertation University of Lund (1919).

15 LINNIK, H. W., "Der Interferenzversuch von Lloyd mit Röntgenstrahlen", *Z. Phys.*, **65,** p. 107 (1930).

16 KIESSIG, H., "Untersuchungen zur Totalreflexion von Röntgenstrahlen", *Ann. Phys.*, **10,** p. 715 (1931).

17 RICHTMEYER, F. K., KENNARD, E. H. and LAURITSON, T., *Introduction to Modern Physics* (New York, McGraw-Hill, 1955).

18 SIEGBAHN, M., *The Spectroscopy of X-Rays* (Oxford, University Press, 1925).

19 MACGILLAVRY, C. H. and RIECK, G. D. (Eds.), *International Tables for X-Ray Crystallography*, Vol. III (Birmingham, Kynoch Press, 1962).

20 BIRKS, L. S., *X-Ray Spectrochemical Analysis* (New York, Interscience, 1959).

21 KIRKPATRICK, P. "The problem of reflection microscopy", from *X-ray Microscopy and Microradiography*, Symposium Proc., Ed. Cosslett, V. E., Engstrom, A. and Pattlee, H. H. (Academic Press, 1957).

22 Hoyle, F. and Bates, D. R., "The production of the E-layer", *Terr. Mag.*, **53,** p. 51 (1948).
23 Burnight, T. R., "Soft x-radiation in the upper atmosphere", *Phys. Rev.*, **76,** p. 165 (1949).
24 Russell, P. C., "Soft x-ray image of the sun", *Nature*, **205,** p. 684 (1965).
25 Bowyer, S., Byram, E. T., Chubb, T. A. and Friedman, H., "Lunar occultation of x-ray emission from the Crab Nebula", *Science*, **146,** p. 192 (1964).
26 Cameron, A. G. W., "Neutron star models", *Astrophys. J.*, **130,** p. 884 (1959).
27 Berman, A. I., "Observatories in space", *Sci. Amer.*, **209,** No. 2, p. 28 (1963).
28 Barkla, C. G., "Polarization in secondary Röntgen radiation", *Proc. Roy. Soc.*, **77,** p. 247 (1906).
29 Compton, A. H. and Hagenow, C. F., "A measurement of the polarization of secondary x-rays", *J. Opt. Soc. Amer.*, **8,** p. 487 (1924).
30 Compton, A. H. and Allison, S. K., *X-Rays in Theory and Experiment*, (New Jersey, Van Nostrand, 1960).

2 | *X-ray Emission Spectra*

X-rays are produced when a fast-moving electron is suddenly stopped by striking the anode or target of an X-ray tube. The X-radiation is of two types, *continuous* and *characteristic*, which differ in their mode of production and in their spectral distribution. It is therefore convenient to consider them separately.

The Continuous Spectrum

It can be deduced from the classical theory of electromagnetism, that X-rays will be produced when an electron is made to change its velocity suddenly [1]. The field surrounding a charge is radial as long as the charge is stationary, or moving uniformly in a straight line at constant velocity. Electromagnetic disturbances are propagated with a velocity c, *in vacuo*. Thus, when the velocity of an electron is suddenly changed, a finite time will elapse before the resultant disturbance has reached a given distance from the new position of the charge. In Figure 2.1 an electron is suddenly accelerated from X to Y for t seconds and then proceeds to Z with an increased, but uniform velocity. During the time T taken for the electron to move from Y to Z, the disturbance will have travelled a distance r along the radius, where $r = cT$. However the radial field centred at X will be undisturbed at points beyond $R = c(T + t)$, because the disturbance will not have reached beyond R in the time $(T + t)$. Consequently the region between the two spheres of radii r and R will contain an electric field oblique to the direction of propagation of the disturbance pulse. The oblique field can be resolved into a component along the original radial direction and a component perpendicular to it. The former corresponds to the original radial electrostatic vector; the latter corresponds to the electric vector (E) associated with the electromagnetic disturbance. The presence of a similar perpendicular magnetic vector (H) can also be deduced.

Thus the pulse emitted is electromagnetic radiation. The wavelength was shown by Nicholas [2] to be inversely proportional to the initial speed of the electrons.

The incident electrons will be stopped by a succession of collisions with the anode atoms. Each collision results in a deceleration of the electron and an emission of electromagnetic radiation. This is the bremsstrahlung process. Using the classical theory it has been possible to predict the energy distribution polar diagram for radiation from a thin target [3], calculations for thick targets present mathematical difficulties owing to secondary effects. Furthermore, the classical theory predicts that when an electron

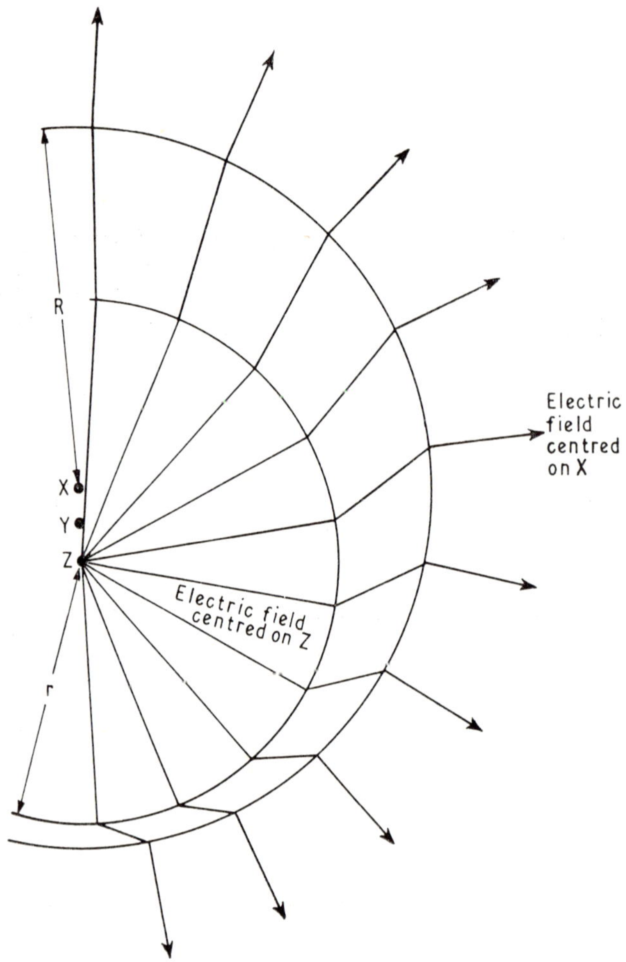

FIGURE 2.1 *Radiation from an Accelerated Electron*

X, electron at original constant velocity, electric field radial and centred on *X*;

X → *Y*, electron accelerating, producing a tangential component to the field;

Y → *Z*, electron at new constant velocity, electric field radial and centred on *Z*.

Whilst the electron is accelerating, it is radiating an electromagnetic wave radially, with a tangential electric field.

collides with a heavy atom such as that of platinum, it will decelerate more rapidly, and therefore emit more X-ray energy per collision, than when the target is made of a lighter material like aluminium; this is confirmed by the quantum theory and by experiment [4].

In fact, the intensity of the X-ray beam is proportional to the atomic number Z. It is proportional also to the number of electrons striking the target per unit time, i.e. the anode current i. A third factor is the tube voltage V_T; the intensity I_{total} is approximately proportional to the square of this, so that

$$I_{\text{total}} \propto iZV_T^2 \text{ (approx)} \tag{2.1}$$

As the rate of energy supplied is equal to iV_T, the efficiency ε of conversion, which is the ratio of X-ray energy to energy supplied to the tube, is given by [4]

$$\varepsilon \propto \frac{iZV_T^2}{iV_T} \text{(approx.)} \tag{2.2}$$

i.e.

$$\varepsilon = kZV_T$$

where k is a constant $= 1{\cdot}1 \times 10^{-9}$ per volt

For a tungsten ($Z = 74$) target operated at 250 kVp, the efficiency is about 0·02, i.e. 2 per cent. The remainder of the energy appears as heat. In practice, the softer components of the X-ray spectrum are removed by filtration (p. 92).

If the tube voltage is increased, the efficiency of X-ray production is increased and the spectral distribution altered to contain more X-rays of shorter wavelengths. It follows that an increase of the tube voltage results in an increase in useful beam intensity greater than that predicted from equation 2.1. Thus, the conditions for a high X-ray exposure rate output are: a high tube voltage, a high tube current, and a target of high atomic number.

The classical electromagnetic theory proved unable to explain some properties of the continuous X-ray spectrum, in particular the fact that for a given tube potential, there is a minimum X-ray wavelength: the *short-wave cut-off*. Therefore, attempts have been made to apply quantum theory methods by Sommerfeld, and others [4]. In any collision if the energy loss of an electron is W, the wavelength of the X-ray photon produced will be λ where

$$W = \frac{hc}{\lambda} = Ve \tag{2.3}$$

Substituting values for hc and e [5]

$$\lambda = \frac{12{,}398}{V} \tag{2.4}$$

where λ is in ångströms (kX units $\times$ 1·00202) and V is in volts. The value of $V\lambda$ is constant and can conveniently be memorized as 12,345.

One electron making, say, ten collisions can emit ten photons of varying wavelengths, so that a stream of electrons will cause the anode to emit many photons with a wide distribution of wavelengths. By varying the tube voltage a different range of photon wavelengths will be obtained.

The spectrum extends from the short-wave cut-off to a limit which in practice is governed by absorption of the longer wavelength X-rays in the apparatus. The spectrum has a single maximum (Figure 2.2). These spectra have been measured by X-ray diffraction methods (Chapter 10). The maximum

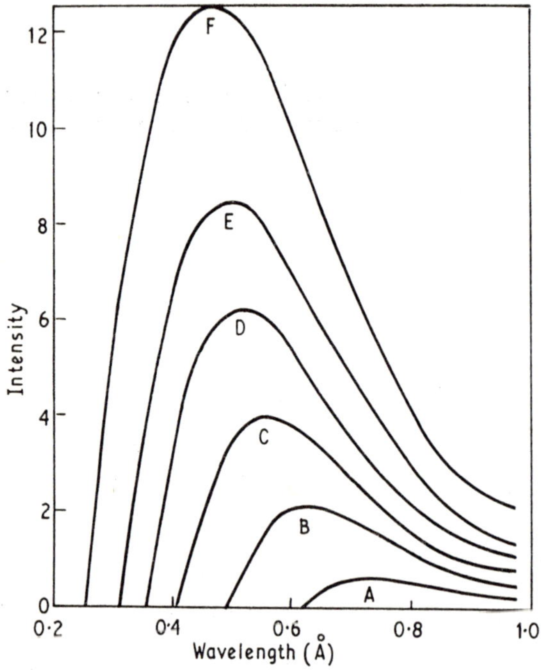

FIGURE 2.2 *Continuous X-ray Spectra at Different Constant Potentials*
Tungsten target. *A*, 20 kV; *B*, 25 kV; *C*, 30 kV; *D*, 35 kV; *E*, 40 kV; *F*, 50 kV.

intensity, or "hump", is dependent on V_{max} and its wavelength is given approximately by

$$\lambda_{peak} = 1{\cdot}4\lambda_{min}$$

The short-wave cut-off is of theoretical interest, as, if this and the corresponding tube voltage are accurately known, it is possible to obtain a precise value for Planck's constant h (equation 2.3). Corrections have to be made for the work functions of the materials of target and cathodes, and for other factors [6]. Duane and Hunt [7] used this principle. They measured the intensity of one particular wavelength for a range of voltages, and by extrapolation found the voltage at which the intensity dropped to zero. Thus they determined λ_{min} and its corresponding V_{max}. The intensities were determined from ionization measurements using an X-ray spectrometer. The ionization–voltage curves are known as "isochromats", because the wavelength is constant.

The Characteristic Spectrum

The bremsstrahlung process is not the only way in which an incident electron may give rise to X-radiation. On striking the target, an electron may lose its energy by removing another electron from a target atom, leaving the atom in an excited and unstable state. The ejected electron will be replaced by another from a higher energy state W_1. The difference in energy between W_1 and the final energy W_2 is emitted as a photon of frequency ν where

$$W_1 - W_2 = h\nu \qquad (2.5)$$

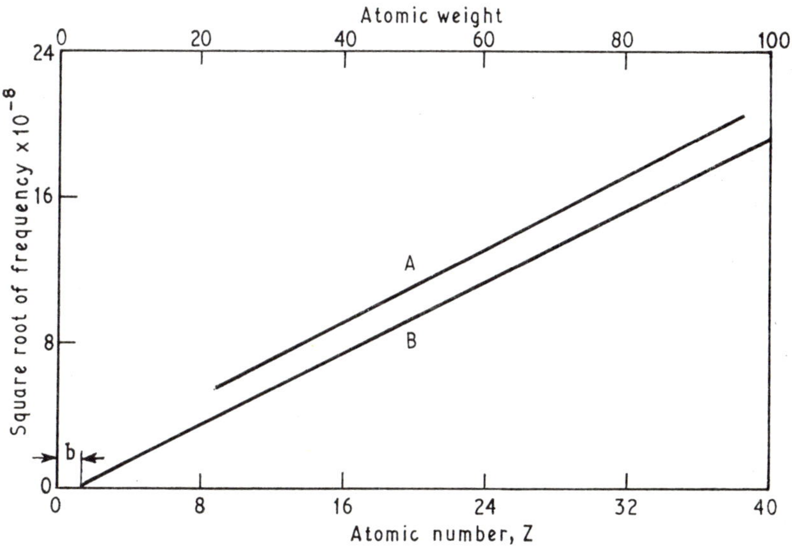

FIGURE 2.3 *Moseley's Curves Showing the Relation Between the Frequencies of X-ray Lines and Atomic Weight and Number (k_α Lines)*
A, $\sqrt{\nu}$ vs. atomic weight; *B*, $\sqrt{\nu}$ vs. atomic number.

(Richtmyer, E. A., Kennard, E. H. and Lauritsen, T., *Introduction to Modern Physics*, 5th ed., p. 364, Fig. 133. New York, McGraw-Hill, 1955)

The atom can return to its stable state in more than one way and by more than one step. Thus, the radiation of the characteristic spectrum consists of X-rays of discrete wavelengths which are characteristic of the target element, and depend on its atomic number.

ATOMIC STRUCTURE

Moseley and Bohr, working independently, made fundamental contributions to the understanding of the characteristic X-ray spectrum. Moseley examined the radiation from 38 elements by using them as targets in an X-ray tube. He discovered that when he plotted atomic number of the target element against the square root of the frequency of the characteristic radiation, he obtained two straight lines (Figure 2.3); he fitted the formula

$$\nu^{\frac{1}{2}} = K(Z - b)$$

where K and b are constants. From the linear relationship, he concluded that the atomic number rather than the atomic weight, characterizes an element. This led to a clarification of the Periodic Table (p. 23), in which four pairs of elements (A and K; Co and Ni; Te and I; Th and Pa) were reversed in order when classified according to their atomic weights. He predicted that certain elements were missing from the Periodic Table, and these have since been discovered.

He identified the two lines as corresponding to the K and L radiations discovered earlier by Barkla and Sadler [8] from absorption measurements of fluorescent radiation. Moseley [9, 10] published his results in 1913 and 1914. He was killed while a young man, in the Dardanelles campaign in 1915.

Bohr [11] had meanwhile applied Planck's quantum theory [12] to optical spectra, using the Bohr model of the atom. In this model the atom can exist in one of a series of discrete energy states. It will emit radiation only when changing from one state to another. The atom has a heavy nucleus with a positive charge of Ze. This charge is neutralized by Z orbital electrons. Not more than two electrons are allowed in the innermost orbit, and these are called K electrons. Up to eight L electrons are allowed in the next orbit; and up to eighteen M electrons in the third, and so on. In an unexcited state, electrons occupy the orbit in which the appropriate energy level is lowest. This is not always the innermost vacant orbit.

The chemical properties of an element are determined mainly by the atomic number and the valency electrons, which are those moving in the orbitals of the highest energy levels.

It follows that, if elements are arranged in order of increasing atomic number, there will be a periodicity in their properties. This was first noted by Mendeleev [13], although he used atomic weights in his periodic table. For example the inert gases, helium ($Z = 2$), neon ($Z = 10$), argon ($Z = 18$), krypton ($Z = 36$), xenon ($Z = 54$) and radon ($Z = 86$), all have their (different) outer shells completely filled, and thus have similar properties.

Using Bohr's treatment, the electrostatic attraction between the heavy positively charged nucleus and an orbital electron is given by

$$\text{Force} = \frac{Ze^2}{a^2} \tag{2.6}$$

where a is the orbital radius.

This attraction is balanced by the centrifugal force mV^2/a of the electron in its orbit, so that

$$\frac{Ze^2}{a^2} = \frac{mV^2}{a} \tag{2.7}$$

and

$$\frac{Ze^2}{a} = mV^2 \tag{2.8}$$

where V is the electron orbital velocity.

Bohr postulated that the only possible orbits are those in which the angular momentum mVa of the electron is an integral multiple n of $h/2\pi$

$$mVa = \frac{nh}{2\pi} \tag{2.9}$$

When an electron jumps from one orbit of energy W_1 and radius a_1 to another of energy W_2 and radius a_2, the frequency of the radiation ν is given by

$$h\nu = W_1 - W_2 \tag{2.10}$$

If W is the total energy, both potential and kinetic, of the electron,

$$W = \frac{mV^2}{2} - \frac{Ze^2}{a} = -\frac{Ze^2}{2a} \quad \text{(from eqn. 2.8)} \tag{2.11}$$

The potential energy term Ze^2/a is negative because it is equal to

$$\int_{\infty}^{a} Ze\frac{e}{r^2}\mathrm{d}r = -\frac{Ze^2}{a} \tag{2.12}$$

Thus the potential energy decreases (becomes more negative) as a decreases.

$$W = -\frac{Ze^2}{2}\cdot\frac{2\pi mV}{nh} \tag{2.13}$$

Dividing equation (2.8) by equation (2.9)

$$V = \frac{2\pi Ze^2}{nh} \tag{2.14}$$

so that

$$W = -\frac{Ze^2}{2}\cdot\frac{2\pi m}{nh}\cdot\frac{2\pi Ze^2}{nh} \tag{2.15}$$

and

$$W = -\frac{2\pi^2 e^4 m Z^2}{n^2 h^2} \tag{2.16}$$

The other orbital electrons exert a screening effect, so that the term $(Z - b)$ should be substituted for Z in the formula. b is called the *screening constant*; its value depends on the type of transition, as well as on the element. It is a complex function of the initial and final energy states. Another consideration is that the orbit is about the centre of mass of the combined system and not about the nucleus alone. For convenience, a correction due to this is included in the value of b. Thus

$$\frac{W_1 - W_2}{h} = \nu = \frac{2\pi^2 e^4 m}{h^3}(Z - b)^2\left(\frac{1}{{n_2}^2} - \frac{1}{{n_1}^2}\right) \tag{2.17}$$

where n_1, n_2 are the quantum numbers referring to W_1, W_2 respectively,
$2\pi^2 e^4 m/ch^3$ is a constant and is equal to R, known as the Rydberg constant,
c = velocity of light.

Hence

$$\frac{\nu}{c} = R(Z - b)^2\left(\frac{1}{n_2^2} - \frac{1}{n_1^2}\right) = \frac{1}{\lambda} \tag{2.18}$$

For a transition of an electron from an L orbit to a K orbit, $n_1 = 2$ and $n_2 = 1$. For a transition from an M to an L orbit $n_1 = 3$ and $n_2 = 2$. For the former transition

$$\frac{1}{\lambda} = \frac{3R}{4}(Z - b_K)^2 \tag{2.19}$$

For the latter transition

$$\frac{1}{\lambda} = \frac{5R}{36}(Z - b_L)^2 \tag{2.20}$$

These correspond to the two straight lines Moseley obtained when plotting $(1/\lambda)^{\frac{1}{2}}$ against Z (p. 23).

Energy Levels

Any transition to a K energy level gives rise to a K line and similarly for L, M, N and so on. The energy levels themselves are subdivided, and therefore the lines have a corresponding fine structure. Thus the K_β doublet arises from transitions from two of the three L energy levels into the K level; the K_α doublet arises from transitions from the M to the K shell.

If an electron drops from the L to the K energy level, and to replace it another electron drops from the M to the L level, then the photon associated with the first transition has an energy of $h\nu_{K_\alpha}$ and the photon associated with the second transition has an energy of $h\nu_{L_\alpha}$. The sum of these two energies is that associated with an electron transition from the M to the K level and is $h\nu_{K_\beta}$. This is expressed in the equation

$$h\nu_{K_\alpha} + h\nu_{L_\alpha} = h\nu_{K_\beta} \tag{2.21}$$

Hence

$$\nu_{K_\alpha} + \nu_{L_\alpha} = \nu_{K_\beta} \tag{2.22}$$

This type of relationship was noticed by Kossel [14] from a study of Moseley's results and is an example of the "combination principle" of spectroscopy. It also lends further support for the idea of stationary energy states.

Wave Mechanical Model

The Bohr model explains a number of X-ray phenomena in terms of simple concepts, and because of this, it still has value. For example, the original Barkla notation is still used in describing the X-ray wavelengths as K, L, M, etc. Shortly afterwards, Bohr developed the Rutherford model of the atom, to include the idea of stationary states, and the rapid development of the study of X-rays during this epoch makes the ideas of the original workers very important. However, the value of the Bohr model in optical spectroscopy is of historical interest only. It fails to explain much of the finer detail of observed optical spectra and some of its fundamental ideas are inadequate. For example, it is difficult to explain why the electrons accelerating in their orbits do not radiate electromagnetic radiation.

The development of wave mechanics led to a new atomic model in which the concept of electron orbits disappeared. An aspect of the relationship between the older Bohr model and the wave mechanical model can be seen by using the de Broglie equation

$$\lambda = \frac{h}{mv} \tag{2.23}$$

which implies that a particle of mass m on moving with a velocity v has an associated wavelength λ. Now, assuming the electron forms a standing wave system round the nucleus with n waves in a circle of radius a,

$$\frac{2\pi a}{\lambda} = n \tag{2.24}$$

Substituting for λ in equation (2.24)

$$\frac{2\pi amv}{h} = n \tag{2.25}$$

or

$$mva = \frac{hn}{2\pi} \tag{2.26}$$

which is the Bohr postulate (equation 2.9).

This picture is an oversimplification because the waves extend throughout space and the explanation above does not postulate any radial stationary state. Furthermore, the waves do not exist along a sharp circumferential line but as a cloud-like mathematical probability function, in which the cloud density represents the probability of the electron being at a particular place. The complete electron distribution function for a given atom is mathematically difficult to derive even for the lighter atoms and approximations have to be used.

Quantum Numbers

Modern quantum theory [15] confirms the earlier ideas of stationary energy states in the atom, but expresses these states in terms of three quantum numbers, one of which corresponds to Bohr's original n. The other two quantum numbers are l and j. The three quantum numbers relate to the possible energy states of the atom.

n is known as the total quantum number. It can have only integral values, and is the main factor describing the binding energy of the electron.

l expresses the angular momentum of the electron. On the Bohr model this momentum is due to its orbital motion. l can have integral values from 0 to $(n - 1)$. When $l = 0$, the angular momentum, but not the energy of the electron, is zero. This can be pictured as an oscillation in a straight line through the nucleus. Such a picture is obviously absurd, and indicates why it is better to think of energy states as electron probability cloud patterns having definite associated energies, rather than to use mental pictures based on classical mechanisms.

j is related to the spin of the electron. The electron spin momentum can be shown to have a quantum number $\frac{1}{2}$, which may be added to, or subtracted from, the l quantum number. This resultant quantum number is called j, so that

$$j = l \pm \tfrac{1}{2} \tag{2.27}$$

The complete atom consists of the nucleus and the number of electrons required to neutralize its positive charge. These electrons are at the appropriate energy levels.

Table 2.1 illustrates the quantum numbers for the K, L and M energy levels.

TABLE 2.1 *K, L and M Energy Levels*

Energy level	n	l	j
K	1	0	$l + \frac{1}{2} = \frac{1}{2}$
L_1	2	0	$l + \frac{1}{2} = \frac{1}{2}$
L_2	2	1	$l - \frac{1}{2} = \frac{1}{2}$
L_3	2	1	$l + \frac{1}{2} = 1\frac{1}{2}$
M_1	3	0	$l + \frac{1}{2} = \frac{1}{2}$
M_2	3	1	$l - \frac{1}{2} = \frac{1}{2}$
M_3	3	1	$l + \frac{1}{2} = 1\frac{1}{2}$
M_4	3	2	$l - \frac{1}{2} = 1\frac{1}{2}$
M_5	3	2	$l + \frac{1}{2} = 2\frac{1}{2}$

As X-ray generation requires large electron energy changes [4], the different electron spin direction energies do not alter the energy of the electron sufficiently to affect the X-ray wavelength. Thus, there is only one K energy level, but it accommodates two electrons.

There are three L energy levels, called L_1 and L_2 and L_3. For all of them $n = 2$, but l can be zero or one (Table 2.1). Similarly, there are seven N levels and so on. Two $j + 1$ electrons may exist in any level; thus the M_2 level has two electrons.

Electron Transitions

Not all transitions are possible, only those governed by *selection rules.*

In general, these are

i the change Δl in l is equal to ± 1

ii the change Δj in j is equal to 0 or ± 1.

The Δl selection rule expresses the *conservation of parity* rule, now recognized to be broken in some nuclear processes. The parity of a system refers to the behaviour of the wave function when the co-ordinates are reversed in sign.

The Δj selection rule arises because the photon is a particle with a spin quantum number of unity; the rule expresses the conservation of angular momentum.

Transitions in which the selection rules are broken may occur under special conditions of excitation, but the intensities of the spectral lines arising from these transitions are never strong [4].

Multiple Ionization

When the more intense and easily resolvable spectral lines, such as the copper K_α doublet, are examined using high resolution spectrometers, the presence of associated weak lines called *satellites* is revealed. Owing to their low intensity, reliable information about them is difficult to obtain, but it is considered that their presence is due to the atom having been in a state of double ionization. For example, an atom may have a vacancy in the K and L states simultaneously, in which case the atom may be said to be in the KL state. Other possible states are the KK, KM, KN, LL, LM and so on. It is possible to show theoretically that a transition from the KL to the LL state would require a slightly different energy change from that from the K to the L state, thus giving rise to satellites. More complex changes may also be possible.

Multiple ionization can be shown to arise from the bombardment of the atoms by a cathode ray stream. However, the intensities thus predicted theoretically by Wentzel [16] and Druyvesteyn [17] do not explain fully the experimental results, unless account is taken of the *Auger effect* [18]. This is a radiationless transition which may leave the atom in a state of double ionization. For example, K_α lines result from an electron leaving the L shell to go to a vacant space in the K shell. The energy of this transition, instead of appearing as a K_α photon, may cause another electron to be emitted from the L shell, leaving the atom in a state of double ionization without the emission of a photon. The electrons ejected by these radiationless transitions were observed by Auger in cloud chamber experiments.

OTHER X-RAY SOURCES

As the wavelength of the x-rays obtained is an inverse function of the rate of velocity change (either acceleration or deceleration), it is important that these rates of change should be large to obtain useful ranges of x-ray energies. Large electron decelerations are relatively easy to produce by collisions between the light electrons and the heavier atoms. This is the basis of the conventional x-ray tube. Large accelerations are very difficult to obtain in the laboratory, but may be produced in space (*see* Chapter 1).

X-rays should also be produced by high-temperature gas explosions, such as probably occur in space. The electrical explosion of metallic conductors in vacuum has been used to investigate x-ray emission from explosions [19]. Analysis of the emitted x-radiation throws light on the basic processes responsible for the emission, namely, the interaction of strong electric fields with the ionized matter of the exploding conductors. The exploding conductor constitutes the source of x-rays, the bulk of the radiation coming from the core of the explosion. The x-ray pulse is very short in duration and its photon energy is about 5 keV, even when the peak voltage applied to the conductor is higher than 150 kVp. This is presumably because the available electrons resulting from an early onset of ionization are decelerated by multiple collisions from the relatively stationary ions of the exploding matter.

Literature References

1 THOMSON, J. J., *Conduction of Electricity through Gases* (Cambridge, Cambridge Univ. Press, 1906).
2 NICHOLAS, W. W., "Continuous spectrum x-rays from thin targets". *Bur. Stand. J. Res. Wash.*, **2,** p. 837 (1929).
3 SOMMERFELD, A., *Atomic Structure and Spectral Lines* (English ed.) (New York and London, Methuen, 1923).
4 COMPTON, A. H. and ALLISON, S. K., *X-Rays in Theory and Experiment* (New Jersey, Van Nostrand, 1960).
5 MACGILLAVRY, C. H. and RIECK, G. D. (Eds.), *International Tables for X-ray Crystallography,* Vol. III (Birmingham, Kynoch Press, 1962).
6 WORSNOP, B. L. and CHALKIN, F. C., *X-Rays* (London, Methuen, 1950).
7 DUANE, W. and HUNT, F. L., "On x-ray wavelengths", *Phys. Rev.*, **6,** p. 166 (1915).
8 BARKLA, C. G. and SADLER, C. A., "Homogeneous secondary Röntgen Radiations", *Phil. Mag.*, **4,** p. 550 (1908).
9 MOSELEY, H. G. T., "The high-frequency spectra of the elements", *Phil. Mag.*, **26,** p. 1024 (1913).
10 MOSELEY, H. G. T., "The high-frequency spectra of the elements", Part II, *Phil. Mag.*, **27,** p. 703 (1914).
11 BOHR, N., "On the constitution of atoms and molecules", *Phil. Mag.*, **26,** p. 1 (1913).
12 PLANCK, M., "Uber das Gesetz der Energieverteilung im normal Spectrum", *Ann. Phys.*, **4,** p. 553 (1901).
13 MENDELEEV, D. I., "On the correlation of the properties and atomic weights of the elements" (transl.), *J. Russ. Chem. Soc.*, **1,** p. 60 (1869).
14 KOSSEL, W., "Uber die Ausbildung der Röntgenserien mit wachsender Ordnungszahl", *Z. Phys.*, **2,** p. 470 (1920).
15 SEMAT, H., *Introduction to Atomic and Nuclear Physics* (London, Chapman and Hall, 1958).
16 WENTZEL, G., "Funkenlinien im Röntgenspectrum", *Ann. Phys.*, **66,** p. 437 (1921).
17 DRUYVESTEYN, M. J., "Das Röntgenspektrum zweiter Art", *Z. Phys.*, **43,** p. 707 (1927).
18 AUGER, P., "Sur les rayons-β secondaires produits dans un gaz par les rayons-x", *Compt. Rend.*, **180,** p. 65 (1925).
19 CHASE, W. G., and MOORE, H. K. (Eds.), *Exploding Wires*", Vol. II (New York, Plenum Press, 1962).

3 | *Absorption and Scattering of X-rays*

When a beam of x-rays traverses any material, gas, liquid or solid, the intensity or energy flux density, I, of the emergent beam is less than that of the incident beam. The beam can be attenuated in the material by several processes (*vide infra*), and the relative importance of each of these depends on the energy of the radiation and the nature of the material. Each of these processes can be denoted by a partial coefficient; together the coefficients constitute the total absorption coefficient (p. 47).

There is some confusion in nomenclature between attenuation and absorption. Photon beams are attenuated during their passage through matter and the degree of attenuation is described in terms of absorption coefficients. The absorption coefficients should not be confused with the absorption of energy, which takes place through the intermediary of directly ionizing particles, and which will be described later in the chapter.

Lambert's Law

Lambert's law describes [1] the manner in which electromagnetic radiation is attenuated in matter. The law implies that equal thicknesses of the same material attenuate equal fractions of the intensity of the respective incident radiation.

The intensity I is defined [2] in terms of the energy fluence F given by

$$F = \frac{\Delta E_F}{\Delta a} \tag{3.1}$$

and ΔE_F is the sum of all the energies, exclusive of rest energies, of all the particles which enter a sphere of cross-sectional area Δa.

$$I = \frac{\Delta F}{\Delta t} \tag{3.2}$$

where ΔF is the energy fluence in the time Δt.
Thus, by Lambert's law,

$$\frac{dI}{I} = -\mu\, dx \tag{3.3}$$

where the loss in intensity is dI in traversing a distance dx; μ is a constant known as the *linear absorption coefficient*. The negative sign indicates that I decreases as x increases.

Integrating over a thickness x_0 of absorber and putting in the condition that $I = I_0$ when $x = 0$,

$$I = I_0 \exp(-\mu x_0) \tag{3.4}$$

The value of the constant μ depends on the energy of the radiation and on the material. The formula ignores any contribution to I from radiation scattered from other parts of the beam, and therefore applies only to narrow beam conditions.

The attenuation of X-rays, in contrast to that of light, is independent of the physical or chemical state of the material. Thus, if a beam of X-rays passes through a vessel filled with hydrogen and oxygen, a certain fraction of the beam will be attenuated. If the gases are combined to make steam, the attenuation will be unchanged, as it will if the steam is condensed to water or the water frozen to ice. This is because, by these changes of state, the alteration in the atoms' electron energy levels is small compared with the energy changes involved in the interaction of X-rays with matter. The value of the linear absorption coefficient depends, however, like the density, on the state of the material. Unlike density, some quantities, such as the numbers of atoms or electrons, do not change with change of state. These can be incorporated into absorption coefficients more fundamental than the linear absorption coefficient.

Atomic Absorption Coefficient

Consider a narrow beam of n photons per unit time, each with energy $h\nu$ incident on material containing A atoms per unit area. A thin slice of this material contains dA atoms per unit area (Figure 3.1).

Let each atom have an associated cross-sectional area μ_a defined so that μ_a is the *effective area* of each atom, in the sense that if a photon is incident on this area μ_a, an interaction will ensue.

In the thin slice, the total area for interaction is therefore

$$\mu_a \, dA$$

and the number of interactions is given by

$$dn = -n\mu_a \, dA \tag{3.5}$$

As $I \propto n$, we have

$$-dI = I\mu_a \, dA \tag{3.6}$$

Integrating and putting in the condition that $I = I_0$ when $A = 0$,

$$I = I_0 \exp(-\mu_a A) \tag{3.7}$$

μ_a is known as the atomic cross-section of the interaction. Its dimensions are $(\text{length})^2$; the practical unit is the *barn*, where 1 barn $= 10^{-22}$ mm^2; μ_a

is also the atomic absorption coefficient (Table 3.1), and may be expressed in terms of cm^2/atom.

Now consider a material having dA' atoms of one element, with an atomic absorption coefficient μ_a'. Similarly, other elements, denoted by double dash, treble dash and so on, are included in the material. Assume that the presence of one atom does not affect the interaction of another.

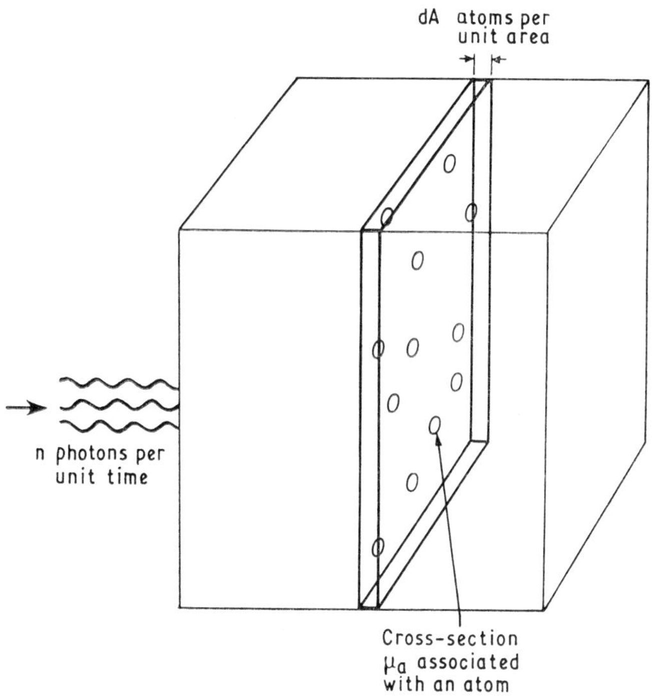

FIGURE 3.1 *Diagrammatic Representation of Atomic Cross-section*

Then, from equation (3.6),

$$dI = -I\mu_a' \, dA' - I\mu'' \, dA'' - \ldots \tag{3.8}$$

Integrating,

$$I = I_0 \exp[-(\mu_a'A' + \mu_a''A'' + \ldots)] \tag{3.9}$$

As the total number of atoms A is equal to $A' + A'' + \ldots$ and μ_a is the effective atomic absorption coefficient of the mixture, then from equations (3.7) and (3.9),

$$\mu_a = \mu_a'f' + \mu_a''f'' + \ldots \tag{3.10}$$

where $f', f'' \ldots$ are the fractional atomic constituents $A'/A, A''/A, \ldots$ of the respective elements.

The values of $\mu', \mu'', \ldots$ are wavelength-dependent, in a manner which depends on the atomic number of the element. Hence, if a mixture is required which is to have the same atomic absorption coefficient as another material, there is no unique solution which applies for all wavelengths. This has important practical considerations in designing, for instance, air-walled chambers (p. 62) or tissue equivalent materials.

Electronic Absorption Coefficient

If Z is the atomic number of the element comprising the slice of material described above, and E is the number of electrons per unit area in the total thickness then in a thin slice, if $\mathrm{d}E$ is the number of electrons per unit area,

$$\mathrm{d}A = \frac{\mathrm{d}E}{Z} \tag{3.11}$$

Substituting for $\mathrm{d}A$ in equation (3.6) and integrating as before, we have

$$I = I_0 \exp(-\mu_e E) \tag{3.12}$$

μ_e is called the electronic absorption coefficient. It has dimensions of $(\text{length})^2$ (Table 3.1) and is equal to μ_a/Z.

As for atomic absorption coefficients, in a mixture of elements,

$$I = I_0 \exp[-(\mu_e'E' + \mu_e''E'' + \ldots)] = I_0 \exp(-\mu_e E) \tag{3.13}$$

where μ_e is the effective electronic absorption coefficient of the mixture and E is the total number of electrons per unit area.

TABLE 3.1 *Inter-relationships of Absorption Coefficients*

Absorption coefficients	*Dimensions*
Atomic μ_a	$(\text{length})^2$
Electronic $\mu_e = \mu_a/Z$	$(\text{length})^2$
Mass $\mu_m = \mu_a N_0/W$	$(\text{length})^2 (\text{mass})^{-1}$
Linear $\mu = \rho\mu_a N_0/W$	$(\text{length})^{-1}$

Hence, equating coefficients,

$$\mu_e = \mu_e'g' + \mu_e''g'' + \ldots \tag{3.14}$$

where $g' = E'/E$, $g'' = E''/E$, . . ., are the fractions of the total number of electrons for each element.

Mass Absorption Coefficient

It is sometimes convenient to consider attenuation in terms of the mass m per unit area of the material. Let the material consist of only one element of atomic mass W grammes.

In a thin slice, the mass per unit area, $\mathrm{d}m$ is given by

$$\mathrm{d}m = \frac{W\,\mathrm{d}A}{N_0} \tag{3.15}$$

where N_0 = Avogadro's constant.

Substituting in equation (3.6) and integrating as before,

$$I = I_0 \exp(-\mu_m m) \tag{3.16}$$

μ_m is known as the mass absorption coefficient (Table 3.1). It has dimensions of (length)^2 (mass)$^{-1}$ and is equal to $\mu_a N_0/W$. For a given element, N_0/W is a constant, being the number of atoms per gramme.

As for the other absorption coefficients, for a mixture of elements,

$$\mu_m = \mu_m' \frac{m'}{m} + \mu_m'' \frac{m''}{m} + \ldots \tag{3.17}$$

where μ_m is the effective mass absorption coefficient and m'/m, m''/m, . . ., are the respective fractional masses due to each element.

Linear Absorption Coefficient

If the single element material has a density ρ and a thickness x_0, then, since $x_0 = m/\rho$,

$$I = I_0 \exp(-\mu_m x_0 \rho) = I_0 \exp(-\mu x_0) \tag{3.18}$$

where μ, the *linear absorption coefficient* is given by

$$\mu = \mu_m \rho \tag{3.19}$$

and has dimensions of (length)$^{-1}$ (Table 3.1). Equation (3.19) should be compared with equation (3.4).

Arguing as before for material consisting of a mixture of elements, each with a linear absorption coefficient μ', μ'', . . ., it can be shown that, if μ is the effective linear absorption coefficient of the mixture and v its volume, then

$$\mu = \mu' \frac{v'}{v} + \mu'' \frac{v''}{v} + \ldots \tag{3.20}$$

where v'/v, v''/v . . ., are the fractional volumes due to each element, and provided that the volumes are conserved, so that

$$v = v' + v'' + \ldots \tag{3.21}$$

Attenuation Processes

Ionizing radiations are attenuated in matter. Fast-moving charged particles lose their kinetic energy in small steps during many interactions. In contrast, X-ray photons may lose either a large fraction or all of their energy in a single interaction.

When an X-ray beam interacts with matter, the field of a photon in the beam and an atom in the material can be considered to form two quantum mechanical systems with an interaction energy. This interaction [3] will cause transitions of the unperturbed system (photon plus atom), which will consist, in general, of a transition of the atom from one quantum state to another, together with an emission or absorption of a photon.

There are four main types of interaction, each occurring with a certain probability which depends on the photon energy and on the nature of the atom. Thus, the total absorption coefficient varies in a complex way with both of these.

In each interaction, energy and momentum are conserved. Equations expressing this conservation take different forms when different components are numerically important.

THE COMPTON PROCESS

When the energy of the photon is large compared with the binding energy of an electron to its atom, the electron may be considered free. A photon interacting with such an electron may give to it some energy and momentum, so that a fast electron and a photon with degraded energy emerge (Figure 3.2).

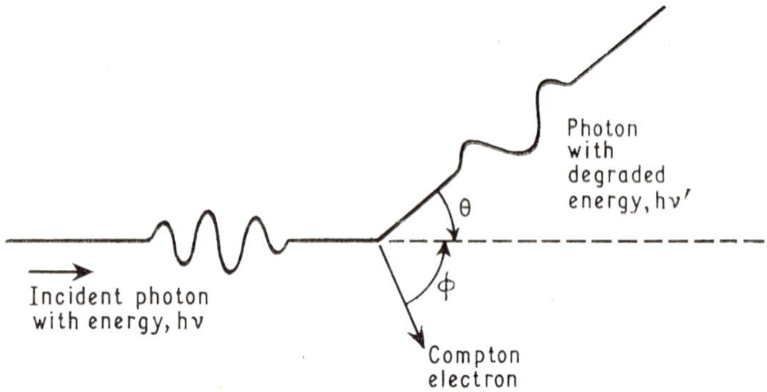

FIGURE 3.2 *The Compton Process*

This is known as the Compton process.

Relevant parameters for photon and electron are shown in Table 3.2.

TABLE 3.2 *Parameters for Photon and Electron*

	Photon	*Electron*
Velocity	c	βc
Frequency	ν	
Energy	$h\nu$	$m_0c^2\left\{\frac{1}{\sqrt{(1-\beta^2)}}-1\right\}$
Momentum	$h\nu/c$	$m_0\beta c/\sqrt{(1-\beta^2)}$
Mass	$h\nu/c^2$	m
		m_0 (rest mass)

Three equations express the conservation of momentum in the direction of the incident photon (equation 3.22), in a perpendicular direction (equation 3.23), and the conservation of energy (equation 3.24):

$$\frac{h\nu - h\nu' \cos\theta}{c} = \frac{m_0\beta c}{\sqrt{(1-\beta^2)}}\cos\phi \qquad (3.22)$$

where β = velocity of electron $\times$ $1/c$.

$$\frac{h\nu'}{c}\sin\theta = \frac{m_0\beta c}{\sqrt{(1-\beta^2)}}\sin\phi \tag{3.23}$$

$$h\nu - h\nu' = m_0c^2\left\{\frac{1}{\sqrt{(1-\beta^2)}} - 1\right\} \tag{3.24}$$

Eliminating ϕ by squaring and adding equations (3.22) and (3.23) and treating the resulting equation with equation (3.24) as simultaneous equations from which β can be eliminated, we have

$$\lambda' - \lambda = \frac{h}{m_0c}(1-\cos\theta) = 0{\cdot}0484\sin^2\tfrac{1}{2}\theta \tag{3.25}$$

This equation implies that the *change in wavelength* between incident and scattered photon depends only on the angle of scatter and is independent of

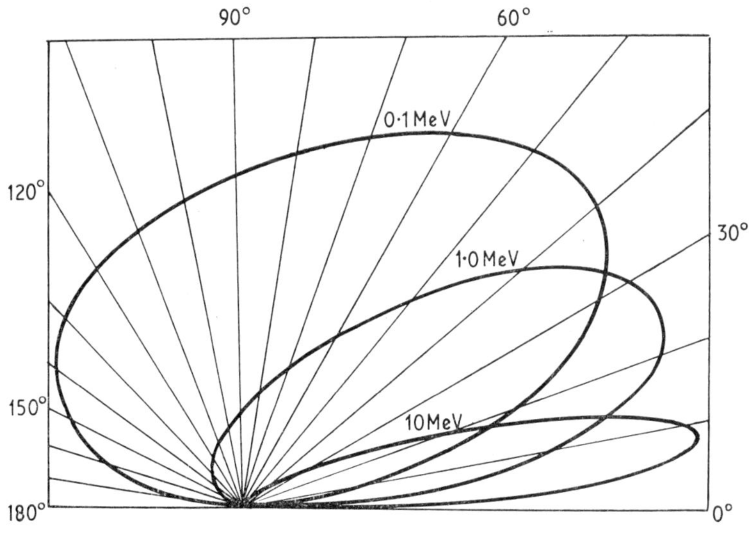

FIGURE 3.3 *Angular Distribution of Compton Scattered Photons*
Energy of incident photons 0·1 MeV, 1·0 MeV, 10 MeV. The radial distance represents the energy scattered through an angle θ.

(Whyte, G. N., *Principles of Radiation Dosimetry*, p. 31, Fig. 2.12. New York, Wiley, 1959)

the wavelength (or energy) of the incident photon. Hence the change in energy between incident and scattered photon increases with increase of incident photon energy.

$\lambda - \lambda'$ is a maximum and has the value of 0·0484 when $\theta = 180°$.

Equation (3.24) may be rearranged:

$$\frac{h\nu}{h\nu'} = 1 + \frac{h\nu}{m_0c^2}(1-\cos\theta) \tag{3.26}$$

The likelihood of a particular angle θ occurring depends on the energy of the incident photon (Figure 3.3). It may be expressed in terms of ${}_s\sigma_e$, the electronic cross-section for Compton scatter. Detailed cross-sections are available [4]. For an angle θ, the electronic scatter cross-section $d\sigma_e$, for a small solid angle $d\Omega$ is given [5] by

$$d\sigma_e = \frac{e^4}{2m^2c^4}\left(\frac{h\nu'}{h\nu}\right)^2\left(\frac{h\nu}{h\nu'} + \frac{h\nu'}{h\nu} - \sin^2\theta\right) d\Omega \tag{3.27}$$

Expressed in words, for a photon of incident energy $h\nu$ and energy $h\nu'$ after scattering, and deviated through an angle θ within the solid $d\Omega$, the probability of an interaction with an electron is given by $d\sigma_e$.

Integrating equation (3.27) over all possible values for the angle of scatter to give the total cross-section per electron σ_e, we have [5]

$$\sigma_e = \frac{2\pi e^4}{m_0{}^2c^4}\left\{\frac{1+\alpha}{\alpha^2}\left[\frac{2(1+\alpha)}{1+2\alpha} - \frac{1}{\alpha}\ln(1+2\alpha)\right] + \frac{1}{2\alpha}\ln(1+2\alpha) - \frac{1+3\alpha}{(1+2\alpha)^2}\right\} \tag{3.28}$$

where $\alpha = h\nu/m_0c^2$.

The fraction of the energy of the incident photons which is converted to the scattered photons, expressed as the Compton scattering cross-section ${}_s\sigma_e$ per electron per square centimetre is given [5] by

$${}_s\sigma_e = \frac{2\pi e^4}{m_0{}^2c^4}\left[\frac{1}{2\alpha^3}\ln(1+2\alpha) + \frac{(1+\alpha)(2\alpha^2-2\alpha-1)}{\alpha^2(1+2\alpha)^2} + \frac{4\alpha^2}{3(1+2\alpha)^3}\right] \tag{3.29}$$

Energy is absorbed in the material from the secondary electron. For this reason, the fraction of the incident photon energy imparted to secondary electrons is called the Compton absorption cross-section ${}_a\sigma_e$ where [6]

$${}_a\sigma_e = \sigma_e - {}_s\sigma_e = \frac{2\pi e^4}{m_0{}^2c^4}\left[\frac{2(1+\alpha)^2}{\alpha^2(1+2\alpha)} - \frac{1+3\alpha}{(1+2\alpha)^2} + \frac{(1+\alpha)(1+2\alpha-2\alpha^2)}{\alpha^2(1+2\alpha)^2} - \frac{4\alpha^2}{3(1+2\alpha)^3} - \left(\frac{1+\alpha}{\alpha^3} - \frac{1}{2\alpha} + \frac{1}{2\alpha^3}\right)\ln(1+2\alpha)\right] \tag{3.30}$$

Now, as energy is conserved

$$h\nu = h\nu' + E \tag{3.31}$$

where E = energy of the secondary electron

and

$$E = m_0c^2\left(\frac{1}{\sqrt{(1-\beta^2)}} - 1\right) \tag{3.32}$$

From equations (3.22), (3.23) and (3.24), it can be calculated that

$$E = h\nu \, . \, \frac{\alpha(1-\cos\phi)}{1+\alpha(1-\cos\phi)} \tag{3.33}$$

The energy distribution in the Compton process may be studied by the relationship of σ_e, ${}_s\sigma_e$ and ${}_a\sigma_e$ (Figure 3.4).

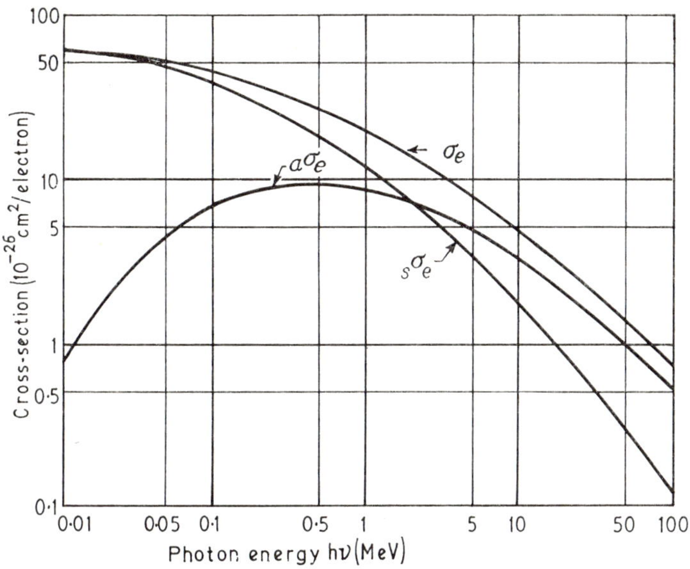

FIGURE 3.4 *Variations of Compton Cross-sections with Photon Energy*
(Whyte, G. N., *Principles of Radiation Dosimetry*, p. 30, Fig. 2.11. New York, Wiley, 1959)

At low energies, the value of σ_e is a maximum of 0·665 barn per electron. It decreases slowly to about 0·006 barn per electron as the photon energy increases to 100 MeV.

At low energies virtually all the energy of the incident photons appears in the scattered photons, the limiting condition being that for Rayleigh scatter (p. 40). However, as the incident photon energy increases, relatively less appears in the scattered photons, until at 100 MeV only about one-fifth of the energy is associated with the scattered photons.

The energy appearing in the secondary electrons is obtained by the difference of the two curves for σ_e and ${}_s\sigma_e$. The fraction

$$\frac{(h\nu')_{\text{average}}}{h\nu}$$

increases as $h\nu$ increases; simultaneously σ_e decreases. Hence ${}_s\sigma_e$ which is equal to

$$1 - \frac{(h\nu')_{\text{average}}}{h\nu}\,\sigma_e,$$

or the fraction of energy in the incoming beam given up to the secondary electrons, is initially low because $h\nu'/h\nu$ is small. ${}_s\sigma_e$ increases with increase of incident photon energy to a flat maximum of about 0·1 barn per electron at 0·5 MeV and then decreases because the decrease in σ_e becomes the deciding factor. At very high energies ${}_a\sigma_e$ approaches σ_e.

It is not always strictly true that the binding energy can be neglected in the Compton process, or that appreciable energy is not imparted to the atomic nucleus. Further, the electron will initially be in some kind of motion and therefore will emit radiation. At low photon energies, all these factors modify the energy of the scattered photon to an appreciable extent. The result is *modified* Compton scatter with associated spread of scattered photon wavelength. Theoretical values [6] of the broadening of modified Compton scatter agree with experimental determinations [7].

As incident and scattered photons have different wavelengths and therefore no constant phase relationship, they are said to be *incoherent*.

Because the incident photon is considered to interact with a free electron in the Compton process, Compton absorption coefficients are independent of the atomic number of the material and dependent on the electron density.

Compton's Experiments

In his original experiments [8] Compton analysed the radiation scattered from a carbon block. He used a crystal analyser and an ionization chamber. The experiments were difficult because the intensity of the scattered radiation was low, and the wavelength difference small. The results were of fundamental importance in that they confirmed the quantum theory of radiation at a time when it was not readily acceptable to many of his contemporaries.

RAYLEIGH OR COHERENT SCATTERING

If a photon interacts with an electron which is not free to move, the relevant value of m_0 in equation (3.25) becomes that of the atom. In this case, $\lambda - \lambda'$ is very small, so that the change in energy between incident and scattered photons is also very small. This is the case of unmodified (Rayleigh) scatter. It is important for low incident photon energies, when the photon has insufficient energy to remove the electron.

Classical theory predicts that when electromagnetic radiation interacts with an electron, the electric field of the radiation together with the charge of the electron will produce an accelerating force. Consequently the accelerated electron will emit radiation (p. 19), and as the electron will oscillate at the frequency of the applied field, the emitted radiation will be also at this frequency. Each scattering electron will be oscillating with a constant phase relationship to any other electron in the material. This scattered radiation has a definite phase relationship to that of the incident beam, and is therefore capable of producing interference (p. 8). Such radiation is said to be *coherent*.

The cross-section of this reaction is small and, particularly for high-energy photons, Rayleigh scattering is negligible compared with other processes. The effect, however, is the basis of X-ray diffraction processes (p. 10).

Classical electrodynamics shows that the electrical and magnetic fields are numerically equal when E is expressed in electrostatic and H in electromagnetic units. The magnitude of these field strengths at a distance r from a charge e with acceleration a is given by

$$E = H = \frac{ae}{rc^2} \sin \phi \tag{3.34}$$

where ϕ is the angle between the electron acceleration direction and r.

The energy per unit volume in an electromagnetic field is equal to

$$\frac{E^2}{8\pi} + \frac{H^2}{8\pi} = \frac{a^2e^2}{4\pi r^2c^4} \sin^2 \phi \tag{3.35}$$

When this expression is summed for all values of ϕ over a sphere of radius r and allowance is made for the wave velocity c, then the energy W radiated per second is given by

$$W = \frac{2}{3}\frac{e^2a^2}{c^3} \tag{3.36}$$

Now, putting m for the mass of the electron,

$$a = \frac{eE}{m} \tag{3.37}$$

and

$$W = \frac{2}{3}\frac{e^4E^2}{m^2c^3} \tag{3.38}$$

From classical theory, the intensity I of the primary beam is given by

$$I = \frac{cE^2}{4\pi} \tag{3.39}$$

Hence

$$W = \frac{8}{3}\pi \cdot \frac{e^4}{m^2c^4} I \tag{3.40}$$

W/I is the *electronic scattering coefficient* σ_e of a free electron, and, of the radiation falling on unit area perpendicular to the beam, a fraction σ_e is scattered by each electron.

Inserting values for e, m and c in equation (3.40), $\sigma_e = 0{\cdot}67$ barn and is independent of wavelength. Now

$$\frac{n\sigma_e}{\rho} = \frac{\sigma}{\rho} \tag{3.41}$$

where ρ is the density of the material and n the number of electrons per unit volume. The term σ/ρ represents the mass scattering coefficient, or the fraction of the beam scattered in unit area of material, and may be experimentally determined. From equation (3.41), n/ρ the number of electrons per gramme is thus known. Knowing the number of atoms per gramme of the material, the method may be used to find the number of electrons per atom.

Both light and heavy [9, 10] atoms have been examined by this method. Calculations may be checked by experiments on the scattering of X-rays by both relatively free atoms in gases and liquids and bound atoms in crystals. Allowance must be made for Compton, or incoherent scattering. Experimental results are in good agreement with the calculations, supporting the modern view of the atom as a charge distribution in space in localized regions.

When the wavelength of the scattering radiation approaches that of an absorption edge, modifications to the theory must be introduced.

It has already been shown (Chapter 1) that the intensity of scattered X-rays at an angle of deviation δ from the direct beam by a particle of charge e and mass m at a distance r from the scatter is given by

$$I_s = \frac{Ie^4}{r^2 m^2 c^4} \cdot \tfrac{1}{2}(1 + \cos^2 \delta) \tag{3.42}$$

When X-rays are scattered by a complex arrangement of bound charges as in an atom, the finite size and the spatial distribution of the charge cause phase relationships to arise between the radiation scattered from the various parts of the atom, resulting in a lower intensity than that obtained by simply adding the contributions from each electron.

For example, a rise in the temperature of a crystal will increase the motion of the atoms, which vibrate with increasing energy in their fixed positions against the restraining elastic forces in the crystal. Consequently, raising the temperature enlarges the effective size of the atom, and reduces the scattered intensity at a given angle.

As the intensity varies inversely as the square of the mass of the particle (equation 3.42) the contribution of the relatively heavy nuclei to the total is neglible compared with that from the electrons. When δ is zero and the scattered beam is parallel to the incident beam, the amplitudes of the scattered waves from all parts of the charge distribution may be added.

THE PHOTO-ELECTRIC EFFECT

The photo-electric effect is that in which incident photons cause the emission of electrons from materials. Such an electron must be bound or attached to a group of atoms and atomic momentum is a factor in conserving momentum.

The photo-electric effect is of historical interest, as its discovery by Hertz [11] in 1887 led Einstein [12] in 1905 to the corpuscular theory of radiation.

Early work on the photo-electric effect was concerned with light. In 1888, Hallwachs [13] found that when ultra-violet light fell on a negatively charged zinc surface, the charge was rapidly lost. As this did not occur with a positively charged surface, it appeared that negatively charged particles were emitted. These were later identified as electrons, first described by Thomson [14].

Suppose the electrons, each of mass m and charge e, have a range of velocities below a certain maximum v. A plate is positively charged to a potential V, of a magnitude just high enough to prevent an electron with a velocity v from escaping. Then

$$\tfrac{1}{2}mv^2 = Ve \tag{3.43}$$

The value of the velocity v depends on the frequency ν of the light and the nature of the surface, and equation (3.43) may be rewritten to express these relationships:

$$\tfrac{1}{2}mv^2 = h\nu - V_0e \tag{3.44}$$

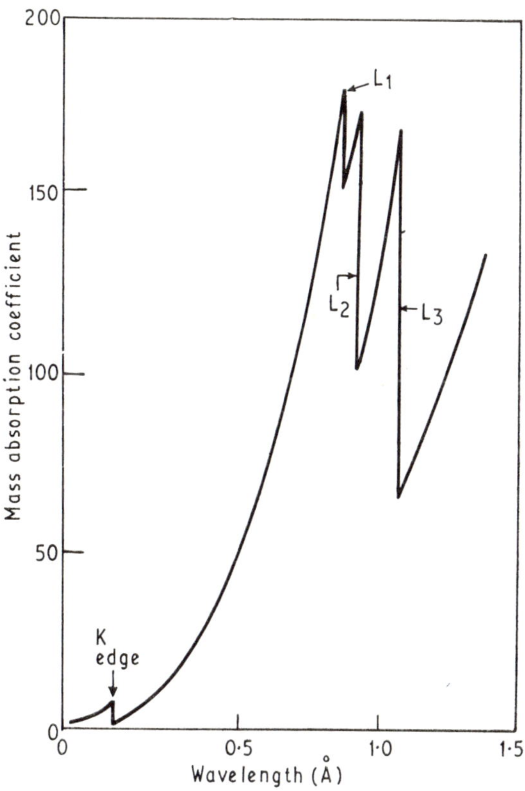

FIGURE 3.5 *Variation in Photoelectric Mass Absorption Coefficient with Wavelength*
K and L absorption edges for lead. K = 0·1405 Å; L_1 = 0·780 Å; L_2 = 0·813 Å; L_3 = 0·950 Å.

(Richtmyer, E. A., Kennard, E. H. and Lauritsen, T., *Introduction to Modern Physics*, 5th ed., p. 378, Fig. 141. New York, McGraw-Hill, 1955)

where V_0e is a surface constant, known as the work function of the material. Thus, there is a minimum frequency ν_0 to cause electron emission given by

$$Ve_0 = h\nu_0 \tag{3.45}$$

These relationships form the basis of a method [15] for measuring Planck's constant h.

The photo-electric effect is of great importance in x-radiation because ν is large, so that $h\nu$ is large compared with $V_0 e$. Also, because x-rays are penetrating, the effect is not limited to the surface, which plays a very minor role.

X-rays have energies sufficient to liberate electrons from the inner atomic orbits. Thus,

$$\tfrac{1}{2}mv^2 = h\nu - \psi \tag{3.46}$$

where ψ is the binding energy of the electron.

Consider the case when a photon ejects an electron from the K shell. This is most likely to occur when the energy of the photon is just greater than ψ, the binding energy. For energies higher than this, the probability of the interaction decreases, although most interactions will be with K electrons.

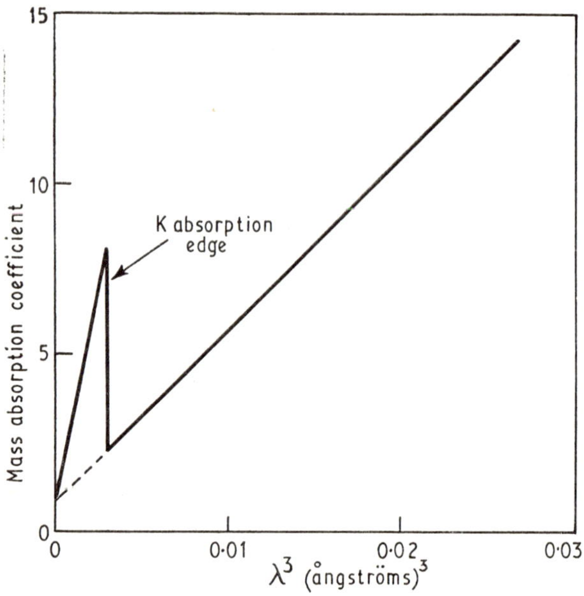

FIGURE 3.6 *Mass Absorption Coefficient for Lead as a Function of Wavelength* (Richtmyer, E. A., Kennard, E. H. and Lauritsen, T., *Introduction to Modern Physics*, 5th ed., p. 379, Fig. 142. New York, McGraw-Hill, 1955)

The energy required to free an electron from the L shell is less than that required to free one from the K shell. Hence, a photon with energy insufficient for the K interaction may achieve an L interaction. Although there are some differences, the rules are similar for all interactions. There is only one K absorption limit, but there are, for example, three L absorption limits [16]. The wavelengths of the absorption limits are just shorter than those of the corresponding emission spectra (p. 23). The discontinuous nature of the interaction probabilities is evident in a graph (Figure 3.5) of the relationship between photon energy and mass absorption coefficient.

When an electron is ejected from a shell, the atom is left in an excited state. Consequently, the vacancy in the shell will be filled by an electron with a higher potential energy. Its excess energy will be emitted as characteristic radiation, which will be unpolarized. The phenomenon is known as *fluorescence*.

It is evident that the intensity and the wavelength of the emission spectra will depend on the atomic number of the absorber. The relationship between mass absorption coefficient and cube of wavelength is a series of straight lines. For a single element, these lines can be extrapolated to converge approximately on the ordinate. Each line corresponds to a particular mode of absorption (Figure 3.6).

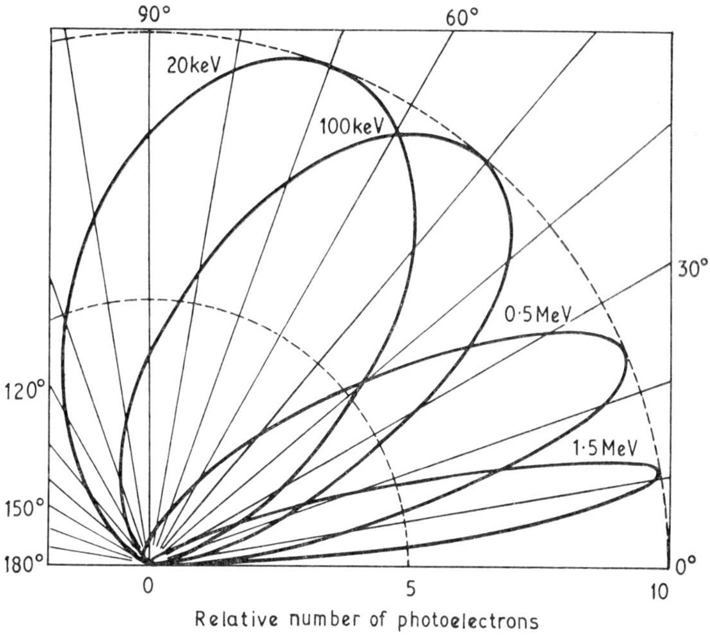

FIGURE 3.7 *Angular Distribution of Photo-electrons for Photon Energies of 20 keV, 100 keV, 0·5 MeV and 1·5 MeV.*
The radial distance indicates the relative number of photoelectrons at different angles.
(Whyte, G. N., *Principles of Radiation Dosimetry*, p. 27, Fig. 2.9. New York, Wiley, 1959)

Each line may be described by an equation of the type

$$\tau/\rho = k\lambda^3 + b \tag{3.47}$$

where τ is the linear photo-electric absorption coefficient, and as the lines meet near a common point, b is approximately constant for one element. The value of k is related to the mode of absorption.

Since τ_m, the photo-electric mass absorption coefficient, is equal to τ/ρ, and also equal to $\tau_a N_0/W$ where τ_a is the atomic photo-electric absorption coefficient, substituting in equation (3.47), we have

$$\tau_a = k\frac{W\lambda^3}{N_0} + b\frac{W}{N_0} \tag{3.48}$$

It is found [16] that kW/N_0 is nearly proportional to Z^4.
Hence, equation (3.48) may be written

$$\tau_a = CZ^4\lambda^3 + b \tag{3.49}$$

where C is a constant. The value of b is usually small compared with the other terms. It is evident from the formula that the photo-electric effect is unimportant for photons of high energy absorbed in elements of low atomic number.

The angular distribution of photo-electrons depends on the photon energies (Figure 3.7). Analysis of photo-electron energies can be made with a magnetic spectrograph in which electrons of one energy are brought to a focus [17].

PAIR PRODUCTION

If a high-energy photon passes near the nucleus of an atom, the photon may vanish and at least part of its energy be transformed into mass: a positive

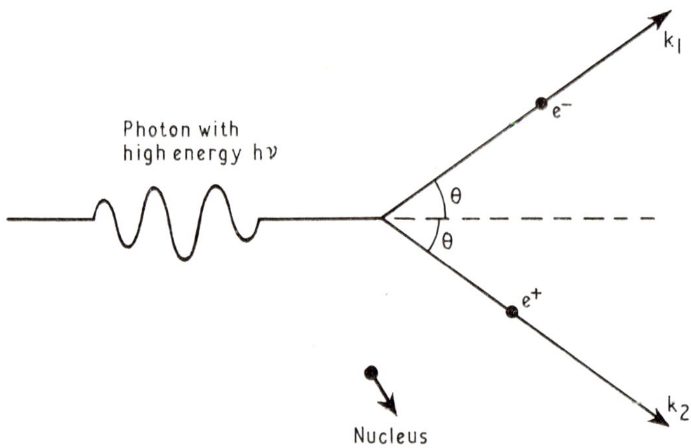

FIGURE 3.8 *Pair Production*

and a negative electron appear. The energy equivalent of the rest mass of an electron is 0·51 MeV, so that this interaction cannot take place unless the photon energy is at least 1·02 MeV. Excess photon energy can be shared between the pair. Thus

$$h\nu = 1{\cdot}02 + k_1 + k_2 \tag{3.50}$$

where k_1 and k_2 denote the kinetic energy of each electron, and are not necessarily equal.

Both negative electrons or *negatrons* and positive electrons or *positrons* lose kinetic energy in many ionizing interactions. After the positron has slowed down, it combines with a negatron to produce two 0·51 MeV gamma rays, which are emitted in opposite directions. This is known as annihilation radiation.

Pair production (Figure 3.8) can occur only in the Coulomb field of a nucleus, because the recoil of the nucleus is necessary to conserve momentum.

The cross-section for this process is therefore dependent on the atomic number of the material and is in fact very nearly proportional to Z^2. If π_a is the atomic cross-section (or atomic absorption coefficient) due to pair production, then π_a/Z^2 is independent of Z, except for the screening effect [5]. Pair production cross-section is very low at 1·02 MeV and then rises smoothly with increase of incident photon energy. Thus pair production is the important absorption process at high energies and for heavy materials.

The pair production absorption coefficients may be written

Atomic absorption coefficient $= \pi_a = \dfrac{\pi}{\rho}\dfrac{W}{N_0} = dZ^2$ cm²/atom
where d is a constant

Electronic absorption coefficient $= \pi_e = \dfrac{\pi}{\rho}\dfrac{W}{N_0 Z} = dZ$ cm²/electron

Mass absorption coefficient $= \pi/\rho$ cm²/g

Linear absorption coefficient $= \pi$ per cm

TRIPLET PRODUCTION

Triplet production can occur when the photon interacts with an electron and not with a nucleus. Three particles result, the original electron as well as the positron and negatron pair. The lower threshold of the reaction is 2·04 MeV. Its cross-section is usually small compared with that of pair production.

Partial Absorption Coefficients

An X-ray beam, on passing through matter, will in general, be attenuated by more than one of the processes described. The total linear absorption coefficient μ may therefore be divided into its constituent parts, σ, γ, τ, π, known as *partial absorption coefficients*. The partial absorption coefficients, or cross-sections, express the probabilities that a photon will undergo a Compton, Rayleigh, photo-electric or pair production interaction respectively.

In general, therefore,

$$\mu = \sigma + \gamma + \tau + \pi \qquad (3.51)$$

This form of equation applies to any of the other absorption coefficients (Figure 3.9) as well as the linear absorption coefficients.

It is often convenient to consider separately the scattered photons and the secondary electrons resulting from attenuation processes, for it is from the secondary electrons that energy is deposited in the absorbing medium.

Thus,

$$\mu = ({}_s\sigma + \gamma + {}_s\pi) + (\tau + {}_a\sigma + {}_a\pi) \qquad (3.52)$$

where the first bracket refers only to scattered photons and the second only to secondary electrons. When attenuation of the primary beam is considered, both scattered radiation and secondary electrons are relevant, and the absorption coefficients include both. Where energy absorbed or dose in the medium is the parameter of interest, only the *true* absorption coefficients should be included, and the scatter contribution is irrelevant.

PARTIAL ABSORPTION COEFFICIENTS OF MIXTURES

Sometimes a material may be required to have the same absorption properties as some other material: an ionization chamber must be "air-walled" (p. 62), or a bolus material has to be matched to soft tissue (p. 171).

From equation (3.17)

$$\frac{\mu}{\rho} = \left(\frac{\mu}{\rho}\right)' h' + \left(\frac{\mu}{\rho}\right)'' h'' + \ldots$$

where a substance having a mass absorption coefficient $\mu_m = \mu/\rho$ has constituent elements with parameters denoted by single dash, double dash,

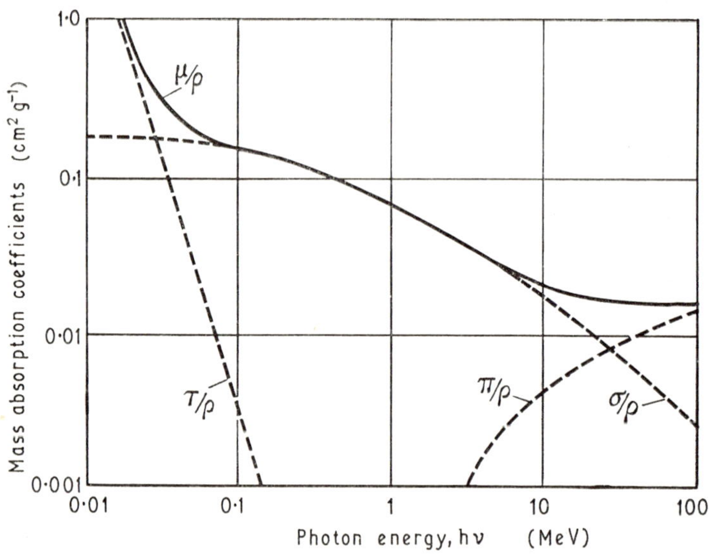

FIGURE 3.9 *Mass Absorption Coefficients For Air*

(After Whyte, G. N., *Principles of Radiation Dosimetry*, p. 38, Fig. 2.17. New York, Wiley, 1959)

etc., and h (with appropriate dashes) denotes the fraction of the total mass due to one of these elements.

Now, considering only the absorption, and omitting the scatter,

$$a\frac{\mu}{\rho} = \frac{h'(\tau' + {}_a\sigma' + {}_a\pi')}{\rho} + \frac{h''(\tau'' + {}_a\sigma'' + {}_a\pi'')}{\rho} + \ldots \quad (3.53)$$

$$= h'\left\{C(Z')^4\lambda^3\frac{N_0}{W'} + {}_a\sigma_e'Z'\frac{N_0}{W'} + \frac{dN_0(Z')^2}{W'} + \frac{bN_0}{W}\right\} + h''\left\{C(Z'')^4\lambda^3\frac{N_0}{W'} + \ldots\right\} + \ldots$$

Therefore, neglecting b which is small compared with the other terms,

$$a\frac{\mu}{\rho} = \frac{h'N_0}{W'}\{C(Z')^4\lambda^3 + {}_a\sigma_e'Z' + d(Z')^2\}$$

$$+ \frac{h''N_0}{W''}\{C(Z'')^4\lambda^3 + {}_a\sigma_e''Z'' + d(Z'')^2\}$$

$$+ \ldots$$

Now the electron density $= n_0$

$$= N_0 \sum \left(\frac{h'Z'}{W'} + \frac{h''Z''}{W''} + \ldots\right)$$

Therefore

$$\frac{{}_a\mu}{\rho} = n_0\{C\lambda^3(\alpha'(Z')^3 + \alpha''(Z'')^3 + \ldots]$$
$$+ [{}_a\sigma_e' + {}_a\sigma_e'' + \ldots]$$
$$+ d[\alpha'Z' + \alpha''Z'' + \ldots]\} \tag{3.54}$$

where $\alpha' = \dfrac{N_0}{n_0}\dfrac{h'Z'}{W'}$ and similarly for double, treble etc. dashes.

A more accurate empirical formula, written [18] in terms of the mass absorption coefficient is

$$\frac{\mu}{\rho} = N_0\left(\sigma_e + 17{\cdot}7\frac{\bar{Z}^{3{\cdot}4}}{E^{3{\cdot}1}} + 1{\cdot}12\frac{\bar{Z}^{1{\cdot}7}}{E^{1{\cdot}7}}\right)10^{-24}\ \mathrm{cm^2\ g^{-1}} \tag{3.55}$$

where E is the photon energy in keV and $Z < 20$.

Consider now restricted photon energy ranges. If photo-electric absorption is present, but photon energies are well above critical absorption energies, the effective atomic number of the mixture is given by

$$\bar{Z}^3 = \alpha'(Z')^3 + \alpha''(Z'')^3 + \ldots \tag{3.56}$$

The index is only approximately equal to 3 (p. 46), and is more often [19] taken as 2·94. However, this value of 2·94 is now considered to be too low and should be replaced [18] by 3·4.

For higher photon energies, where pair production is the only important process

$$\bar{Z} = \alpha'Z' + \alpha''Z'' + \ldots \tag{3.57}$$

From these two equations, it follows that there is no unique value of effective atomic number of a mixture which holds for all photon energies. For most materials, the photon energy range in which photoelectric absorption is important is separated from that energy range where pair production occurs (Figure 3.9). Thus, it may be sufficient to design a material with an effective atomic number for either one of these ranges. If this is not the case, constituent materials may be selected bearing in mind that the effective atomic number for the two energy regions will differ least when the constituent atomic numbers have similar values (Table 3.3). in the "matched" materials.

The relative importance of Compton cross-section varies with photon energy (Figure 3.9). The Compton process depends on the electron density $n_0 = Z/A$. This quantity varies only very slowly with Z, so that Compton absorption is virtually independent of effective atomic number. The value of n_0 (Table 3.3) can be calculated from the composition of the material.

TABLE 3.3 *Electron Density and Effective Atomic Numbers*
(From Hine, G. J. and Brownell, G. L., *Radiation Dosimetry*, New York, Academic Press, 1956)

Material	ρ(g/cm^3)	η_0(electrons/g)	$\bar{Z}$ *Photo-electric absorption*	$\bar{Z}$ *Pair Production*
Air	0·001293	$3{\cdot}03 \times 10^{23}$	7·64	7·36
Water	1·00	$3{\cdot}34 \times 10^{23}$	7·42	6·60
Muscle	1·00	$3{\cdot}36 \times 10^{23}$	7·42	6·60
Subcutaneous fat	0·91	$3{\cdot}48 \times 10^{23}$	5·92	5·2
Bone	1·85	$3{\cdot}00 \times 10^{23}$	13·8	10·0

The Energy Absorption Process

When X-rays are attenuated in matter, energy is not absorbed from scattered photons unless and until they undergo a further absorption process. The secondary electron is the agent through which energy is deposited in the absorbing material.

This high-speed electron loses its energy in many interactions which may take different forms: chemical reactions, elastic scattering and the emission of bremsstrahlung, excitation and ionization. In general the greater part of the energy of the electron is dissipated in the creation of ion pairs, and this energy ultimately appears locally as heat. This is not necessarily the case for chemical energy nor for energy emitted as bremsstrahlung.

In general, a high-speed electron loses a relatively small amount of energy at each interaction, so that the deflexions along its path are slight, the path becoming more tortuous as the electron slows down.

ENERGY LOSS FROM THE SECONDARY ELECTRONS

An electron ionizes by the direct action of its electric field on the electrons of the atom. At energies above about 1 MeV, the electric field becomes disturbed [20], contracting in the direction of motion and expanding in a transverse direction. Consequently, there is a slight decrease in mean free path (and an increase in ion density), as the electron energy increases above 1 MeV. The relationship becomes more marked as the energy increases further. At somewhat lower electron energies, the mean free path decreases with electron velocity, (equation 3.58).

The collision energy loss $-\mathrm{d}E_c$ is given by [20]

$$-\left(\frac{\mathrm{d}E_c}{\mathrm{d}l}\right) = \frac{2\pi e^4 NZ}{m_0 v^2}\left[\ln\frac{m_0 v^2 E}{2I^2(1-\beta^2)} - \ln 2\{2\sqrt{(1-\beta^2)} - 1 + \beta^2\} + \{1-\beta^2\} + \tfrac{1}{8}\{1-\sqrt{(1-\beta^2)}\}^2\right] \tag{3.58}$$

where N = number of absorbing atoms per unit volume
Z = atomic number of absorbing material
v = velocity of the electron
I = average ionization energy of the absorber
e = electron charge
E = kinetic energy of electron.

Thus, the collision energy loss per unit path length is proportional to NZ, the electron density. There is also a loss $\mathrm{d}E_r/\mathrm{d}l$ in electron energy due to radiation. This is approximately proportional [20] to E^2Z.

The total loss

$$\frac{\mathrm{d}E_c}{\mathrm{d}l} + \frac{\mathrm{d}E_r}{\mathrm{d}l} = \frac{\mathrm{d}E_s}{\mathrm{d}l}$$

known as the *stopping power, S, of a material for a charged particle.* It is more usually expressed [2] as the *mass stopping power*, S/ρ.

Although $\mathrm{d}Es$ denotes energy lost by ionization, electronic excitation and radiation, for some purposes it is desirable to exclude bremsstrahlung losses from stopping power. In these cases, S/ρ must be multiplied by an appropriate factor [2] less than unity.

The energy lost from the electron may not all be locally absorbed. That part which is so absorbed is called the *linear energy transfer L* given by [2]

$$L = \mathrm{d}E_e/\mathrm{d}l \tag{3.59}$$

where $\mathrm{d}E_e$ is the average energy locally imparted to a medium by a charged particle of specified energy, in traversing a distance $\mathrm{d}l$. The term "locally imparted" may refer either to a maximum distance from the track or to a maximum value of discrete energy loss by the particle. In either case, the limits chosen should be specified [2].

An electron released by the creation of an ion pair, may itself have energy sufficient to ionize. It is then known as a delta ray, and its track appears as a spur off the main one.

As the mean free path of an electron is small, often less than a micron, at least a part of the energy lost will be absorbed within the attenuating material. When the range of the secondary electron is small compared with the physical dimensions of the material, virtually all its energy is so absorbed. The range, when expressed in terms of the mass traversed, for example, grammes per square centimetre, is insensitive to the nature of the absorbing material. Thus, the range–energy curve for air (Figure 3.10) may be used for other substances.

The energy of the electron is continually, but irregularly decreasing along its track, with corresponding increases in the value of linear energy transfer (LET). Furthermore, an X-ray set will produce photons with a range of energies; one of these photons can, after a Compton interaction, result in an electron with any initial energy within a certain range. A further complication is the production of δ-rays along the track.

The result is that initial energies of secondary and tertiary electrons from an X-ray beam form a spectrum very much more complex than that of the X-ray photon energies. In spite of this, it is sometimes useful to specify the mean LET of a certain radiation in a certain medium. But a knowledge of the distribution of LET is of more interest. Its most useful form is a spectrum of the fraction of the absorbed energy per unit interval of LET, plotted

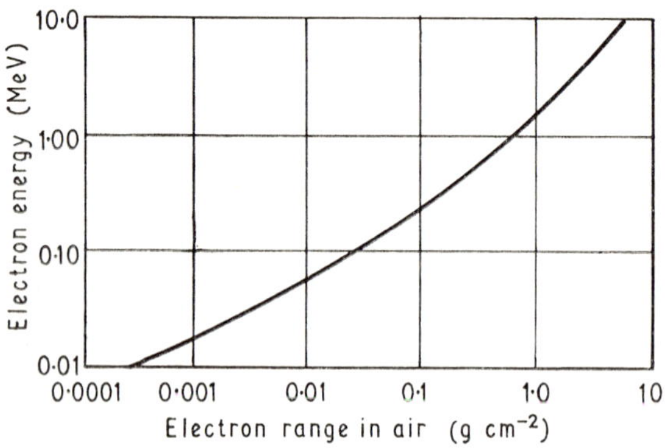

FIGURE 3.10 *Approximate Maximum Range–Energy Curve for Electrons in Air*
From calculations by Nelms.
(*Physical Aspects of Irradiation*, Recommendations of the ICRU, NBS Handbook 85. Washington, D.C., NBS, 1964)

against LET. Such a curve may be obtained experimentally or by calculation [21].

LET is conveniently expressed in terms of keV/mμ, i.e. keV/nm.

The energy absorbed in a mass of material, rather than along the track of an ionizing particle is often the relevant consideration. This is the absorbed dose D, given by

$$D = \frac{\Delta E_D}{\Delta_m} \tag{3.60}$$

where ΔE_D is the energy imparted by ionizing radiation to the matter in a volume element and Δm, the mass of the matter in that volume.

The special unit of absorbed dose [2] is the rad, where

$$1 \text{ rad} = 100 \text{ erg/g} = \frac{1}{100} \text{ J/kg} \tag{3.61}$$

Absorbed dose is a macroscopic concept, and can be applied only when a statistically large number of ionizations occur in the volume element. Therefore, it breaks down for very small volumes. The critical volume depends on the dose rate: it is smaller for high dose rates than for lower ones.

The effect of a certain radiation dose often depends on the LET of the radiation. It would be inappropriate to add doses with different effectiveness and the term *dose equivalent* (DE) has been introduced [2] where

$$\mathrm{DE} = D \times \mathrm{QF} \times \text{any other modifying factors.}$$

Dose equivalents can be added to give a meaningful measurement of radiation energy absorption. It is measured in *rems*. QF is a quality factor which is correlated [22] with LET (Table 3.4). It is used in radiation protec-

TABLE 3.4 *Relationship between LET and QF*
From Neary, G. J., *Phys. Med. Biol.*, 7, p. 419, 1963.

LET keV *per micron of water*	*QF*
3·5 or less	1
3·5 to 7	1 to 2
7·0 to 23	2 to 5
23 to 53	5 to 10
53 to 175	10 to 20

tion and is unity for X-radiation of all energies. The equivalent term for use in radiobiology is *radiobiological equivalent* (RBE) and is evaluated with more precision than is the QF.

A distinction must be drawn between the energy absorbed in a volume element (absorbed dose D), and the energy removed from the beam in that same volume element. The latter has been named [2] *kerma*, K. It is derived from initial letters of "kinetic energy released in material" and is given by

$$K = \frac{\Delta E_k}{\Delta m} \tag{3.62}$$

where ΔE_k is the sum of the initial kinetic energies of the charged particles liberated by the indirectly ionizing particles in a volume element of mass Δm. Kerma is closely related to the röntgen (p. 55) and to the "first collision dose" used in neutron dosimetry.

Kerma and absorbed dose are both measured in terms of energy per unit mass. The dose absorbed in a volume element may include only part of the energy liberated from charged particles originating outside the volume element. On the other hand, kerma is the sum of the kinetic energies of all the charged particles originating in the volume element. In special cases, such as the uniform irradiation of an infinite medium, absorbed dose and kerma may be numerically equal. In practice, they often differ widely, particularly for high energy radiation [23].

An important quantity in radiation dosimetry is W, the average energy required to produce an ion pair. The value of W in air is 33·73 ± 0·15 eV [2]. An electron with an initial energy of, say, 0·1 MeV can therefore produce several thousand ion pairs. Such an electron will have a range of about 0·017 g/cm² or 13 cm in air. After ionization, the atom is left in an excited state, and the return to ground state will be accompanied by the emission of ultra-violet light.

Literature References

1 LONGHURST, R. S., *Geometrical and Physical Optics* (London, Longmans, 1960).

2 INTERNATIONAL COMMISSION on RADIOLOGICAL UNITS AND MEASUREMENTS, *Radiation Quantities and Units*, ICRU Report 10(*a*), 1962. NBS Handbook 84, (Washington, D.C. Nat. Bur. Stand., 1962).

3 HEITLER, W., *The Quantum Theory of Radiation* (Oxford, Clarendon Press, 1954).

4 KLEIN, O., and NISHINA, Y., "Uber die Streuung von Strahling durch freie Elektronen nach der neuen relativistischen Quantendynamik von Dirac", *Z. Phys.*, **52,** p. 853 (1928).

5 WHYTE, G. N., *Principles of Radiation Dosimetry* (New York, Wiley, 1959).

6 HINE, G. J. and BROWNELL, G. L., *Radiation Dosimetry* (New York, Academic Press, 1956).

7 KIRKPATRICK, H. A. and DUMOND, J. W. M., "Shape of the modified Compton line for hydrogen and Ceylon graphite scatterers", *Phys. Rev.*, **54,** p. 802 (1938).

8 COMPTON, A. H. and ALLISON, S. K., *X-rays in Theory and Experiment* (New Jersey, Van Nostrand, 1960).

9 FERMI, E., "Eine statistiche Methode zur Bestimmung einiger Eigenschaften des Atoms and ihre Anwendung auf die Theorie des periodischen Systems der Elemente", *Z. Phys.*, **48,** p. 73 (1928).

10 HARTREE, D. R., "The wave mechanics of an atom with a non-Coulomb central field, Part I: Theory and Methods", *Proc. Camb. Phil. Soc.*, **24,** p. 89 (1927–28).

11 HERTZ, H., "Uber einen Einfluss des ultravioletten Lichtes auf die electrische Entladung" (XIV), *Ann. Phys. u. Chem.*, **31,** p. 983 (1887).

12 EINSTEIN, A., "Uber einen die Erzeugung und Verwandlung des Lichtes betreffenden heuristischen Gesichtspunkt", *Ann. Phys.*, **17,** p. 132 (1905).

13 HALLWACHS, W., "Uber den Einfluss des Lichtes auf electrostatisch geladene Körper", *Ann. Phys. u. Chem.*, **33,** p. 301 (1888).

14 THOMSON, J. J., "Cathodc rays", *Phil. Mag.*, **44,** p. 293 (1897).

15 RICHARDSON, O. W., and COMPTON, K. T., "The photo-electric effect", *Phil. Mag.*, **24,** p. 575 (1912).

16 RICHTMYER, F. K., KENNARD, E. H. and LAURITSEN, T., *Introduction to Modern Physics* (New York, McGraw-Hill, 1955).

17 ROBINSON, H. and RAWLINSON, W. F., "The magnetic spectrum of the β-rays excited in metals by soft x-rays", *Phil. Mag.*, **28,** p. 277 (1914).

18 WEBER, J. and VAN DEN BERG, D. J., "The effective atomic number and the calculation of the composition of phantom materials", *Brit. J. Radiol.*, **42,** p. 378 (1969).

19 MAYNEORD, W. V., "The significance of the roentgen", *Acta Un. int. Cancr.*, **2,** p. 271 (1937).

20 BETHE, H. A. and ASHKIN, J. A., *Experimental Nuclear Physics*, Ed. SEGRE, E. (New York, Wiley, 1953).

21 INTERNATIONAL COMMISSION ON RADIOLOGICAL UNITS AND MEASUREMENTS, *Physical Aspects of Irradiation*, Recommendations of the ICRU, Report 10(*b*), 1962. NBS Handbook 85 (Washington, D.C., Nat. Bur. Stand., 1964).

22 NEARY, G. J., "The significance of the rem and parameters on which it depends", *Phys. Med. Biol.*, **7,** p. 419 (1963).

23 BOAG, J. W., "Radiological quantities and units, their inter-relationships and conditions of use", *Phys. Med. Biol.*, **7,** p. 409 (1963).

4 Measuring Methods and Instruments

In the measurement of X-rays, the objective is usually the determination of the energy absorbed in a material. This depends on both the output from the X-ray set and on the nature of the absorbing material. It may be convenient to measure output and then relate it to absorbed energy or dose in any particular material.

Output is measured in terms of exposure, which is the ionization produced in air under specified conditions. Although definition and terminology have changed over the years, measurement of exposure has proved to be both useful and capable of high sensitivity. It is supplemented by more direct measurements of absorbed dose and by various radiation-sensitive devices, each of which has its own particular advantages for comparative measurements.

Exposure

The energies of X-ray photons and their rate of emission from an X-ray tube depend on the voltage across the tube and the current through it, as well as on the design features such as the target material and geometry and the inherent filtration of the tube. The energy spectrum of photon emission is complex, and although a specification in these terms would describe completely the radiation, it is not generally practicable nor essential. The "quality" of the radiation is often given as its HVL (p. 89), but this does not specify a spectral distribution entirely unambiguously. It has been shown [1] that the approximate spectral distribution can be derived from two of the three parameters: first HVL, second HVL and kVp of the tube.

The radiation exposure X is defined [2] in terms of the ionization it produces in air:

$$X = \frac{\Delta Q}{\Delta m} \tag{4.1}$$

where ΔQ is the sum of the electrical charges of one sign produced when all the electrons (negatrons and positrons), liberated by photons in a volume element of air whose mass is Δm, are completely stopped in air. The special unit is the röntgen (R), where

$$1\text{R} = 2{\cdot}58 \times 10^{-4}\ \text{coulomb/kg} \tag{4.2}$$

It is numerically identical with the old röntgen, defined as 1 e.s.u. of charge per 0·001293 g of air.

When exposure rate, rather than integrated exposure, is of interest

$$1 \text{ R/sec} = 2{\cdot}58 \times 10^{-4} \text{ A/kg} \tag{4.3}$$

As the mass of air in an ionization chamber is usually a small fraction of a gramme, it is evident that ionization currents, too, will be small; several orders of magnitude less than one micro-ampere are usual.

The average energy W required to produce an ion pair in air [3] is $33{\cdot}73 \pm 0{\cdot}15$ eV. An electron with an initial energy of, say, 0·1 MeV can therefore produce several thousand ion pairs. Such an electron will have a range of about 0·17 g/cm^2 or 13 cm of air.

MEASUREMENT OF THE RÖNTGEN

It is convenient to measure ionization in a defined volume rather than in a defined mass of air. Therefore, before conversion to röntgens, corrections [4, 5, 6] must be made for temperature and pressure of the air. As water and dry air molecules have different stopping powers, a humidity correction [3] also is required.

The Free-air Ionization Chamber

The free-air ionization chamber is one without a confining wall to the air volume in which the ionization is measured.

Measurements of x-ray exposure in terms of the röntgen demands that the ionization associated with all the corpuscular emission should be measured. Now the range of secondary electrons (Table 4.1) is, in general, larger

TABLE 4.1 *Approximate Range of Electrons in Air*

Electron energy (keV)	*Range in air* (cm)
1	$3{\cdot}9 \times 10^{-3}$
10	$1{\cdot}9 \times 10^{-1}$
100	13
1000	$9{\cdot}3 \times 10^{2}$

than the dimensions of the air volume, the so-called *sensitive volume*, in which the exposure is being measured. Indeed for x-ray energies above about 3 MeV, the range of secondary electrons is so long that measurement of exposure in röntgens is impracticable and the röntgen is not defined above this energy. It is difficult also to measure at energies below a few keV [2]. Absolute measurement of the röntgen may be made using a free-air ionization chamber for energies between a few keV and about 500 kVp. Such measurements are carried out in national standardizing laboratories.

Consider a sensitive volume whose dimensions are less than the maximum range of the secondary electrons (Figure 4.1). Some electrons A, originating inside the sensitive volume, will cause some ionization outside this volume, and this ionization will be lost to the measurement. Now surround the sensitive volume with air exposed to the same intensity of radiation and of dimensions greater than the maximum range of the secondary electrons; the loss in ionization referred to above will now be exactly compensated for by a gain from secondary electrons *B* originating outside the sensitive volume, but causing some ionization inside it. This is an illustration of the concept of electronic equilibrium.

A free-air ionization chamber (Figure 4.2) has a collimated beam of x-radiation passing between two parallel plates. The edges of the beam between the plates are somewhat diffuse and therefore the defining area of the beam is taken to be that at the collimating diaphragm *D*. The cross-sectional area at *D* will be appropriate, and electronic equilibrium will be

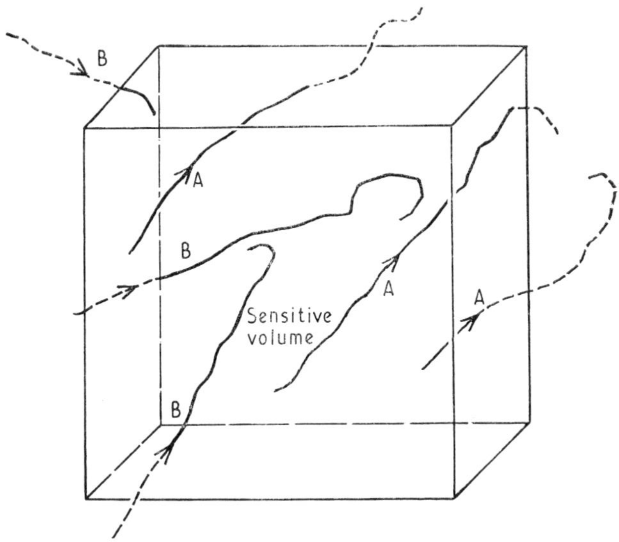

FIGURE 4.1 *Concept of Electronic Equilibrium*
The irregular lines are electron tracks. The part of the track outside the sensitive volume is shown dotted.
The ionization from those tracks (*A*) emerging outside the sensitive volume (but originating inside it) is balanced by the ionization from the tracks (*B*) entering the sensitive volume (but originating outside it.)

maintained if the beam is not significantly attenuated during its passage from *D* through the chamber and if *D* is at a distance from the centre of the chamber great enough to prevent any secondary electrons from the collimating diaphragm reaching the collector plate. In addition, the plates must be far enough [7, 8, 9] from the x-ray beam to ensure that the secondary electrons have spent all their ionizing energy before they are collected.

In the arrangement [10] shown in Figure 4.2, negative ions formed by free electrons attaching themselves rapidly to oxygen molecules, are attracted to the positive (high voltage) plate and positive ions travel to the collector plate. Immediately before the measurement, the slider *R* is moved to *A* and the switch *S* closed; thus the voltage across the capacitor *C* and that across the electrometer are both zero. The switch *S* is opened as the exposure commences. Positive ions collected cause the collector plate voltage to rise and the electrometer to start to deflect; it is, however, returned to its null position by moving the slider *R* towards *B*. This adjustment is continuous throughout the exposure, so that the voltage *V* across *C* is always equal and opposite to that across the potentiometer *RA*. Thus, at the termination of the measurement, the position of the slider *R* is a measure of the voltage across the capacitor *C* and hence also of the charge collected. At the same time the collector plate is kept at earth potential.

The length *PQ* of the sensitive volume is defined by the lines of force of the electrostatic field between the plates. The collector plate is closely surrounded

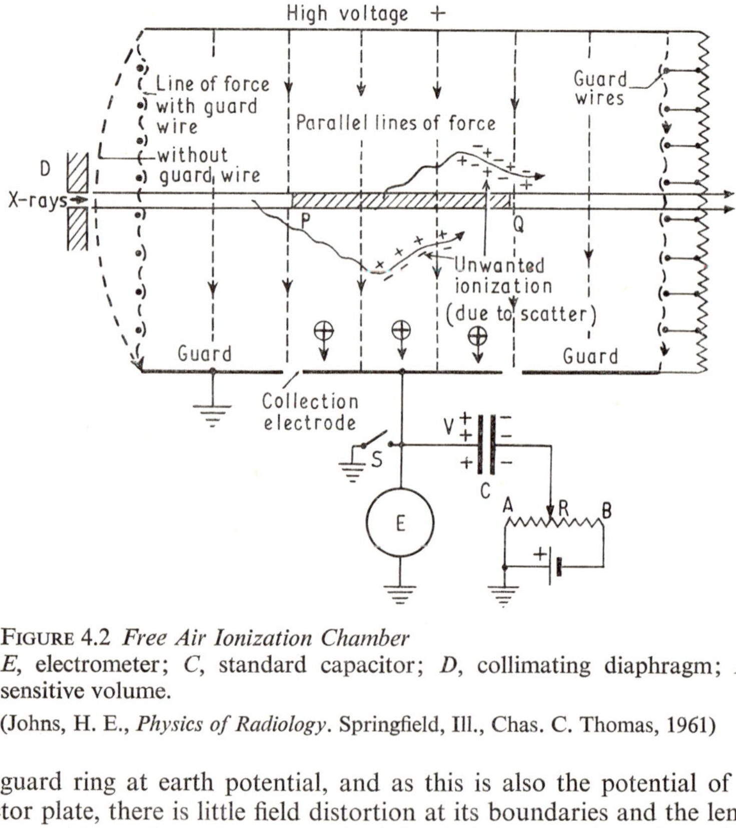

FIGURE 4.2 *Free Air Ionization Chamber*
E, electrometer; *C*, standard capacitor; *D*, collimating diaphragm; *PQ*, sensitive volume.

(Johns, H. E., *Physics of Radiology*. Springfield, Ill., Chas. C. Thomas, 1961)

by a guard ring at earth potential, and as this is also the potential of the collector plate, there is little field distortion at its boundaries and the length of the sensitive volume is accurately defined, being equal to that of the collector plate. Field distortion in parallel plate and other ionization chambers can be further reduced by guard wires placed between the plates [11, 12, 13].

The ionization measured will be in terms of the röntgen, only if all the associated corpuscular emission contributes, of if there is exact compensation for losses. Besides losses due to the long range of secondary electrons which have already been considered, there are a number of other ways in which ionization may be lost.

Recombination

Negative ions and electrons each tend to recombine with positive ions. The number dn of ions per unit volume which combine in the time dt is proportional to the square of the number n of ions present. Thus

$$\frac{dn}{dt} = \alpha n^2 \tag{4.4}$$

where α is defined as the coefficient of recombination. Integrating,

$$\frac{1}{n} - \frac{1}{n_0} = \alpha t \tag{4.5}$$

where n_0 is the initial number of ions and n the number at time t. The value of α is much lower for electrons than for negative ions [14].

The effect of recombination may be seen by lowering the voltage across a parallel plate ionization chamber (Figure 4.3). The ions will then take longer

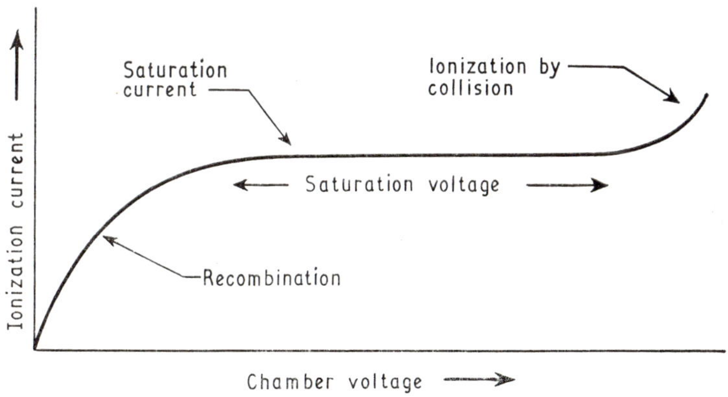

FIGURE 4.3 *Variation of Ionization Current with Chamber Voltage*

to cross the chamber and consequently have a greater chance of recombining. This will result in a decrease in the ionization current. Dust or other suspended particles increase the rate of recombination, probably because the particles carry charges; Owens [15] found that when tobacco smoke was blown between the parallel plates of an ionization chamber the current diminished to a small fraction of the saturation current, and it recovered as the smoke settled or disappeared. Local high ionization density along the track of densely ionizing radiation may cause recombination at comparatively high plate voltages. This can be an important source of recombination [16].

At the other extreme, if the voltage is increased too much, electrons will acquire an energy high enough to ionize neutral molecules with which they collide. This is known as ionization by collision (Figure 4.3), and is a random process between ions from different tracks.

For measurement of the röntgen, the voltage should be that for saturation current.

Formation of Negative Ions

As negative ions are more likely than electrons to combine with positive ions, the presence of many negative ions may lead to significant recombination at a chamber voltage high enough for saturation current for electrons.

The number of collisions necessary for an electron to form a negative ion varies with the type of gas; it is about 10^3 for the halogen gases, about 10^4 for oxygen and water vapour and 10^6 or more for argon, hydrogen, nitrogen and carbon monoxide [17]. Thus, the rate of negative ion formation depends on the mean free path of the electron. Negative ion formation is not generally a serious source of inaccuracy in x-ray measurements, because the density of gas molecules is low and hence the mean free path large. However, at high pressures the mean free path may be so short that the large number of collisions may result in an appreciable amount of negative ion formation with consequent heavy recombination. For example, at a gas pressure of 20 atmospheres, the voltage required for saturation current is appreciably lower for an ionization chamber filled with argon than for one filled with air [18].

Diffusion

In the absence of an electric field, positively and negatively charged particles move, in general, at random. However, if the charge distribution is uneven a *diffusion current* will flow: a positive ion will be repelled from a local concentration of positive ions and attracted to one of negative ions; conversely for a negative ion. If an ionization chamber has walls which are conducting, some ions will come into contact with the walls and give up their charges, so that the concentration of ions will decrease towards the walls; this pressure gradient for the ions will result in a steady diffusion drift towards the walls. Such diffusion currents will be at the expense of the measured ionization current, although in an ordinary ionization chamber, the loss by diffusion is usually negligible compared with that by recombination [19].

Diffusion is proportional to the product of temperature in degrees Kelvin and ionic mobility k, where k is the ionic velocity per unit electric field producing it; in the same circumstances, therefore, diffusion is greater for electrons than for heavy ions.

Leakage Current

Ionization currents in the measurement of the röntgen are small, and therefore special precautions must be taken so that leakage currents do not invalidate the measurement.

Leakage currents can arise across the insulators supporting either plate. However, a leakage current from the high voltage plate (Figure 4.2) to earth does not affect the measurement, but merely provides a minute drain on the power supply. But a leak across the collector plate insulator to earth provides a path in parallel with the measuring system. Consequently the insulator supporting the collector plate must be of high quality. Even so, slight contamination, moisture or exposure to radiation may materially decrease the resistance of the insulator. For this reason, an ionization chamber should have a guard ring held very closely at the same potential as the collector plate [20], placed so that there is no direct path across the insulator surface to the earthed case [16]. In the arrangement shown in Figure 4.2, the collector plate is held very close to earth potential and the guard rings are earthed, so that the potential difference between the collector plate and its surroundings is small, and, consequently, so is any leakage current.

Walled Chambers

The free-air ionization chamber used for standardization has a large overall size, and is usually unsuitable for routine use, although Greening [21] has described a free-air chamber for measuring under clinical conditions the output of X-ray sets of 10 to 50 kVp. The radiation intensity must be measured in a defined volume, and this implies that, except in free-air chambers, the ionization chamber must have a confining outer wall. This outer wall forms one electrode of the chamber and therefore must have a conducting inner surface. The collecting electrode is placed inside the wall, and is highly insulated from it.

Wall Thickness

Consider a large volume of air exposed to a uniform X-ray intensity, and a very thin-walled ionization chamber within it. The range of secondary electrons originating within the chamber will, in general, be larger than the chamber dimensions and the considerations described earlier in the chapter and illustrated in Figure 4.1 (p. 57) will apply. Electronic equilibrium will be achieved provided that the surrounding air volume has dimensions greater than the maximum range of the secondary electrons.

These conditions are not achieved in practice, and therefore it is the custom to make the wall of the chamber thick enough for electronic equilibrium to be established within the wall [3]. The uniformity of X-ray intensity is then relevant only over the external dimensions of the chamber. A further advantage is that electrons scattered from intervening objects cannot penetrate to the chamber gas, although photons so scattered can produce ionization within the chamber, and it is therefore advisable to remove scattering materials from the neighbourhood.

In such a thick-walled chamber, nearly all the electrons contributing to the ionization current arise within the chamber wall, and it cannot be assumed that a measurement will necessarily give the exposure in röntgens. A number of corrections may have to be made.

A wall thick enough for electronic equilibrium will attenuate the primary photons [22] and a correction must usually be made for this. Gray [23] has shown that, with such chamber walls, the ionization is a measure of the photon flux at the centre of the chamber, and the photon flux in the absence of the chamber is the quantity normally required.

The distance separating chamber and X-ray source is unlikely to be long enough for electronic equilibrium to be established in the intervening air. Consequently, in a uniform X-ray flux, as the front wall of the chamber is thickened, the number of secondary electrons in the chamber will increase, and so will the ionization current, until electronic equilibrium is achieved. However, as the X-ray flux is not uniform, and is higher nearer the X-ray source, this increase can be slightly modified by some secondary electrons arising from the chamber wall instead of from air which is in an area of higher X-ray flux.

In accurate determinations of the röntgen, a narrow beam of X-rays is used in order to reduce complications arising from scatter in broad beams [24]. When the chamber wall is thick, multiple absorption processes can increase the measured ionization. For instance, a Compton photon arising in the front wall of the chamber and scattered in the forward direction can be absorbed within the inner part of the chamber wall to give a secondary electron which enters the cavity. It is possible that this phenomenon can lead to an additional wall thickness having an apparently *negative absorption.* Thus, a 4 mm Perspex cap is sufficient for electronic equilibrium at 2 MV. If it is added to a chamber used for 250–290 kVp x-rays (2·5 mm Cu HVL to 3·4 mm Cu HVL) the effective absorption is negligible, but it is −1·1 per cent for radiation of 170 kVp (1·0 mm Cu HVL) and −1·9 per cent for

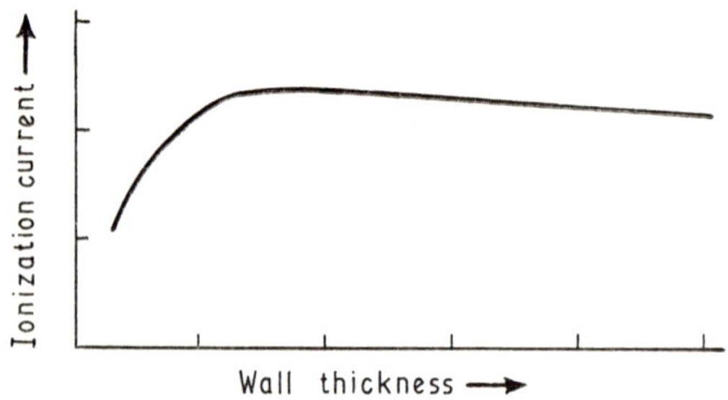

FIGURE 4.4 *Variation of Ionization Current with Chamber Wall Thickness*

radiation of 145 kVp (0·5 mm Cu HVL) [24]. Thus, the same thickened cap may be used with accuracy for X-rays generated at voltages ranging from 250 kVp to 2 MV. This is an exception, justified by a knowledge of all the variables, to the general recommendation that the chamber wall should be only thick enough for both electronic equilibrium and adequate mechanical strength [3]. For 4 MV X-rays, this [25] is about 1·5 g/cm^2.

In general, as the chamber wall is thickened, the ionization current will increase until electronic equilibrium is reached and then it will decrease slowly because of attenuation within the walls (Figure 4.4). The peak will not necessarily occur at the wall thickness required for electronic equilibrium. Indeed, a peak may not be present [26] if, for instance there are scattering materials in the vicinity. The accurate measurement of exposure using a thimble chamber demands an analysis of the variables contributing to the ionization, and if this is done, agreement between it and a standard free-air chamber measurement can be achieved [27].

Material of Chamber Wall

If exposure is to be measured in röntgens using a thick-walled chamber, the absorption coefficient for each absorption process of the X-ray photons, and the stopping power and energy spectrum for the secondary electrons, should be the same for the material of the wall as for air. The chamber is then *air-walled*; it should have the same composition as air. In fact, it is not practicable to make a chamber of exactly the same composition as air. The inclusion of elements not present in air make it impossible to design a chamber which is air-walled for all X-ray photon energies, but a close approximation may be achieved which is valid for a range of energies (Chapter 3). A chamber may be tested for its air-walled properties by filling it with air at different pressures. The pressure should range from a value so low that virtually all the electrons arise in the chamber wall to a value so high that most of the electrons arise in the gas cavity. If the ionization measured shows no discontinuity, the chamber is air-walled.

The wall material should consist only of elements having atomic numbers close to those present in air. Carbon or various plastic materials are commonly used. Sometimes the inner electrode may be of aluminium and the outer wall of a different material designed to give an overall effectively air-walled chamber. Non-conducting materials may be coated with colloidal graphite to make the electrodes conducting.

Walled chambers should be calibrated for a range of energies covering those for which they are used. Such a calibration enables the measuring instrument used with its chamber to read exposure in röntgens for a particular energy of incident radiation, but correction factors must usually be applied for other energies. If the effective atomic number of the wall is lower than that of air, the chamber will give too low an exposure at lower than calibration X-ray energies, when the photoelectric effect is more important, and vice versa.

More than 90 per cent of the total ionization from 250 kVp X-rays arises from photon absorption in an inner wall skin 0·03 mm thick. In spite of the small contribution from photoelectric absorption at this energy, any steel dust remaining on the wall from the manufacture of the chamber may materially increase the ionization current. At 2 MV, the photoelectric absorption is much smaller still and electrons arise from deeper in the chamber wall, and this type of contamination has little effect [27]. Consequently, two chambers which behave identically at 2 MV may not do so at 250 kVp owing to metal dust contamination.

Thin-walled chambers for low-energy X-rays may have an inflated balloon [28] as the outer wall. A balloon of carbon loaded with latex rubber is conducting and very nearly air equivalent. High sensitivity is obtained as the balloon can have a large volume (over 3 litres). The excess pressure for inflation is only 6 mm Hg, so that the filling gas is very nearly atmospheric air. For soft X-rays of 0·03 mm Al HVT the sensitivity is about 77 per cent of that for X-rays of 1·0 mm Al HVT. Polar characteristics are good.

Insulators

High-quality insulating materials must be used in ionization chambers so that leakage current is small compared with ionization current. However, ionizing radiations induce conductivity in many insulating materials [29]. After the end of exposure, recovery is slow, often taking some hours, the recovery time depending on the nature of the insulator, the dose rate and the duration of radiation. In selecting an insulator, more weight may have to be given to recovery time than to the value of the static conductivity. Polystyrene is recommended for continuous irradiation, and amber for intermittent use (Fowler, J. F., personal communication).

The problem is different for chambers used to measure low doses, as in protection work. Here it is important to produce insulators with clean surfaces, often under mass production [30] conditions.

It may be necessary to keep the surface of insulators dry, clean and unscratched; they may be built in desiccated compartments or coated with, for example, ceresin wax.

Condenser Chambers

Ionization chambers may be used to measure the exposure rate as ionization current, or they may measure total exposure during the time for which they are exposed. In the latter case the chamber is initially charged, and the loss of charge due to induced conductivity of the chamber gas is measured after exposure; this type is known as a condenser chamber. The measuring system may, or may not, be part of the instrument; sometimes it is permanently attached to the chamber, in which case it will form part of the condenser system.

If a condenser chamber is charged to a voltage V, and during exposure loses a voltage $\mathrm{d}V$, then if the loss of charge is $\mathrm{d}Q$,

$$\mathrm{d}Q = C\,\mathrm{d}V + V\,\mathrm{d}C \tag{4.6}$$

where C is the combined capacity of chamber and measuring instrument during measurement, and $\mathrm{d}C$ is the change in this capacity due to the change in charge, when the changes are small. Now $\mathrm{d}Q$ is proportional to the exposure in röntgens (R). Hence unless C is constant, a calibration of the instrument in terms of R and $\mathrm{d}V$ will not be linear; this is often evident in string electrometers.

In systems where the capacity is constant [21, 32, 33, 34, 35, 36, 37, 38], $\mathrm{d}Q$ will be proportional to $\mathrm{d}V$, and it may be convenient to express the sensitivity in volts (lost) per röntgen. As the volume and not the mass of air is constant, the sensitivity is accurately constant only under conditions of standard temperature and pressure.

Cable-connected X-ray Dosemeters

A versatile instrument[1] incorporating an ionization chamber connected by cable to a battery-operated valve electrometer [39, 40] can measure exposure from 0·45 mR to 150 mR full scale or exposure rate 0·045 mR/sec to 15 R/sec. full scale.

Other ionization meters, e.g. the Ionex[2], are available, some with very small chamber volumes, necessary because it has been shown that large chambers give anomalous percentage depth doses at small depths [41, 42].

In this type of dosemeter, the polarizing voltage is often taken to the wall of the chamber, making it necessary to surround the chamber with an earthed screen. Including the inner electrode, three connexions are necessary. A simpler system is to place the battery between the central electrode and the input of the measuring instrument [43]. The wall of the chamber can then be earthed.

Sub-standard X-ray Dosemeters

The output of an X-ray set may be accurately measured with a sub-standard dosemeter such as that designed by Farmer [44]. The ionization chamber is in the form of a thimble chamber on the end of a light Dural stem. It is permanently connected to the grid of the electrometer valve of the measuring instrument. It may be recharged and measured remote from the chamber position, as the cable is long. The response is energy-independent for HVL of 0·1 mm Cu to 3 mm Cu. Models are available with full scale readings of 60 R, 120 R, 600 R or 1,200 R. Valve drift during radiation exposure is compensated for in the d.c. amplifier; readings are repeatable to 0·2 per cent, and the instrument has proved to be stable and reliable in operation.

The Victoreen Condenser *r*-meter[3] has a somewhat similar thimble chamber available in several sizes, but it is built on a heavy metal stem which can cause scatter and shadow effects [45].

Parallel-plate Monitoring Chambers

In radiotherapy, it is desirable to have a continuous indication of X-ray output, and this can be provided by a parallel-plate monitoring chamber placed either between the X-ray tube and the filter or between the filter and the patient [26], and with a meter on the control desk. Two parts of the filter may be insulated and used to form the monitor walls [46]. These monitors are often not very stable in performance but are useful in detecting faults; they are not intended as calibrated dosemeters.

Safety considerations have led to an increasing practice of measuring the dose rate from diagnostic X-ray sets and Reisma [47] has designed a parallel plate ionization chamber to measure the exposure of an X-ray beam of from 2 mm Al HVL to 7 mm Al HVL.

Special difficulties arise in measuring very low-energy X-radiation [20] owing to absorption in the chamber walls; in addition, most absorption at those energies is by the photoelectric process, and as this absorption varies as the cube of the atomic number, any small difference in effective atomic number of air and wall material will have a large effect. However, a chamber designed empirically has been made to be substantially energy independent between 0·02 mm AL HVL and 1·0 mm AL HVL (grenz rays) [48].

In this energy range "open-air" chambers are the only ones free from wall effect [49]; these have guard wires to make the electrostatic field uniform over the length of the collector electrode. The wall correction for nylon chambers is, however, not high for X-rays generated at 10 to 15 kVp. Graphited-Cellophane walls may be used also [50].

Extrapolation Chambers

Ionization chambers have a finite thickness. If the exposure is varying rapidly with distance, chambers may be too thick to measure the dose at a place of interest. This is often the case at the surface of the skin or of a phantom, where back-scattered radiation decreases rapidly with distance above the surface.

Extrapolation chambers have been designed to evaluate exposure in such cases [51]. One plate of the chamber is usually on the surface and the second at a variable distance above it. A guard ring is included. The ionization current is measured for a number of spacings between the plates, and a graph plotted showing the relationship between ionization current per unit volume and plate separation. This curve is extrapolated to zero separation to give the ionization current density, and hence the exposure at the surface.

Extrapolation chambers are unsuitable for clinical use, because there is seldom the opportunity to take a series of readings for one determination; they are laboratory instruments [52].

Absorbed Dose

When material receives a certain exposure X, the dose absorbed is proportional to X and also to a factor which depends on the properties, particularly the effective atomic number, of the material in a manner appropriate to the particular energy of the radiation.

MEASUREMENT OF THE RAD

It often happens that a particular dosemeter is made of material different from the material of interest, and consequently the dose absorbed in the two is different. It may well be, too, that the measured effect results from only a small fraction of the energy absorbed in the device, so that a calibration must be included.

Two methods do not suffer from these limitations. In the first, the energy absorbed is assumed to appear eventually as heat, and the resulting temperature rise measured. This is the calorimetric method. In the second, the ionization is measured in a gas cavity in the material, and the Bragg–Gray principle shows that, under certain conditions, the ionization in the cavity can be used to measure the energy absorbed in the material, if certain other parameters are known.

The Calorimetric Method

Although a small fraction of radiation energy absorbed may sometimes give rise to chemical changes, it is usually justifiable to assume that all the energy absorbed appears finally as heat.

The magnitude of radiation dose of practical interest ranges from a fraction of a rad up to a few megarads. One calorie is equivalent to $4{\cdot}2 \times 10^7$ ergs, and this is the quantity of heat required to raise the temperature of 1 g of water by 1 °C. Consequently, if water absorbs a dose of 1 Mrad, the rise in temperature will be only about 2·4 °C, and proportionately for other doses. Thus, the calorimetric method [53, 54] of measuring dose is technically difficult. However, changes in temperature down to 10^{-3} °C and less can be measured by modern methods [55], and accurate measurements down to 100 rad have been made using a carbon calorimeter [56]. Dosage from x-radiation of below 30 keV has been measured by calorimetry [57].

A limitation of the calorimetric methods is that it measures dose in a small isolated mass and not in a segment of an extended medium [3].

Bragg–Gray Principle

Under certain circumstances, the ionization in a small gas cavity surrounded by an absorbing medium may be used as a measurement of the energy absorbed in that medium. Thus, an ionization chamber measurement may be used to evaluate absorbed dose.

Consider such a small cavity. If it is small enough, the number, energy and direction of secondary electrons arising in the absorbing medium but traversing the cavity will be the same as they would have been before the introduction of the cavity; furthermore, the electrons arising from absorption processes of incident photons in such a small cavity will be negligible in number compared with those arising in the surrounding medium [58].

Fano [59] has shown that "in a medium of given composition exposed to uniform primary radiation intensity, the flux of secondary radiation is also uniform and independent of the density of the medium as well as of density variations from point to point". Thus, provided medium and cavity are matched, the latter need be small enough only to maintain uniform primary radiation intensity.

The secondary electrons will lose energy at a rate in the gas cavity different from that at which energy would have been lost in the medium.

Let $(S/\rho)_g$ = the loss of energy d*Es* of the electrons divided by the product of the path length traversed d*l* and the density of air ρ_g, known as the *mass stopping* of air.

$$\left(\frac{S}{\rho}\right)_g = \left(\frac{1}{\rho}\right)_g \frac{\mathrm{d}Es}{\mathrm{d}l} \tag{4.7}$$

Let $(S/\rho)_m$ = the *mass stopping power* of the medium for the same ionizing particle, similarly defined.

Let $\dfrac{(S/\rho)_m}{(S/\rho)_g} = S =$ the *relative stopping power*

Then the absorbed dose E in the medium is given by

$$E = \frac{(S/\rho)_m}{(S/\rho)_g} \times \text{(Energy lost by the secondary electrons in crossing the cavity)}$$
$$= S \times W \times J \tag{4.8}$$

where J = the number of ion pairs formed
W = the energy required to form one ion pair.

In evaluating equation (4.8), J may be measured by using the gas cavity as an ionization chamber, with a chamber wall material accurately matched to the medium. W may be obtained from published tables [3].

S may also be obtained from published tables [3] for monoenergetic electrons of energies between 0·001 MeV and 10 MeV for water, tissue and a number of elements. The energy of the charged particle at the point of interest may well differ materially from its initial energy, but it is the former which is relevant to the calculation. This point is discussed in detail in the Report of the International Commission on Radiological Units and Measurements [3], where values of mean mass stopping power are tabulated.

Tissue equivalent chambers filled with tissue equivalent gas have been used to measure the absorbed dose from X-rays generated at 30 kVp to 210 kVp [60]. Burlin [61] has examined the limits of the cavity ionization theory.

An exposure of one röntgen produces, in air, an absorbed dose of 0·869 rad [3]. Hence D_m, the absorbed dose in rads in a medium exposed to X röntgens is given by

$$D_m = 0{\cdot}869 \times \frac{(\mu/\rho)_{\mathrm{med}}}{(\mu/\rho)_{\mathrm{air}}} = fX \qquad (4.9)$$

where X is the measured cavity ionization exposure in röntgens, $(\mu/\rho)_{\mathrm{med}}$, $(\mu/\rho)_{\mathrm{air}}$ are the mass absorption coefficients of medium and air respectively, and f is the relevant factor in rads/R.

For X-radiation in the megavoltage region, the exposure is not measured (p. 56). Instead, it is usual to measure the 2 MV "röntgen" in an ionization chamber at 50 mm deep in a water phantom (Chapter 6). This measurement can be converted into rads by using the appropriate factors [62].

Solid State Dosemeters

These dosemeters measure the energy absorbed in the solid material, which is sometimes similar to that of the irradiated material of interest. They are,

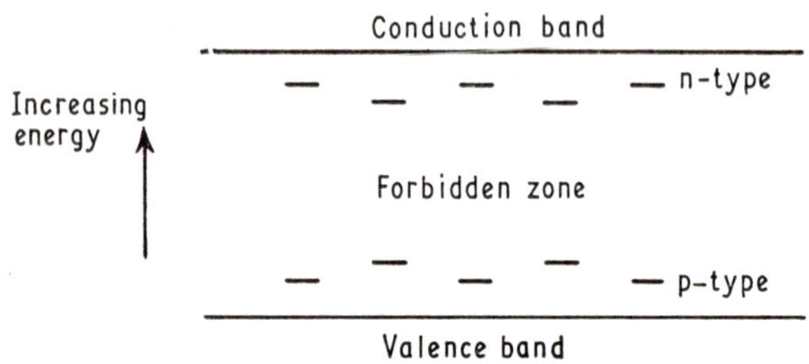

FIGURE 4.5 *Solid State Devices: Energy Levels*

in general, sensitive devices; an energy of 3·5 eV is required for the creation of an electron-hole pair in silicon and this is only one-tenth of that required for an ion pair in air (although in solid state dosemeters there may be other energy losses, for example, by excitation); the stopping power of a solid state dosemeter is much higher than a gas medium.

Crystals of semiconductor dosemeters may be considered as having at least three energy bands (Figure 4.5) arising from broadening the discrete energy levels associated with the constituent atoms [63, 64]. The valence band is normally completely filled with electrons and the conduction band is normally empty. These two bands are separated by a forbidden band, in which electrons cannot normally exist. The presence of an impurity, however, or a crystal imperfection, can create energy levels at isolated points in the forbidden band [17]. The energy required to raise an electron to the conduction band is about 2·5 to 20 eV for an insulator and about 0·1 to 2·5 eV for a semiconductor [63]. The latter contains impurities which create energy levels within the forbidden zone, so that when the temperature is too low for thermal vibrations to raise electrons from the valence to the conduction band, it may be high enough to raise them from an impurity level just below the conduction band into this band (*n*-type semiconductor), or from the valence band to an impurity level just above the latter band (*p*-type semiconductor). Thus, solid state dosemeters are sensitive to the amount of impurity present and to temperature.

Energy absorbed during exposure to ionizing radiations may raise electrons from the valence to the conduction band, leaving the atoms in an excited state.

SCINTILLATION COUNTERS

In certain materials, the atoms quickly return to ground state with the emission of light, when the materials are known as scintillators. The light

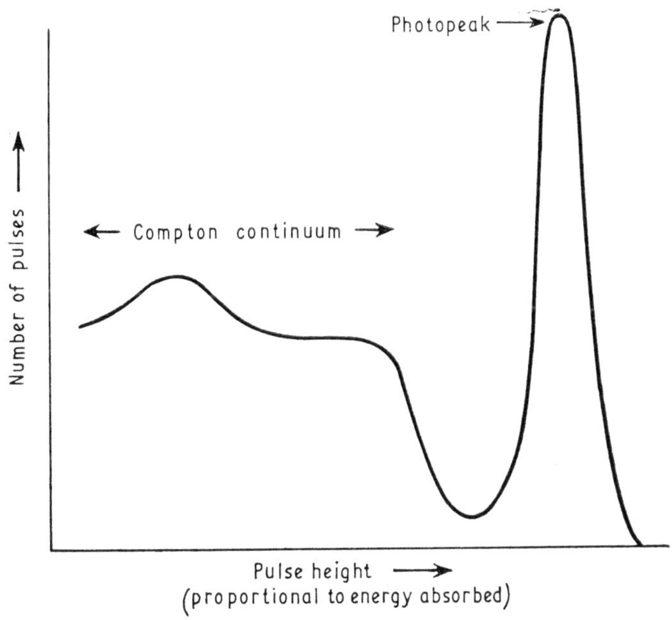

FIGURE 4.6 *Typical Spectrum for Monoenergetic Photons, using a Scintillation Counter*

can be measured with a photomultiplier, the amount of light emitted per event being a measure of the energy absorbed in the scintillator. Thus, after a photoelectric attenuation process, the amount of light emitted can be used to measure the energy of the incident photon. After Compton absorptions, the Compton electrons extend over an energy spectrum and the amount of light emitted reflects this range. The scintillations are converted into pulses and can be sorted to show the energy absorption spectrum [65] (Figure 4.6).

Sodium iodide (thallium activated) is the most common solid scintillator. Provided the X-ray photon energy is below the megavoltage region, crystals of sodium iodide are available large enough to give very high efficiencies and, because of multiple absorption processes, Compton counts are often small [66, 67] compared with the photoelectric absorption peak.

For higher X-ray energies, other methods are available for reducing that part of the spectrum due to Compton absorption. Organic phosphors have a lower effective atomic number than inorganic ones such as sodium iodide and therefore the former type will absorb X-radiation less by the photoelectric process than will inorganic crystals. Each phosphor may be made to "look at" the same source simultaneously, and the output of each succeeding amplifier subtracted (Figure 4.7). If the degree of amplification in each

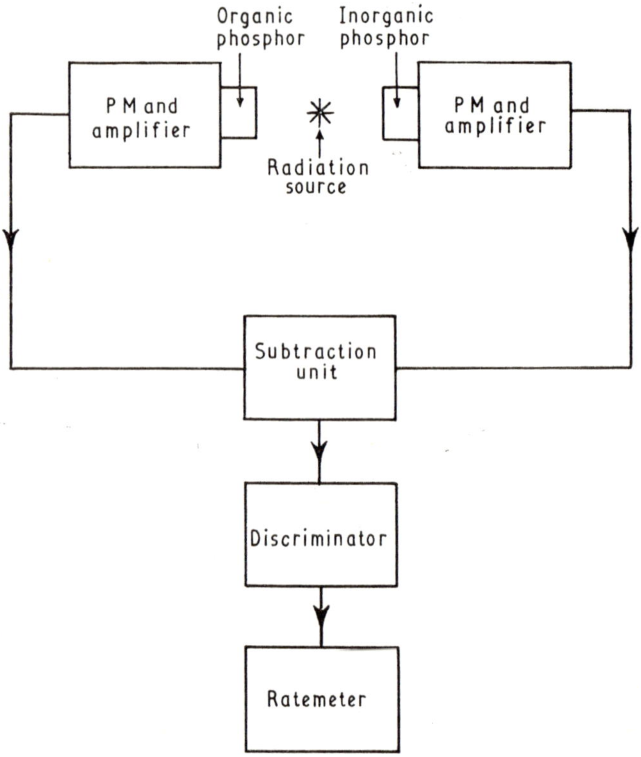

FIGURE 4.7 *Method of Discriminating in Favour of Photo-peaks in Scintillation Counting*

system is adjusted so that the Compton spectra (Figure 4.6) are of equal amplitude, the net output will be that due only to photoelectric absorption [68].

Alternatively, the exposed phosphor may be surrounded by another of the same material, and each viewed by their own photomultipliers. Photons absorbed in the central phosphor by the Compton process may be deflected into the annular phosphor, where they are likely to be further absorbed. If, therefore, the two outputs are fed into an anti-coincidence unit, these Compton absorptions will not be passed by the unit and mainly photoelectric processes will be counted.

In such cases, scintillation counting can be used to examine the X-ray photon spectrum of the incident radiation. Statistical fluctuations in the various stages, and particularly in the photomultiplier amplification, result in a rather low, typically 6–10 per cent, photo-peak discrimination.

The intrinsic efficiency of the scintillator may be defined as

$$\eta = h\nu / W \tag{4.10}$$

where $h\nu$ = energy of each luminescent light quantum,

W = average energy required for the production of each light quantum.

For the best scintillators W is comparable to the energy required for the production of one ion pair in air (15–60 eV compared with 34 eV [20].

THERMOLUMINESCENT DOSEMETERS

After absorption of energy from ionizing radiation and the raising of electrons to the conduction band, some of the electrons may fall into traps at such a depth that the probability is very low that they will be released at ambient temperatures. If the material is heated, however, this probability increases and the electrons may subsequently return to ground state with the emission of light. The amount of light emitted is a measure of the radiation energy absorbed. The phenomenon is known as radio-thermoluminescence or TLD [69, 70].

As the probability is not zero for any particular trap to be emptied at a particular temperature, a constant heating rate is necessary if the light output at a certain temperature is to be used as the dose measurement. Alternatively, the integrated light output can be used, as this is independent of the heating rate. Care has to be taken to eliminate infra-red radiation from the measuring system, as the equipment is inevitably heated at the same time as the thermoluminescent powder. Lithium fluoride [71, 72] is often used as a TLD material. If low doses have to be measured, read-out must be in an inert atmosphere to eliminate triboluminescence. Each batch of powder must be separately calibrated, and although it can be re-used after heat treatment, crystal size variations may lead to changes in sensitivity.

The efficiency of LiF is energy-independent between 100 keV and 1,000 keV, and shows only a slow rise of about 40 per cent for 40 keV X-ray photons.

As only a few milligrammes are required for a measurement, very small dosemeters are available, suitable for finger-tip monitoring or measurement of dose at bone interfaces.

Manganese-activated lithium fluoride is energy-dependent, the response increases with decrease of energy below about 500 keV. However, with suitable shielding, a dosemeter incorporating this material may be made substantially energy-independent for X-radiation from 40 keV to 1·25 MeV. This material shows a linear response to total exposure from a few milliröntgens to 2×10^5 röntgens, and the response is independent of exposure rate between 10 mR/min and 7,000 R/min. A disadvantage of manganese-activated lithium fluoride is that is shows some loss in thermoluminescence if it is stored after irradiation [73].

TLD materials incorporated in an inert base such as Teflon avoid the possibility of triboluminescence and are easier than powders to handle and dispense.

GLASS DOSEMETERS

If silver-activated phosphate glass is exposed to ionizing radiation, electrons are liberated. These free electrons are trapped by Ag^+ ions of the glass, reducing the ions to atomic silver centres [17]. If this glass is then irradiated with ultra-violet light of about 365 nm, it will fluoresce and the orange fluorescence, measured on a photomultiplier, may be used to measure exposures down to 50 mR [3, 74]. This is known as radiophotoluminescence. Small rods 1 mm in diameter and 6 mm long are suitable for measuring higher doses such as those met with in radiotherapy [75]. They are known as *Fluorods*. Glass dosemeters are versatile, as the effective atomic number can be designed to suit the application [3, 75]. Measurements are reasonably independent of exposure of the glass to light or to temperature between $-70°$ and $+100°C$. They are energy-dependent, but this can be reduced, as for conduction detectors, by suitable shielding [76].

After exposure to high doses, the glass loses its transparency and the optical density may be used to measure exposure in the region of 10^3 to 10^8 R [77]. The thickness of glass can be selected: thin microscope slides 180 μ thick have been used to measure 35 keV X-rays [78] within the exposure range $1{\cdot}5 \times 10^3$ to $1{\cdot}4 \times 10^7$ R.

RECTIFYING JUNCTION COUNTERS

Semiconductors may be of *n*-type where the charge carriers are electrons, or *p*-type where the carriers are holes. If two semiconductors, one of each type, are placed in contact, each set of carriers tends to diffuse into the other region. When a reverse bias is applied, the free charge-carriers are removed from either side of the junction, forming a depletion layer [79]. This depletion layer behaves as a solid state ionization detector, so that if it is irradiated, electron-hole pairs are formed and can be measured. It is technically difficult to obtain large depletion depths, which are proportional to (resistivity $\times$ voltage)$^{1/2}$. For *p*-type silicon, the resistivity is about 10^4 Ω-cm and the voltage about 100. Such a counter will have a depletion depth of only about 300 μ and so is inefficient for the measurement of X-rays. Germanium (Li-drifted) has a higher atomic number and hence greater absorption, but it has an efficiency also limited by the obtainable depth. It also has the disadvantage of having to be at liquid nitrogen temperature.

The advantage of this type of counter is the very high resolving power (an order of magnitude better than for scintillation counters), which makes them attractive for X-ray spectroscopy.

Radiation Polymerization Dosemeters

These dosemeters have a visible end point, that of a change from liquid to solid state, and therefore require no auxiliary measuring equipment, and they can be read by untrained personnel. They consist of a suitable monomer designed to gel [80] at a certain dose. For instance, ethylene glycol dimethacrylate gels after a dose of $5{\cdot}0 \times 10^4$ rads and divinyl benzene after $4{\cdot}2 \times 10^6$ rads, and other materials have been developed to cover the dose range from 300 to 10^7 rads. They are energy-independent for X-rays with a HVL above 1 mm Cu, but are dose-rate-dependent except for high dose rates above 2,500 rads/min.

The dosemeters are small in volume (0·5 ml) and expendable and have been developed for use with food irradiation projects. If their dose-rate-dependence could be matched to that of some useful radiation effect such as one in radiotherapy or radiobiology, they would provide a simple indication that a certain effective dose had been delivered.

Chemical Methods

These measure energy absorbed in the chemical system, which, if an aqueous solution is used, may well be similar to that absorbed in a biological material.

When water is irradiated, free radicals are formed and a number of reactions can take place, for example

$$H_2O \rightarrow H_2O^+ + e^- \tag{4.11}$$

$$H_2O^+ \rightarrow H^+ + OH \tag{4.12}$$

$$H^+ + e^- \rightarrow H \tag{4.13}$$

Many substances act as *acceptors* for these free radicals and the resulting chemical products are a measurement of the radiation dose. In general, oxidizable solutes will be oxidized by the OH radicals and reducible solutes will be reduced by the H radicals [3, 17, 81].

Conditions for a suitable chemical system are:

1. The radiation indicator must react readily with the system.
2. For work in the biological field its effective atomic number should be similar to that of water.
3. The indicator should be stable in air; otherwise the reaction must be carries out anaerobically.
4. The irradiation products, similarly, should be stable in air and capable of being estimated in very high dilutions.
5. The yield should be independent of the concentration of the acceptor.

The yield may be expressed as a G value: the number of molecules produced or converted per 100 eV absorbed [17].

FERROUS SULPHATE DOSEMETER

This is sometimes known as a Fricke dosemeter, after the author who first studied it [82]. When a 0·8 N sulphuric acid solution of ferrous sulphate is exposed to radiation, the ferrous ions are oxidized to ferric and the concentration of these may be measured volumetrically, calorimetrically, spectrophotometrically [3] by the absorption of the solution at 304 to 305 mm, or by a photometric method using 1:10 phenanthreline [83]. The yield for X-rays has been measured as $13{\cdot}4 \pm 0{\cdot}6$ (8 kVp X-rays) to 16·3 (30 MeV X-rays) ions per 100 eV [3, 84]. The G value is approximately constant [85] between 10 and 20 kVp. Doses of from 10^3 to 10^6 rads may be measured, if care is taken to avoid oxygen depletion in measuring doses over 50,000 rads. As with all chemical dosemeters, a control experiment with an unirradiated solution should be carried out [86]. The yield is independent of radiation energy for X-ray photons of greater than 100 keV, but tends to fall off at higher LET values.

A more sensitive method of measurement has been developed using ^{59}Fe as a tracer [87]. After irradiation the ferric ions are separated from the ferrous by extraction of the ferric thiocyanate with isoamyl alcohol, and the activity of both organic phase and water counted. It is suggested that by this method it should be possible to measure radiation doses of less than 100 rads. This range (0–100 R) has been attained by combining a tracer method with ion-exchange separation [88].

CERIC SULPHATE DOSEMETER

In this system, the ceric ion in 0·1–0·8 N sulphuric acid is reduced to cerous. The *G*-value is several times lower than for the oxidation of ferrous sulphate solutions, but unlike the latter, it is independent of oxygen concentration. Hardwick [89] has measured the yield for this system as well as for the ferrous sulphate dosemeter with particular reference to the energy of the ionizing electron, and has found $G = 5{\cdot}2$ ions changed per 100 eV for 14 kVp X-rays and 5·5 for 50 kVp X-rays. For energetic γ-ray photons in the region of 1 MeV, $G = 3{\cdot}2$ ions changed per 100 eV. Doses between 10^5 and 10^8 rads can be measured [3].

Very high standards of purity of reagents and cleanliness of glassware are necessary for reproducible results [3].

Colorimetric Dosemeters

Colour changes in various dyes can be used to measure radiation quantitatively by optical absorption changes, and they have the special property of being able to show dose variations over an area or volume of interest [75].

METHYLENE BLUE DOSEMETER

An aqueous solution of 3·2 per cent polyvinyl chloride, 0·04 per cent methylene blue, 7·0 per cent lead nitrate, 17 per cent ethylenediamine tetra-acetic acid and 8·5 per cent glycerol is dried on a glass slide to give a clear blue film [90]. This is sealed from the atmosphere and decolorized by exposing to light. The slide is then exposed to irradiation and if stored at $-10°$C, no colour will develop. At room temperature, the full colour develops in five to ten minutes, and the optical density is proportional to the radiation exposure.

As exposures of the order of one röntgen can be measured with this dosemeter, it might be used for the visual indication of areas of radiation hazard, should it prove sufficiently stable over long periods of time. However, results are difficult to reproduce [75].

Proportional and Geiger Counters

Consider an ionization chamber run under conditions of saturation current, so that each ion is collected before recombination occurs. If the voltage is raised further, electrons will acquire sufficient energy to ionize by collision neutral gas molecules, so that the ionization current increases (Figure 4.8). A heavily ionizing particle will lead to more ionization by collision than one sparsely ionizing. Consequently the ionization current depends on the character as well as the number of initial ionizing events. The ratio of the number of ions collected to the number of initiating ions, or the *gas amplification factor* can be several orders of magnitude and may be as high as 10^6, so after further amplification it is simple to detect and count each ionizing event. In a certain voltage region the size of each pulse is proportional to the energy liberated in the counter by the original ion, and this proportionality gives the *proportional* counter its name. Analysis of the pulse sizes may be used to measure the energy spectrum of the incident radiation and to distinguish

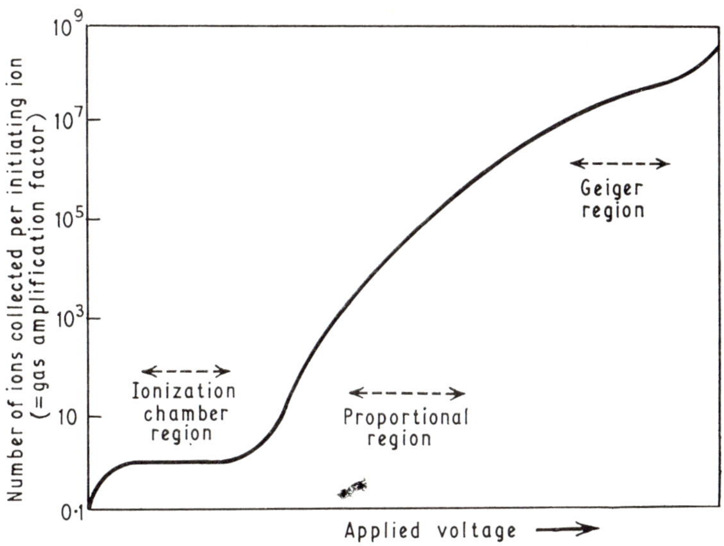

FIGURE 4.8 *The Variation of Gas Amplication Factor with Applied Voltage*

between more or less heavily ionizing types of radiation. Proportional counters are filled at a low pressure with a suitable gas, so that there is a long mean free path and a low probability of recombination.

If the voltage is increased above that for the proportional region, there will be more ionizations by collision (Figure 4.8). The resulting electrons will pass quickly to the anode, leaving a sheath of positive ions surrounding its length. A characteristic of a tube operating in this *Geiger* region is that the discharge spreads along the length of the wire, so that the size of each resulting pulse is independent of the nature of the initial ionizing event; the build-up of ionization is known as a *Townsend avalanche*. The gas amplification factor may be up to 10^{11} and depends somewhat on the applied voltage. A Geiger counter usually consists of a cylindrical outer case forming the cathode and a thin central wire of, for example, 0·03 mm diameter tungsten, forming the anode.

Free electrons are essential for the Townsend avalanche, and therefore the filling gas must not readily form negative ions. Argon and neon are often used, and more rarely hydrogen and nitrogen. The gas pressure is usually 70–200 mmHg and is a major factor in determining the operating voltage. Under ideal conditions at voltages within the Geiger region each ionizing event causes one Townsend avalanche and one pulse.

During a Geiger discharge, the electrons acquire energy high enough to ionize by collision only when they are very near the anode wire. The positive ion sheath they leave behind drifts towards the cathode and during its passage may cause photon or electron emission in the gas, or on reaching

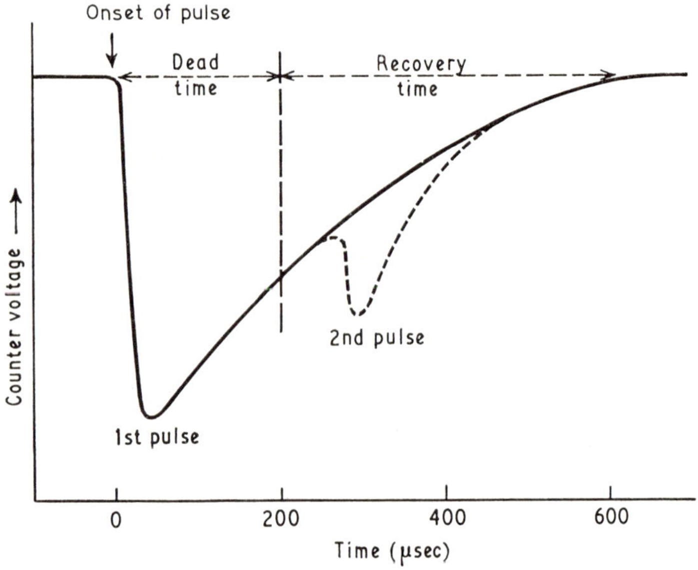

FIGURE 4.9 *Geiger Counter Voltage as Affected by Townsend Avalanche*

the cathode owing to the difference of ionization potential and work function. These photons may initiate a second Townsend avalanche, and so on. This is known as *multiple pulsing* and will result in a number of counts being recorded for each ionizing event, unless the discharge is *quenched*; initiation of a discharge may be made to trigger a negative pulse, which is applied to the anode, so that the tube voltage is made insufficient for a Townsend avalanche. After a few hundred microseconds this pulse terminates and the tube is ready to count again. Alternatively, or more often additionally, the filling gas contains a quenching agent; such tubes are known as *self-quenching*, and are of two types: organic- and halogen-quenched.

Organic-quenched tubes may contain argon at about 90 mmHg pressure and ethyl formate or ethyl alcohol at, for example, 10 mmHg pressure as a quenching agent. Argon molecules have a higher ionization potential than ethyl alcohol or formate molecules and consequently positive ion charge is transferred to the quenching agent during migration of the sheath towards the cathode. On reaching the cathode, alcohol or formate ions lose their energy by dissociating rather than, like argon ions, by causing further ionization, so that further discharge does not take place. These tubes have a finite life of about 10^{10} counts, after which insufficient alcohol molecules remain for quenching.

Halogen-quenched counters contain, for example, 0·1 per cent chlorine and act in a similar fashion except that dissociated molecules recombine. There is therefore no limitation to the life of these tubes.

The passage of a Geiger pulse lowers the tube voltage, which subsequently rises towards its original value (Figure 4.9), and the tube will be inoperative until the voltage is restored to a value high enough for initiation of a Townsend avalanche. This time depends to some extent on the nature and position of the next ionizing event and is therefore somewhat variable. An external quenching pulse with a sharp lagging edge makes the resolving time definite

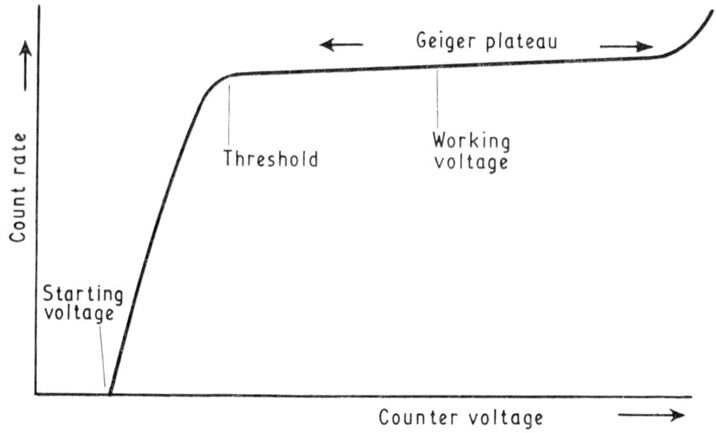

FIGURE 4.10 *Characteristic Curve for a Geiger Counter*

between two pulses, at the expense of an increase in its value. In either case, when random pulses are being counted, there is a chance, which increases with count rate, that some initiating events will arrive too soon after their predecessors to be counted.

If T = resolving time in seconds and n = observed count rate in reciprocal seconds (sec^{-1}), then the counter will be inoperative for nT second in each second, so that the effective counting time is $(1 - nT)$ second in every second. Therefore, the true count rate n_0 is given by

$$n_0 = \frac{n}{1 - nT}\ sec^{-1} \tag{4.14}$$

The ideal of one initiating event to one pulse is not attainable, and as the voltage is increased, so is the probability of multiple pulsing and consequently the count rate. The slope of the Geiger plateau (Figure 4.10) is expressed as percentage increase in counts per volt (or per 100 volts), and different designs of counter have maximum specified slopes varying between 0·02 and 0·15 per cent per volt. Desirable characteristics are a long plateau and a small slope, and both these depend to some extent on the external electronic circuit as well as on the counter [91]. A typical circuit will consist of a constant voltage supply of up to 2,000 V for an organic-quenched counter and up to 1,000 V for a halogen-quenched counter, a pre-amplifier or probe unit close to the counter, a main amplifier and either a scaler to count individual pulses or a rate meter reading in pulses per second. Any quenching pulse is usually supplied from the probe unit. It is possible to dispense with a pre-amplifier or probe unit when only a short cable connects the Geiger counter to the scaler or ratemeter. The capacity associated with a long cable tends to degrade the pulse size and shape and in an extreme case will prevent the counting of any pulses on the plateau.

A low-energy X-ray photon, of the order of ten or twenty keV is likely to be absorbed in the filling gas, and the resulting ionization will trigger the tube. This, however, is unlikely for higher-energy photons and, in this case, most counts in the G.M. tube arise from absorption of photons in the tube wall.

The intrinsic efficiency ε may be defined as the number of secondary electrons entering the sensitive volume of the tube for each photon striking the tube. Then

$$\varepsilon = \tau R_\tau + \sigma R_\sigma + 2\pi R_\pi \tag{4.15}$$

where τ, σ, π are the partial linear absorption coefficients of the wall material for the photoelectric, Compton and pair-production processes respectively and R_τ R_σ, R_π are the corresponding ranges of the secondary electrons in the wall material [17], provided that the wall thickness is not less than the maximum range of the secondary electrons but not great enough to attenuate the X-rays appreciably. As these two conditions are incompatible, this formula tends to overestimate ε. The value of ε is usually lower than 1 per cent for photon energies up to 1 MeV, but rises with increase of energy [92]. Efficiencies can be increased by making the cathode area large, using a construction of fine mesh or a plate and fin [17, 93] and by using appropriate materials for the cathode.

Geiger counters measure directly neither exposure nor absorbed dose, but are useful for selected measurements of X-rays. Special counters[4,5] have been designed to measure soft X-radiation in X-ray spectrometers; they have filling gas and tube dimensions chosen for high counting efficiency and a window thickness of a few milligrammes per square centimetre.

Because of their high sensitivity compared with ionization chambers, Geiger counters of almost any type may be used to locate areas of unwanted high radiation levels near X-ray sets, and the magnitude of the exposure rate can subsequently be measured by an ionization chamber, film badge or other appropriate means. However, with careful design, radiation monitors[6] using Geiger counters may be substantially ($\pm$40 per cent) energy-independent from 80 keV to 2,000 keV.

High-current Geiger tubes[4,5] have been designed to measure X-ray exposure rates. These have current output up to about 50 μA; when they are connected to a microammeter in series with the counter, this forms a very simple circuit. They can measure exposure rates up to about 6 R/min and have resolving times of up to 50 μsec. They are halogen-quenched and operate at about 500 V.

Photographic methods of measuring X-radiation are described in Chapter 7.

QUANTUM COUNTING EFFICIENCY

The quantum-counting efficiency E of a detector is defined [94] in terms of the fraction of the incident X-radiation absorbed in the active part of the detector, i.e. not the windows, etc. For crystallographic X-ray wavelengths it is mainly due to the photoelectric effect and is therefore dependent upon absorption edge effects in the gas of the counter (Kr or Xe), the iodine of the scintillation crystal, or the silver and bromine absorption edges in the film emulsion.

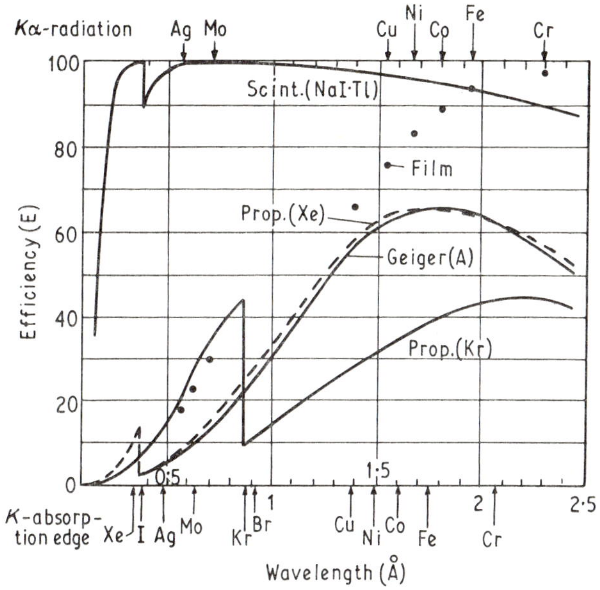

FIGURE 4.11 *Calculated Quantum-counting Efficiency E as a Function of Wavelength λ for four Counter Tubes*

(*International Tables for X-ray Crystallography*, Vol. III, p. 148, Fig. 3.1.3.4. Kynoch Press)

Figure 4.11 shows the response of the scintillation, proportional and Geiger counters compared with that of film for a range of characteristic radiations. The quantum counting efficiency for a given wavelength is decisive in the selection of a counter tube in circuits used in X-ray crystallography and X-ray spectro-chemical analysis.

Literature References

1 GREENING, J. R., "The derivation of approximate X-ray spectral distribution and an analysis of X-ray 'quality' specifications", *Brit. J. Radiol.*, **36**, p. 363 (1963).

2 INTERNATIONAL COMMISSION ON RADIOLOGICAL UNITS AND MEASUREMENTS, *Radiation Quantities and Units*, Recommendations of the ICRU, Report 10*a*, 1962. NBS Handbook 84 (Washington, D.C., Nat. Bur. of Stand., 1962).

3 INTERNATIONAL COMMISSION ON RADIOLOGICAL UNITS AND MEASUREMENTS, *Physical Aspects of Irradiation*, Recommendations of the IRCU, Report 10(*b*). NBS Handbook 85 (Washington, D.C., Nat. Bur. Stand., 1964).
4 BARNARD, G. P., ASTON, G. H. and MARSH, A. R. S., *Effect of Variation in the Ambient Air on the Calibration and Use of Ionization Dosemeters* (London, H.M.S.O., 1960).
5 FOWLER, J. F., "A simple indicator for the temperature and pressure correction", *Brit. J. Radiol.*, **29,** p. 462 (1956).
6 VAN STEKELENBURG, L. H. M., "An air density correction meter", *Brit. J. Radiol.*, **29,** p. 464 (1956).
7 KEMP, L. A. W., and HALL, S. M., "A precise investigation of some aspects of parallel-plate free air chamber geometry", *Brit. J. Radiol.*, **27,** p. 226 (1954).
8 WYCKOFF, H. O., ASTON, G. H. and SMITH, E. E., "A comparison of x-ray standards", *Brit. J. Radiol.*, **27,** p. 325 (1954).
9 WYCKOFF, H. O., *et al.*, "Intercomparison of national Roentgen and gamma ray exposure—Dose standards", *Acta Radiol.*, **1,** p. 57 (1963).
10 JOHNS, H. E., *The Physics of Radiology* (Springfield, Ill., Chas. C. Thomas, 1961).
11 KEMP, L. A. W., "Guarded-field ionization chambers: a new principle applied to both free-air and thimble versions", *Nature*, **178,** p. 1250 (1956).
12 BARBER, B., "The basic data of a guarded-field ionization chamber: further analytical investigations", *Phys. Med. Biol.*, **6,** p. 389 (1962).
13 KEMP, L. A. W. and BARBER, B., "The construction and use of a guarded-field cavity ionization chamber for the measurement of supervoltage radiation", *Phys. Med. Biol.*, **8,** p. 149 (1963).
14 RUTHERFORD, E., *Radioactivity* (Cambridge, University Press, 1905).
15 OWENS, R. B., "Thorium radiation", *Phil. Mag.*, **48,** p. 360 (1899).
16 WILKINSON, D. H., *Ionization Chambers and Counters* (Cambridge, University Press, 1950).
17 PRICE, W. J., *Nuclear Radiation Detection* (New York, McGraw-Hill, 1958).
18 TAYLOR, D. and SHARP, J., "Nuclear particle radiation detectors, Part I. Ion chambers and ion chamber instruments", *Proc. Instn. Elect. Engrs.*, Part II, **98,** p. 174 (1951).
19 CROWTHER, J. A., *Ions, Electrons and Ionizing Radiations* (London, Edward Arnold, 1949).
20 HINE, G. J. and BROWNELL, G. L., *Radiation Dosimetry* (New York, Academic Press, 1956).
21 GREENING, J. R., "A compact free air chamber for use in the range 0–50 kV", *Brit. J. Radiol.*, **33,** p. 178 (1960).
22 MAYNEORD, W. V. and ROBERTS, J. E., "An attempt at precision measurements of gamma rays", *Brit. J. Radiol.*, **10,** p. 365 (1937).
23 GRAY, L. H., "Radiation Dosimetry", Part II. *Brit. J. Radiol.*, **10,** p. 721 (1937).
24 BARNARD, G. P., AXTON, E. J. and MARSH, A. R. S., "A study of cavity ion chambers for use with 2 MV x-rays: equilibrium wall thickness: wall-absorption correction", *Phys. Med. Biol.*, **3,** p. 366 (1959).
25 BARNARD, G. P., "Standardization of ionization chambers for megavoltage radiation", *Phys. Med. Biol.*, **9,** p. 287 (1964).
26 INTERNATIONAL COMMISSION ON RADIOLOGICAL UNITS AND MEASUREMENTS, ICRU Report, 1959. NBS Handbook 78. (Washington, D.C., Nat. Bur. Stand., 1961).
27 BARNARD, G. P. *et al.*, "On the congruity of N.P.L. exposure standards", *Phys. Med. Biol.*, **9,** p. 333 (1964).

28 KEMP, L. A. W., and READ, L. R., "An inflated balloon ionization chamber for low-energy, low-level x-ray exposure measurements", *Phys. Med. Biol.*, **13,** p. 451 (1968).

29 FOWLER, J. F., "X-ray induced conductivity in insulating materials", *Proc. Roy. Soc.* A, 1956, p. 464.

30 LYSONS, H., "The use of polytetrafluorethylene insulators in integrating ion chambers", *J. Sci. Inst.*, **27,** p. 334 (1950).

31 FARMER, F. T., "Condenser ionization chambers for measurement of x-ray dose in body cavities", *Brit. J. Radiol.*, **18,** p. 148 (1945).

32 SIEVERT, R. M., "Eine Methode zur Messung von Roentgen-, Radium-, und Ultrastrahlung nebst einigen Untersuchungen über die Anwendbarkeit derselben in der Physik und Medizin", *Acta Radiol.*, Suppl. 14 (1932).

33 SIEVERT, R. M., "Über die Anwendung der Kondensator-kammer für sowohl Roentgen wie γ-Strahlenmessungen: zugleich ein Beitrag zu den Vergleichen der biologischen Wirkungen dieser beiden Strahlenarten". *Acta Radiol.*, **15,** p. 193 (1934).

34 CEDERLAND, J., "A three-electrode ionization condenser chamber", *Brit. J. Radiol.*, **33,** p. 105 (1960).

35 FARMER, F. T., "An electrometer for measurement of voltage on small ionization chambers", *Proc. Phys. Soc.*, **54,** p. 435 (1942).

36 SUTHERLAND, W. H., "A dynamic electrometer for small condenser chambers", *Brit. J. Radiol.*, **22,** p. 255 (1949).

37 BEWLEY, D. K., "Loss of charge in BD 11 chambers due to mechanical shock", *Brit. J. Radiol.*, **26,** p. 601 (1953).

38 DAY, M. J. and FARMER, F. T., "Effects of mechanical shock on type BD 11 ionization chambers", *Brit. J. Radiol.*, **26,** p. 601 (1953).

39 STUART, S. and OSBORN, S. B., "A wide range direct reading dosemeter", *Phys. Med. Biol.*, **3,** p. 255 (1959).

40 OSBORN, S. B. and BURROWS, R. G., "An ionization chamber for diagnostic radiation", *Phys. Med. Biol.*, **3,** p. 37 (1958).

41 WOOD, R. G., SUTHERLAND, W. H. and COHEN, M., "Some factors influencing the experimental determination of percentage depth doses for medium energy x-rays", *Brit. J. Radiol.*, **36,** p. 266 (1963).

42 OLIVER, R. and KEMP, L. A. W., "An investigation into some factors affecting x-ray dose distribution and its measurement", *Brit. J. Radiol.*, **22,** p. 33 (1949).

43 SVENSSON, G. and BJARNGARD, B., "Ionization chamber system with emitted wall for intra-cavitary use", *Phys. Med. Biol.*, **9,** p. 465 (1964).

44 FARMER, F. T., "A sub-standard x-ray dosemeter", *Brit. J. Radiol.*, **28,** p. 304 (1955).

45 ADAM, G. D., "On the use of thimble chambers in phantoms", *Radiol.*, **78,** p. 77 (1962).

46 THOREUS, R., "Monitoring filters in Roentgen therapy", *Acta Radiol.*, **45,** p. 414 (1956).

47 REISMA, K., "*Dosemeters for X-Ray Diagnosis*" (Eindhoven, Philips Technical Library, 1962).

48 OOSTERKAMP, W. J. and PROPER, J., "Free-air thimble chambers for grenz-ray dosimetry", *Acta Radiol.*, **37,** p. 33 (1952).

49 TAYLOR, L. S., "The measurement of x- and gamma-radiation over a wide energy range", *Brit. J. Radiol.*, **24,** p. 67 (1951).

50 VENNART, J., "Some physical measurements in the grenz ray region", *Brit. J. Radiol.*, **27,** p. 324 (1954).

51 WEYL, C., WARREN, S. R. and O'NEILL, D., *Radiologic Physics* (Springfield, U.S.A., Charles C. Thomas, 1941).

52 GLASSER, O. *et al.*, *Physical Foundations of Radiology* (London, Pitman Medical Publ., 1961).
53 LAUGHLIN, J. S. and BEATTIE, J. W., "Calorimetric determination of energy flux of 22·5 MeV x-rays", *Rev. Sci. Inst.*, **22,** p. 572 (1951).
54 LAUGHLIN, J. S. *et al.*, "Calorimetric evaluation of the roentgen for 400 kV and 22·5 MeV Roentgen rays", *Amer. J. Roent.*, **70,** p. 294 (1953).
55 KEENE, J. P. and LAW, J., "A determination of the G-value of ferrous sulphate for ^{60}Co radiation using calorimetric dosimetry", *Phys. Med. Biol.*, **8,** p. 85 (1963).
56 BEWLEY, D. K., "The measurement of locally absorbed doses of megavoltage x-rays by means of a carbon calorimeter", *Brit. J. Radiol.*, **36,** p. 865 (1963).
57 GREENING, J. R., RANDLE, K. J. and REDPATH, A. T., "The measurement of low-energy x-rays. II. Total absorption calorimetry", *Phys. Med. Biol.*, **13,** p. 359 (1968).
58 GRAY, L. H., "Radiation dosimetry. Part I", *Brit. J. Radiol.*, **10,** p. 600 (1937).
59 FANO, U., "Notes on the Bragg-Gray cavity principle for measuring energy dissipation", *Rad. Res.*, **1,** p. 237 (1954).
60 GROSS, W. *et al.*, "Experimental determination of the absorbed dose from x-rays in tissue", *Rad. Res.*, **18,** p. 326 (1963).
61 BURLIN, T. E., "The limits of validity of cavity ionization theory", *Brit. J. Radiol.*, **35,** p. 343 (1962).
62 BARNARD, G. P., "Dose-exposure conversion factors for megavoltage x-ray dosimetry", *Phys. Med. Biol.*, **9,** p. 321 (1964).
63 WILLIAMS, I. R. and WILLIAMS, M. W., *Basic Nuclear Physics*, (London, George Newnes, 1962).
64 SHAW, D. F., *An Introduction to Electronics* (London, Longmans, 1962).
65 CORMACK, D. V. *et al.*, "Spectral distributions of 280 kVp x-rays", *Brit. J. Radiol.*, **31,** p. 565 (1958).
66 WANG, P. K. S., RARIDON, R. J. and TIDWELL, M., "X-ray spectrum from a beryllium window tube. 1. Scintillation spectrometry", *Brit. J. Radiol.*, **30,** p. 70 (1957).
67 PALMER, R. C., WILLIS, W. V. and DAVIES, D. K., "X-ray spectrum from a plastic window tube", *Phys. Med. Biol.*, **6,** p. 401 (1962).
68 PEIRSON, D. H. and IREDALE, P., "The gamma-ray scintillation spectrometer", *Brit. J. Appl. Phys.*, **8,** p. 422 (1957).
69 MCDOUGALL, D. J. (Ed.), *Thermoluminescence of Geological Materials* (London, Academic Press, 1968).
70 BJARNGARD, B., "$CaSO_4$(Mn) thermoluminescence dosimeters for small doses of gamma radiation", *Rev. Sci. Inst.*, **33,** p. 1129 (1962).
71 CAMERON, J. R. *et al.*, "Radiation dosimeter utilizing the thermoluminescence of lithium fluoride", *Science*, **134,** p. 333 (1961).
72 KARZMARK, C. J., WHITE, J. and FOWLER, J. F., "Lithium fluoride thermoluminescence dosimetry", *Phys. Med. Biol.*, **9,** p. 273 (1964).
73 SCHULMAN, J. H. *et al.*, "New thermoluminescence dosemeter", *Rev. Sci. Inst.*, **31,** p. 1263 (1960).
74 YOKOTA, R., NAGAJIMA, S. and SAKAI, E., "High-sensitivity silver-activated phosphate glass needles for the simultaneous measurement of thermal neutrons, γ- and/or β-rays", *Health Phys.*, **5,** p. 219 (1961).
75 FOWLER, J. F., "Solid state dosimetry", *Phys. Med. Biol.*, **8,** p. 1 (1963).
76 SCHULMAN, J. H. *et al.*, "Dosimetry of x-rays and gamma rays by radiophotoluminescence", *J. Appl. Phys.*, **22,** p. 1479 (1951).
77 SCHULMAN, J. H., KLICK, C. C. and RABIN, H., "Measuring high doses by absorption changes in glass", *Nucleonics*, **13,** p. 30 (1955).

78 BRUSTAD, T. and HENRIKSEN, T., "Glass dosimetry of soft X-rays", *Brit. J. Radiol.*, **31,** p. 163 (1958).

79 DEARNALEY, G. and NORTHROP, D. C. *Semiconductor Counters for Nuclear Radiations* (London, Spon, 1966).

80 HOECKER, F. E., "Characteristics of the radiation polymerization dosimeter", *Health Phys.*, **8,** p. 381 (1962).

81 DAY, M. J. and STEIN, G., "Chemical measurement by ionizing radiations", *Nature*, **164,** p. 671 (1949).

82 FRICKE, H. and MORSE, S., "The action of X-rays on ferrous sulphate solutions", *Phil. Mag.*, **7,** p. 129 (1929).

83 MILLER, N., "Quantitative studies of radiation-induced reactions in aqueous solutions. 1. Oxidation of ferrous sulphate by X- and γ-radiation", *J. Chem. Phys.*, **18,** p. 79 (1950).

84 CHAPIRO, A. (Ed.), *Radiation Chemistry of Polymeric Systems* (New York, Interscience, 1962).

85 LAW, J. and REDPATH, A. T., "The measurement of low-energy X-rays. III. Ferrous sulphate 'G' values", *Phys. Med. Biol.*, **13,** p. 371 (1968).

86 HALE, E. J. and OLIVER, R., "A pitfall to avoid in ferrous sulphate dosimetry", *Brit. J. Radiol.*, **34,** p. 397 (1961).

87 RUDSTAM, G. and SVEDBERG, T., "Use of tracers in chemical dosimetry", *Nature*, **71,** p. 648 (1953).

88 CAL, O. S., "The use of ion-exchange columns in the measurement of doses below 1,000 rad with the Fricke dosemeter", *Int. J. Appl. Rad.*, **13,** p. 304 (1962).

89 HARDWICK, T. H., "The effect of the energy of the ionizing electron on the yield in irradiated aqueous systems", *Disc. Faraday Soc.*, **12,** p. 203 (1952).

90 OSTER, G. and BROYDE, B., "A sensitive chemical dosemeter for ionizing radiations", *Nature*, **184,** p. 545 (1959).

91 MACINTOSH, I. W., *Geiger Muller Tubes*, Mullard Technical Report, Mullard Ltd., X-ray Division, Mitcham Junction, Surrey, 1958.

92 BRADT, H., "Empfindlichkeit von Zählrohren mit Blei, Messing und Aluminiumkathode fur γ-Strahlung im Energieintervale 0·1 bis 3 MeV", *Helv. Phys. Acta.*, **19,** p. 77 (1946).

93 EVANS, R. D. and MUGELLE, R. A., "Increased gamma-ray sensitivity of tube counters and the measurement of thorium content of ordinary materials", *Rev. Sci. Inst.*, **7,** p. 441 (1936).

94 *International Tables for X-ray Crystallography*. Vol. III, p. 147 (Birmingham, Kynoch Press).

5 *Calibration of X-ray Sets*

The object of calibrating an x-ray set is to provide information on the quantity and quality of x-ray exposure, under all conditions of interest. The geometrical distribution of the radiation depends on the condition, position, size and angle of the target, as well as on beam-limiting devices. Exposure rates and energy of the radiation are determined by the target material and tube design and by the operating current and voltage; the output may be modified by the addition of metal filters.

Safety aspects of x-ray installations must be investigated (Chapter 14), and any faults rectified before proceeding to calibration.

Position of Focal Spot

The position of the focal spot must be known for calculations of the magnification of an image. It is required also in calculating field size and central axis exposure rates at positions other than those investigated during the calibration.

For tube voltages up to 300 kVp, the height of the focal spot above a reference plane AB (Figure 5.1), may be obtained by photographing the focal spot through two pinholes. The recommended [1] separation s of the pinholes is 7 mm. The distance $(x - h)$ of focal spot above pinholes should be at least 100 mm. The distance h of pinholes above film should be at least $2(x - h)$ for focal spot sizes up to 2·5 mm, and at least $(x - h)$ for larger focal spots [1]. An approximate value of x must be used in arranging for the satisfaction of these conditions. A more precise value may then be calculated from the equation

$$\frac{s}{s'} = \frac{x - h}{x} \tag{5.1}$$

where s' is the measured distance between pinhole images on the photographic film.

In a second method, the magnification m is measured of some object at different distances d above a reference plane. A linear relationship holds between $1/m$ and d, and if these quantities are plotted on a graph, the line may be extrapolated to give d_f, the distance of the focal spot above the plane of reference, when $1/m$ is zero.

A third method is valid only if the absorption of the x-rays in air is negligible and if there is no appreciable scatter contributing to exposure measurement. The latter condition is unlikely if an applicator with an end plate is used [2]. In this method, exposure rates x/t are measured on the central beam axis for different apparent focal distances d. Because the conditions are such that the inverse square law holds, $\sqrt{(t/x)}$ will be proportional to the real focal distance d_f, which can be found by extrapolating the line relating $\sqrt{(t/x)}$ and d, to the point where $\sqrt{(t/x)}$ is zero.

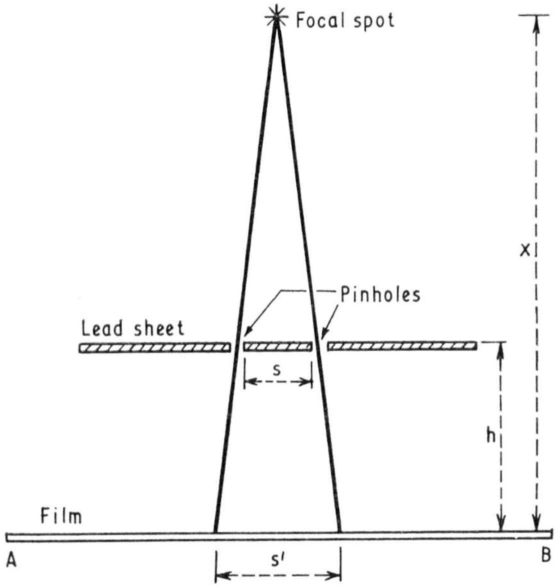

FIGURE 5.1 *Two-pinhole Method for Determining Focal Distance*

Size of Focal Spot

A knowledge of the size of the focal spot is required in calculations of penumbra widths and in assessing questions of resolution.

Focal spot size may be calculated from a pinhole photograph (Figure 5.2). The pinhole must be small compared with the dimensions of the focal

TABLE 5.1 *Recommended Pinhole Dimensions* (see Figure 5.3)
(From I.C.R.U. Report 10 f, *Methods of Evaluating Radiological Equipment and Materials*, NBS Handbook 89, Washington D.C., NBS, 1963)

Focal spot size	*Nominal pinhole diameter*	D(mm)	L(mm)
Below 1·0 mm	0·030 mm	0·030 ± 0·005	0·075 ± ·010
1·0–2·5 mm	0·075 mm	0·075 ± 0·005	0·350 ± ·010
Above 2·5 mm	0·100 mm	0·100 ± 0·005	0·500 ± ·010

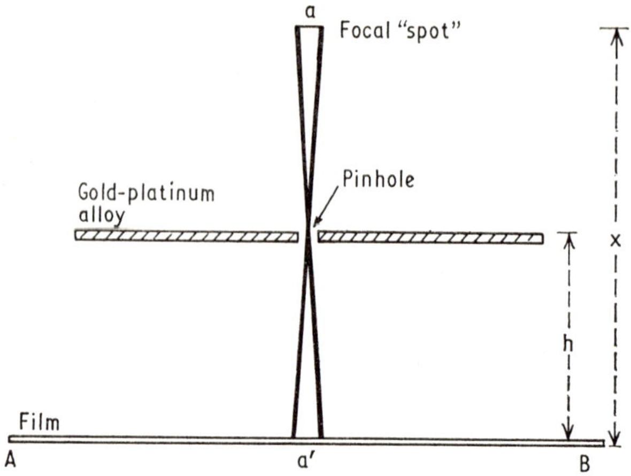

FIGURE 5.2 *Pinhole Method for Determining Focal Spot Size*

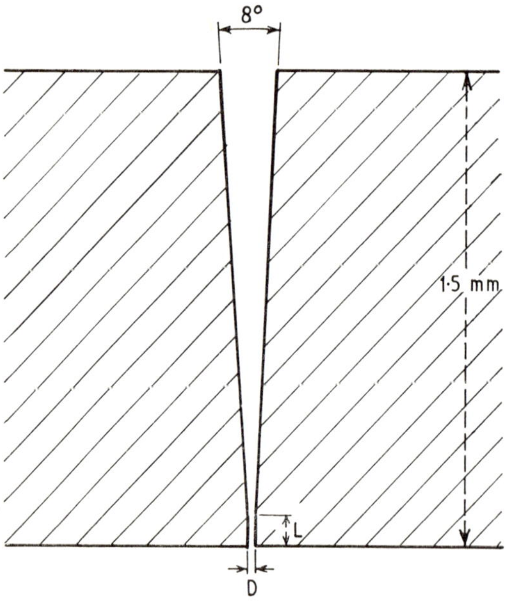

FIGURE 5.3 *Pinhole Dimensions*

(ICRU Report 10*f*, *Methods of Evaluating Radiological Equipment and Materials*, NBS Handbook 89. Washington, D.C., NBS, 1963)

spot (Table 5.1) and of a particular shape (Figure 5.3). It should be manufactured[7] from a 90:10 gold–platinum alloy [1], although platinum has also proved satisfactory. The pinhole should be on the central axis of the beam; its position and that of the film should be governed by the same conditions as when using the two pinholes. The image density should be between 0·8 and 1·2 and the size of the image on the film measured with a magnifying glass having a built-in scale with 0·1 mm divisions.

In a line-focus tube, the lengthwise distribution of energy in the focal spot usually diminishes gradually to zero at the ends; to allow for this, the measured length of the image should be multiplied by a factor, tentatively given [1] as 0·7. The length must be divided by the enlargement (magnification) factor $h/(x - h)$ (Figure 5.2).

Angle of Focal Spot

The focal spot of a line-focus tube is set at right angles to the emergent beam, so that the apparent size of the focal spot, *BC* or *GH* (Figure 5.4) will vary

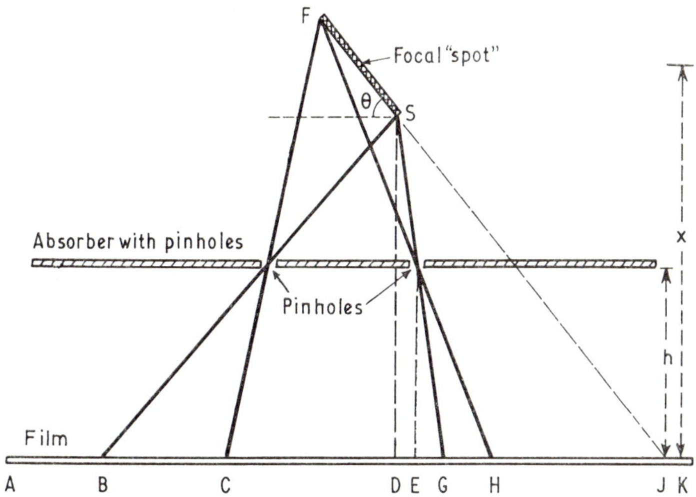

FIGURE 5.4 *Pinhole Method for Determining Angle of Focal Spot*

with the position from which it is viewed. This difference in apparent size can be exploited [3] to find the angle θ, between focal spot and film *AK*. A pinhole photograph is taken, using two pinholes, each small compared with the focal spot dimensions. The pinholes are in a sheet of absorbing material, placed parallel to the film and a distance h above it.

X-ray Exposure Rates

Measurement of exposure rates from X-ray tubes were at one time confined largely to radiotherapy work, but the introduction of more stringent safety standards for workers and for patients undergoing X-ray diagnosis has led to increasing use of exposure measurements in industry and diagnostic X-ray departments in hospitals.

A particular exposure rate will be specific for a certain kilovoltage, tube current, added filtration and target distance of a particular set. Measurements of exposure should be carried out with minimal extraneous scatter present.

The dosemeter should have been recently calibrated for the full range of radiation qualities encountered.

For low-energy X-radiation, say of HVL (p. 89) 0·01 mm to 1·0 mm Al, chambers may be made of thin nylon, polystyrene, paper or other suitable solid material, or of mesh; Jennings [4] has compared depth dose measurements using different types of chamber in this energy range. Special problems arise due to focal distance, which is unlikely to be large compared with chamber dimensions [2], and due to wavelength-dependence of the chamber response (Chapter 4).

For higher-energy radiations generated at up to 400 kVp, an instrument such as the Baldwin-Farmer substandard dosemeter (Chapter 4), is suitable, with the chamber wall appropriately thickened, if necessary. The field area should not be so large that it results in undue scattered radiation, nor should unnecessarily large parts of the chamber insulator be irradiated. But the whole effective volume of the chamber must "see" the whole focal spot, and hence some scattered radiation is inevitable. For radiation of 1·8 mm Cu HVL, the relative exposure rate may increase [5] some 7 per cent as the area of the beam increases from 100 to 10,000 mm^2, and it is recommended that a standard field size of 100 mm × 100 mm should be adopted. In order to minimize scatter from the collimating system, measurements taken at 50 mm depth in a water system are preferable [2] to measurements in air, as a basis for depth-exposure measurements.

It is difficult to measure exposure when X-ray photons have energies higher than a few MeV. Nevertheless, since 1956 the National Physical Laboratory has undertaken the calibration of secondary standard dosemeters with 2 MV X-rays. These are based on measurements of ionization in a graphite cavity ionization chamber in terms of the 2 MV X-ray "röntgen". Studies have since been extended [6, 7] to cover X-rays from 4, 6, 8 and 15 MV linear accelerators. The consistency of the 2 MV X-ray "röntgen" with that based on a free-air ionization chamber and for orthovoltage radiation has been checked [8] using a graphite cavity chamber for both energy ranges.

A "Code of practice for the dosimetry of 2 to 8 MV X-ray and caesium-137 and cobalt-60 gamma-ray beams" describes [9] an accepted method for measuring the dose in rads at a specified point in an absorbing medium. The dose is obtained from ionization measurements made with a dosemeter previously calibrated for 2 MV X-rays at the National Physical Laboratory. The build-up cap of the chamber should be of 4·6 mm Perspex. Measurements should be made on the central beam axis at a depth of 50 mm inside a water phantom of adequate size. The dose D in rads is given by

$$D = RNC_\lambda \tag{5.2}$$

where R = ionization chamber instrument reading corrected to dry air at 22°C, 760 mmHg

N = quoted N.P.L. 2 MV factor to röntgens for dry air at 22°C, 760 mmHg

C_λ = overall conversion factor, which is a function of the radiation quality and the chamber plus Perspex cap (Table 5.2).

For any quality, if the dosemeter has no switch for commencing and terminating the exposure, precautions must be taken to avoid errors due to the slow increase in tube voltage while running up, or to a shutter opening and closing. If such errors cannot easily be eliminated, their effects may be minimized by subtracting two exposure readings, each for different times, the difference in exposure readings being that for the difference in times [10].

When an X-ray set is to be used under a variety of conditions, exposure rates for all these should be experimentally investigated. For instance, the target distance may be varied; it may be that scatter from the collimating system and air absorption are negligible, when exposure rate will be inversely proportional to the square of the target distance. Should this be established, exposure rates for intermediate distances may be accurately interpolated.

TABLE 5.2 *Values of Conversion Factors for Two Dosemeters with 4·6 mm Perspex Cap*
(From Hospital Physicists' Association, *Phys. Med. Biol.*, 9, p. 457, 1964)

X-ray equivalent generating voltage (MV)	C_λ	
	Baldwin-Farmer dosemeter	*Victoreen dosemeter*
2	0·945	0·940
4	0·939	0·934
6	0·933	0·929
8	0·926	0·923

When it is proposed to alter tube running conditions or add filters, exposure rates should be separately measured. Over a limited range, the tube current may be found to be proportional to the exposure rate.

Day-to-day exposure rates may vary with temperature, supply voltage and ageing of the tube, as well as with the development of minor faults and ageing of circuit components. Consequently, a parallel-plate monitor in the beam is useful as a continuous check on exposure rate. In addition, accurate exposure or absorbed dose rate calibrations should be repeated every one or two weeks [11] under reproducible conditions, for which a jig is convenient. Only one set of conditions of kilovoltage, filter and so on, need be investigated for these routine calibrations.

Half-value Layer or Half-value Thickness

When describing X-radiation, it is important to know the effective energy as well as the exposure rate. The energy is determined primarily by the tube voltage, which is often continuously varying in a manner dependent on the circuit (Chapter 13). The value of neither the maximum nor the average tube voltage is sufficient to determine the spectrum of photon energies, which is further modified by the target material and beam filtration.

A measure of the energy of the radiation may be obtained from its half-value layer (HVL) or half-value thickness (HVT). These are synonyms, the quantity being [12] the thickness or the surface density of a layer of specified material which attenuates the beam to such an extent that the exposure rate is reduced to one half, under narrow beam conditions. The material used for attenuation should be in sheets thick enough to measure and maintain, but which has attenuation coefficients varying rapidly with photon energy, with no discontinuities. Certain standards of purity must be maintained. Jones [13] considers that there are advantages in using aluminium for all energies, but other materials are acceptable [2] (Table 5.3).

TABLE 5.3 *Materials for HVL Measurements*
(From ICRU Report 10b, *Physical Aspects of Irradiation*, NBS Handbook 85, Washington, D.C., NBS, 1963)

Material	*Acceptable level for HVL measurements*
Aluminium	10 kV — 10 (or more) MV
Copper	35 kV — 8 MV
Lead	350 kV — 3·5 MV

In an experimental determination of HVL, readings should be made under conditions where no significant amount of scatter is recorded [14]; the area of the beam should be just large enough to irradiate the measuring chamber, and the absorbers should be placed about half-way between tube and chamber. Dimensions of a suitable measuring arrangement may be obtained from published curves [2].

Steps must be taken to counteract the effects of fluctuations in output; there are three main methods. First, an automatic device may be installed to stabilize the output [15]. The second alternative is to use a comparator technique: the Kemp comparator is admirably suited to this type of measurement. In this method, a monitor ionization chamber is exposed simultaneously with the measuring chamber; the comparator measures, by a null method, the charge ratio of the two chambers [16, 17, 18]. The third alternative is less elegant than, but in principle similar to, the second: two chambers are exposed simultaneously, one under standard conditions to monitor the output of the X-ray set and the other with appropriate absorbers between it and the tube; the two chamber readings are compared. In these and other repetitive measurements, the fitting of jigs to hold the chambers increases both accuracy and speed of working.

Photons of homogeneous energy are attenuated according to the equation

$$X = X_0 \, e^{-\mu d} \tag{5.3}$$

where X_0 is the exposure when no absorber is present
X is the exposure with an absorber of thickness d and linear absorption coefficient μ.

Hence, a graph showing the relationship between log X and d will be linear.

However, in general, photons of a range of energies will be present, and those of lower energy will be preferentially absorbed, as they have a higher value of μ. The relationship between log X and d will be curvilinear (Figure 5.5, *lower curve*).

The calibration correction of the ionization chamber generally varies with the energy of the radiation, that is with its HVL, and therefore when each experimental point on the uncorrected curve is multiplied by its appropriate calibration factor, a corrected curve of slightly different shape will result (Figure 5.5, *upper curve*). The correct curve will be approached after a number of successive approximations, although in practice one corrected

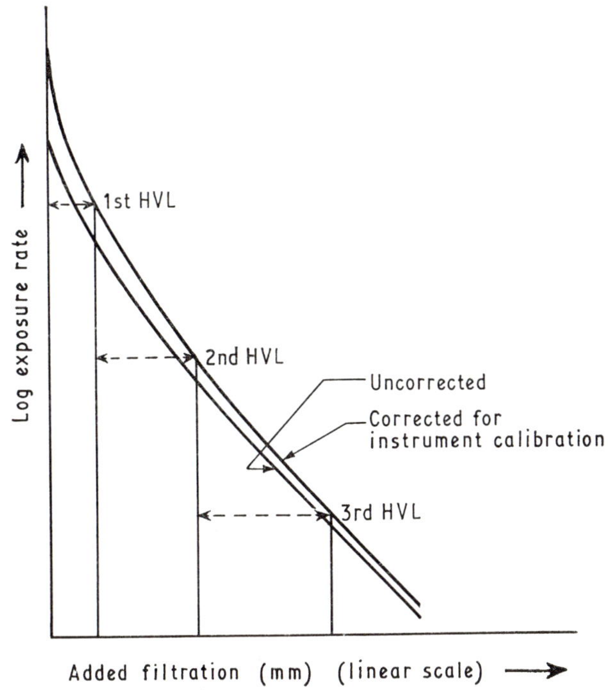

FIGURE 5.5 *Measurement of HVL*

curve is enough. From this curve, values of HVL can be read, both for the tube output and for more heavily filtered beams.

Particularly for radiotherapy, the HVL has proved to be a useful measure of X-ray energy, but it is not specific for the spectrum of energies present. It is possible to imagine two beams of X-rays, one of homogeneous and one of heterogeneous energies, yet both having the same HVL. In practical terms, the former approximates to a heavily filtered beam and the latter to one lightly filtered.

Some information about the degree of heterogeneity can be obtained by comparing the value of the first HVL (Figure 5.5), with that of the second, the ratio being known as the homogeneity coefficient [12]. Greening [19] has examined the degree to which X-ray spectral distributions may be derived from a knowledge of one or more of the three parameters: first HVL, second HVL and tube voltage. In radiotherapy, for voltages up to 2 mV, both the kV or MV and the (first) HVL should be specified [12]. Above 2 MV, the quality of the radiation is not significantly modified by the filtration and only the MV need be stated.

Inherent and Added Beam Filtration

At low energies, the window of the tube may act as an unwanted beam filter, absorbing useful soft X-rays. X-ray tubes are available with an inherent filtration of only 1 mm beryllium; they are operated at 2 kVp to 50 kVp or higher [20, 21, 22]. Exposure rates as high as 1,000 R/sec are obtainable. If the low-energy photons are unwanted, an added filter of 0·025 mm Al will absorb them to an extent great enough to reduce the exposure rate four-fold when the tube is run at about 5 kVp.

X-ray sets for crystallography are operated typically at about 40 kVp. The anode is at right angles to the electron beam, because only narrow beams are required.

When X-radiation generated at 200 kVp to 500 kVp is used for deep radiotherapy, the softer components of the beam must be absorbed in tube filters, in order to increase the tumour:skin dose ratio.

For megavoltage radiation, the X-ray beam is usually transmitted through the target, which acts as a filter. Photoelectric absorption coefficients are small in this energy range, and filters are not useful to harden the beam. However, low atomic number materials, which show a slight preferential absorption for lower energy photons, are used to improve the uniformity of radiation in a plane perpendicular to the central axis of the beam. Megavoltage X-ray beams may be contaminated [2] by electrons scattered from diaphragms, and if these cannot be prevented by distance from reaching the skin or other medium, filters of moderate atomic number materials such as tin or brass, may be used to absorb the electrons.

Size and Uniformity of Field

The maximum size of field obtainable may be measured on a photographic film, exposed on a plane perpendicular to the central axis of the tube. If the target–film distance is known, maximum field sizes at other distances may be calculated.

For a reflection target, variation of in-air dose across the field is, in general, greatest in a direction parallel to the tube axis and least in a direction perpendicular to this axis. With a proper choice of target angle and with a good design of tube shield, the variation should not exceed about 10 per cent across a 250 mm diameter field at 500 mm FSD [23]. In general, the HVL is higher towards the anode side of the tube: the *heel* effect [2]. This is due to the internal filtration at the target. If the target is too steeply inclined, the area of highest exposure rate may be towards the cathode [20] (Figure 5.6). If the target angle is too small, distortion of the target face during the tube life may result in the uniformity of dose distribution deteriorating [24].

Transmission targets give x-ray fields with dose rates higher at the field centre than at the periphery. The dose distribution may be examined radiographically or using detectors of small size [25]. Beam-flattening filters can correct for non-homogeneity of dose rate over the field area [26, 27], but as they achieve their purpose by absorbing radiation in the central region of the field, the dose rate is reduced, sometimes by as much as one-third [1, 27].

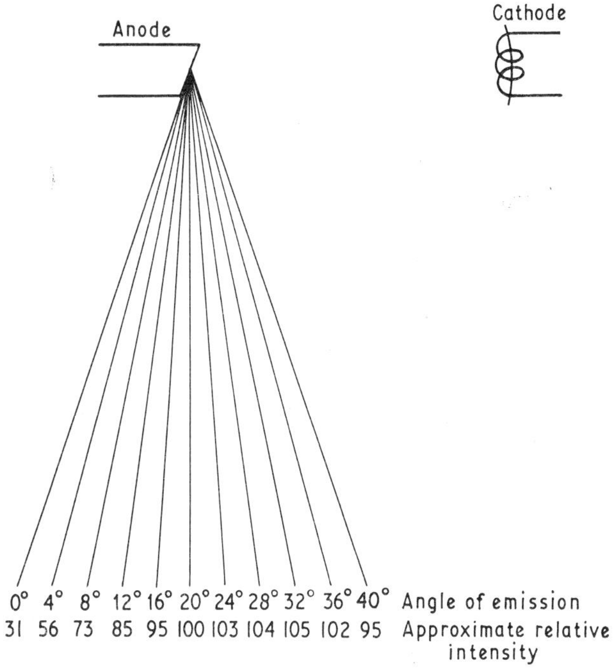

FIGURE 5.6 *Relative Intensity Across an X-ray Beam with a Tube Target Angle of 20°*

(Jacobi, C. T. and Hagen, D. E., *X-Ray Technology*. St. Louis, The C. V. Mosby Co., 1960)

Beam-limiting Devices

Beams should be as small as possible for the purpose in hand (Chapter 14). Beam limiters are of three types: the cone or applicator, the variable diaphragm, and the pinhole system. The tube window can be considered also as a beam limiter.

A cone or applicator can be fixed on to the tube housing. It is usual to have available a range of sizes. The device serves four purposes: to locate the beam, to fix the focal distance, to reduce the penumbra and to limit the beam size, although the collimator effecting this may or may not be incorporated in the device.

A limiting diaphragm (Figure 5.7), at the tube end will reduce scatter; it may be made of lead or of some other radio-opaque material. It should transmit a beam slightly larger than the size required, so that this is determined by the dimensions of the cone or applicator near the skin, with consequent limitation of penumbra width. Hence, the sides must be made of radio-opaque material; as the beam strikes the sides obliquely, the thickness effective for absorption is much greater than the thickness of the sides.

The applicator face (Figure 5.7), should be open-ended for soft X-rays up to about 150 kVp, as any applicator end plate might absorb the primary

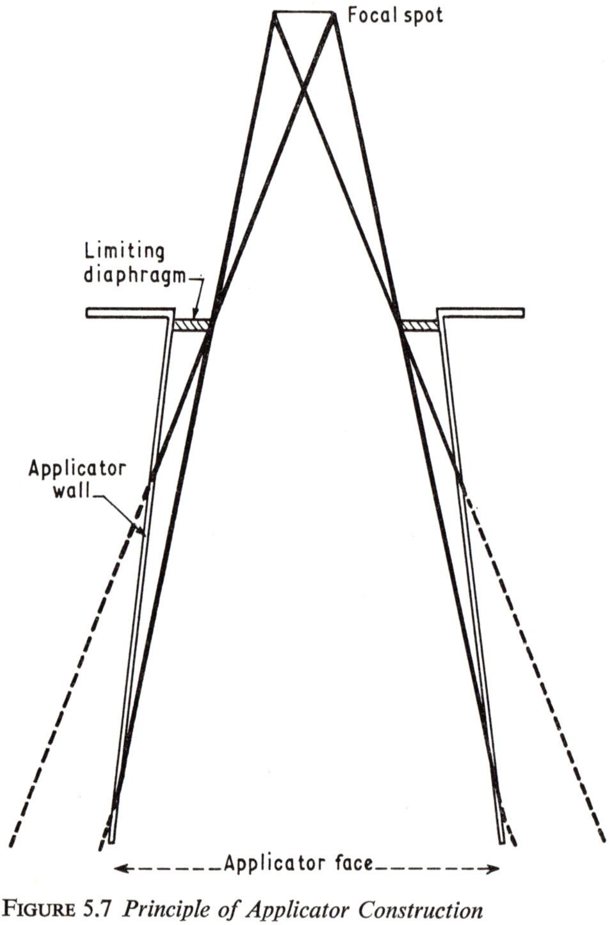

FIGURE 5.7 *Principle of Applicator Construction*

beam appreciably; an open end is advantageous also for megavoltage radiation, to eliminate forward Compton scatter from the face [28]. At intermediate energies of a few hundred kVp, it is usual to have a closed end, often of Perspex for ease in setting-up. The face is useful for marking the field centre and for absorbing scatter from the sides of the applicator, but an unsuitable material or thickness may materially affect the output at or near the surface, introducing a deviation from the inverse square law relating focal distance and exposure rate, as well as modifying the percentage depth dose and the backscatter factor in tissue [12, 29]. It is therefore advantageous to use as a reference, the dose rate at 50 mm depth. Reliable depth doses are thus obtained at the price of some uncertainty of surface dose.

It is necessary to check that a particular applicator transmits a beam of its nominal size, for all orientations between it and the tube, and that there are no leaks of radiation from the applicator.

Variable diaphragms consist most simply of two pairs of parallel plates (Figure 5.8), made of lead or another radio-opaque material. For megavoltage radiation, the weight of material necessitates more complex geometrical arrangements. Diaphragm leaves to give irregular shapes are possible.

Visual indication of the X-ray field is given by means of a light device, which usually marks also the field centre and focal distance. It is important

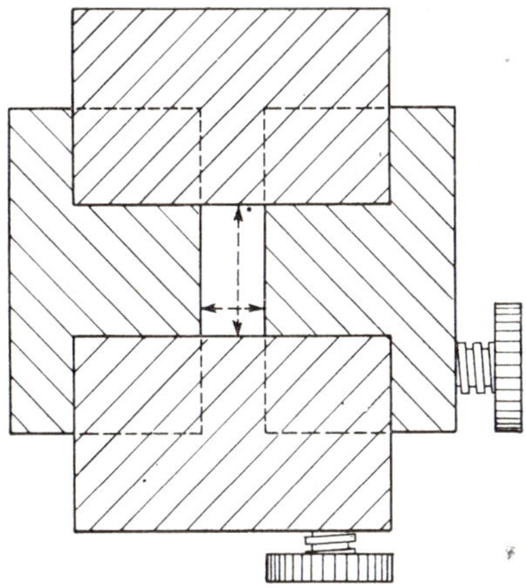

FIGURE 5.8 *Variable Diaphragm*

to ensure that light and X-ray fields are coincident, and necessary to check that they remain so. The examination can be made by setting up a particular field using the light device, marking the field edges on a photographic film or a fluorescent screen. An X-ray exposure is then made, and the light and X-ray fields examined for coincidence. If a fluorescent screen is used, very careful safety precautions must be taken to avoid exposure of the operator.

Variable diaphragms have the advantage that they can be used to produce fields of any size, within the limits of the machine. They can be permanently attached to the tube housing, thus eliminating handling of heavy parts. They are convenient for use with moving fields, where applicators are inappropriate because the focal-skin distance is continually changing as tube or patient rotates. On the other hand, as the beam is limited at some distance above the skin, penumbra are wider than with applicators.

COLLIMATION AND BEAM LIMITING IN CRYSTALLOGRAPHY

To collimate means to render parallel (the word is really derived from the Latin *collineare*, not *collimare*), and the ideal collimator would produce a parallel beam of X-rays. X-ray cameras have devices called collimators, but the use of this word is not strictly appropriate as they are there merely to limit the beam and not to render the beam parallel.

The real object of collimators in X-ray camera design is to produce intense lines or spots of high resolution with a minimum possible background. The X-ray source is usually the focal spot of the X-ray tube, and this forms an essential part [30] of the collimator and camera geometry. There is usually an entrance aperture which may be designed to view the whole of the focal spot where high intensity is required or a small part of the spot when a fine beam is required. The exit aperture of a collimator will diffract X-rays, and this may complicate the X-ray pattern received on the film from the specimen, so there is often a guard aperture to collect X-ray scatter from the collimator. Collimators are usually made of lead, but this is not the ideal material because of its poor mechanical properties and the best collimators are made of platinum. A good test to see if a collimator is functioning properly in an X-ray powder camera is to run the camera without a specimen. No scatter should be received on the film.

In the powder diffractometer, the X-ray beams are defined by special slits, which allow the beam to diverge on to the flat specimen and converge on to the counter, to conform with the focusing geometry. Collimators for single-crystal diffractometers are of the pinhole type.

Literature References

1 International Commission on Radiological Units and Measurements, *Methods of Evaluating Radiological Equipment and Materials*. Recommendations of the ICRU, Report 10*f*, 1962. NBS Handbook 89 (Washington, D.C., Nat. Bur. Stand., 1963).

2 International Commission on Radiological Units and Measurements, *Physical Aspects of Irradiation*, Recommendations of the ICRU, Report 10*b*, 1962. NBS Handbook 85 (Washington, D.C., Nat. Bur. Stand., 1963).

3 Stern, B. E. and Walford, G. V., "The target angle of an X-ray tube", *X-Ray Focus*, **8**, p. 30 (1967).

4 Jennings, W. A., "A survey of depth dose rates for X-rays from 6 to 75 kVp", *Brit. J. Radiol.*, **26**, p. 481 (1953).

5 Farr, R. F., "The specification of Roentgen ray output and quality", *Acta Radiol.*, **43**, p. 152 (1955).

6 Barnard, G. P., "Standardization of ionization chambers for megavoltage X-rays", *Phys. Med. Biol.*, **9**, p. 287 (1964).

7 Barnard, G. P., Marsh, A. R. S. and Hitchman, D. G. I., "Studies of cavity ionization chambers with megavoltage X-rays", *Phys. Med. Biol.*, **9**, p. 295 (1964).

8 Barnard, G. P. *et al.*, "On the congruity of N.P.L. exposure standards", *Phys. Med. Biol.*, **9**, p. 333 (1964).

9 Hospital Physicists' Association, "Code of practice for the dosimetry of 2 to 8 MV X-rays and caesium-137 and cobalt-60 gamma ray beams", *Phys. Med. Biol.*, **9**, p. 457 (1964).

10 International Commission on Radiological Units and Measurements, ICRU Report, 1959. NBS Handbook 78 (Washington, D.C., Nat. Bur. Stand., 1961).

11 Bull, S. M. and Burlin, T. E., "Routine calibration of diagnostic x-ray machines", *Brit. J. Radiol.*, **35,** p. 796 (1962).

12 International Commission on Radiological Units and Measurements, *Clinical Dosimetry*, Recommendations of the ICRU, Report 10(*d*), 1962. NBS Handbook 87 (Washington, Nat. Bur. Stand., 1962).

13 Jones, D. E. A., "The suitability of materials used for the measurement of half-value-thickness for x-ray beams". *Brit. J. Radiol.*, **34,** p. 801 (1961).

14 Hospital Physicists' Association, "A code of practice for x-ray measurements", *Brit. J. Radiol.*, **33,** p. 55 (1960).

15 Clayton, C. G. and Farmer, F. T., "Stabilization of x-ray therapy apparatus by ionization current", *Brit. J. Radiol.*, **22,** p. 669 (1949).

16 Kemp, L. A. W., "A new circuit with special applications in gamma-ray and x-ray dosimetry", *Brit. J. Radiol.*, **18,** p. 107 (1945).

17 Kemp, L. A. W., "A method of calibrating an ionization current comparator without reference to radiation", *Brit. J. Radiol.*, **24,** p. 211 (1951).

18 Kemp, L. A. W., "A review of the theory, calibration techniques and applications of an ionization current comparator, with an investigation of its capabilities as a precision instrument", *Amer. J. Roent.*, **71,** p. 853 (1954).

19 Greening, J. R., "The derivation of approximate spectral distributions and an analysis of x-ray quality specifications", *Brit. J. Radiol.*, **36,** p. 363 (1963).

20 Jennings, W. A., "Physical aspects of the Roentgen radiation from a beryllium window tube operated over the range 2–250 kVp for clinical purposes", *Acta Radiol.*, 33, p. 435 (1950).

21 Jennings, W. A., "A theoretical study of the radiation outputs and qualities from a beryllium-window tube operated at low kilovoltages from 10–50 kVp", *Brit. J. Radiol.*, **26,** p. 193 (1953).

22 Jennings, W. A., "A survey of depth dose rates from x-rays from 6 to 75 kVp", *Brit. J. Radiol.*, **26,** 481 (1953).

23 Farr, R. F., "The distribution of dose-rates across the field of an x-ray therapy tube", *Brit. J. Radiol.*, **28,** p. 364 (1955).

24 Beetlestone, A. and Thurmer, G., "Some considerations of focal spot sizes", *Brit. J. Radiol.*, **31,** p. 168 (1958).

25 Koschel, K. W. and Clarke, K. H., "An instrument for the rapid measurement of x-ray field flatness", *Brit. J. Radiol.*, **36,** p. 538 (1963).

26 Greene, D., "Size of the focal spot in a betatron", *Brit. J. Appl. Phys.*, **8,** p. 293 (1957).

27 Miller, C. W., "Recent developments in linear accelerators for therapy. I. The continuing evolution of linear accelerators", *Brit. J. Radiol.*, **35,** p. 182 (1962).

28 McEvedy, M. B., "Open-ended applicators for high energy gamma-ray therapy", *Brit. J. Radiol.*, **30,** p. 168 (1957).

29 Oliver, R. and Kemp, L. A. W., "An investigation into some factors affecting x-ray dose distribution and its measurement", *Brit. J. Radiol.*, **22,** p. 33 (1949).

30 Bernal, J. D., "Universal x-ray photogoniometer", *J. Sci. Instr.*, **5,** p. 241 (1928).

6 Dosimetry

When material is exposed to X-radiation, the dose absorbed depends on two main factors. The first concerns the output or exposure from the X-ray generator, and is obtained from its calibration (Chapter 5). The second

TABLE 6.1 *Conversion Factors from Röntgens to Rads for Typical Radiation Spectra* (From Whyte, G. N., *Principles of Radiation Dosimetry*, New York, Wiley, 1959)

Radiation	*Filter* (mm)	*HVL* (mm)	rads/R *Water*	*Muscle*	*Bone*
100 kV X-rays	0·18 Al	5·5 Al	0·91	0·94	3·1
150	0·075 Cu	0·2 Cu	0·92	0·94	2·7
200	0·20 Cu	0·5 Cu	0·94	0·95	2·0
250	0·17 Cu + 3·0 Al	1·0 Cu	0·95	0·95	1·76
250	0·9 Cu + 3·0 Al	2·0 Cu	0·96	0·96	1·42
280	—	2·5 Cu	0·97	0·96	1·22
400	—	4·2 Cu	0·97	$0{\cdot}96_5$	1·11
^{137}Cs γ-rays (0·66 MeV)	—	—	$0{\cdot}97_5$	$0{\cdot}96_5$	0·93
^{60}Co γ-rays (1·25 MeV)	—	—	$0{\cdot}97_5$	$0{\cdot}96_5$	0·93

TABLE 6.2 *Calculated Conversion Factors for Baldwin-Farmer Chamber fitted with 4·6 mm Perspex Cap*
(From Barnard, G. P., *Phys. Med. Biol.*, **9**, p. 321, 1964)

X-Radiation	*rads in water per 2 MV "röntgen"*
2 MV	0·9453
4·2	0·9380
5·6	0·9344
7·6	0·9277
13·3	0·9146

factor on which dose depends is the nature of the material, as this influences the absorption properties of the X-rays in question.

A particular material, when exposed to X-rays of a certain energy, may be said to absorb so many rads per röntgen (Table 6.1), or for megavoltage X-rays, so many rads in water per 2 MV "röntgen" (Table 6.2).

Scatter Factor

The ratio of the exposure at a reference point in a phantom to the exposure at the same point in space under similar conditions of irradiation in the absence of the phantom is known [1] as the scatter factor. For up to 400 kV radiation [1], consider an x-ray set calibrated and found to give an exposure rate of P röntgens per minute at a point A on the central ray (Figure 6.1*a*). Now consider absorbing material of depth at least 100 mm, placed to fill the beam and some distance outside it, so that its cross-section is at least

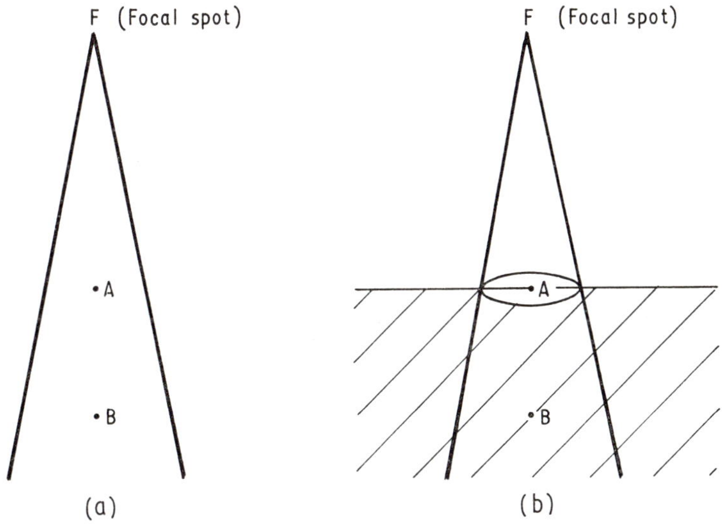

FIGURE 6.1 *Exposure in Material*

300 mm × 300 mm. The point A is arranged to be on the surface of the material (Figure 6.1*b*). Then if P' is the new exposure rate at A,

$$P' = P + S \tag{6.1}$$

where S is the exposure rate at A from secondary radiation back-scattered from the absorbing material. In this case, the scatter factor P'/P is called [1] the *back-scatter factor* (BSF), and

$$\text{BSF} = 1 + S/P \tag{6.2}$$

The value of the BSF depends on the amount and direction of Compton scatter. It rises with x-ray energy to a maximum when the primary beam has a HVL of about 0·6 mm Cu, and falls to unity for 10 MeV photon beams [2].

For all energies, S is by definition zero for infinitely small fields. For larger fields, the value of S depends on both the area and the shape of the field [3, 4]. The value of the BSF increases with the dimensions of the field until they are so large that scattered radiation originating at the periphery contributes little to the exposure rate at the centre (Figure 6.2). Circular fields have higher BSFs than elongated fields of the same area, where the amount of scatter reaching the field centre is limited by the longer length of the field.

Dose in Material

Consider a point B in air (Figure 6.1a) and assume that absorption in, and scatter from, the air is negligible. Then exposure rate X/t at B is

$$(X/t)_B = P \times \frac{AF^2}{BF^2} \tag{6.3}$$

When absorbing material is present, as in Figure 6.1(b), then

$$(X/t)_B = P \times \frac{AF^2}{BF^2} \times e^{-\mu AB} + S \tag{6.4}$$

where μ = linear absorption coefficient of the radiation in the material
S = exposure rate at B from the scattered radiation.

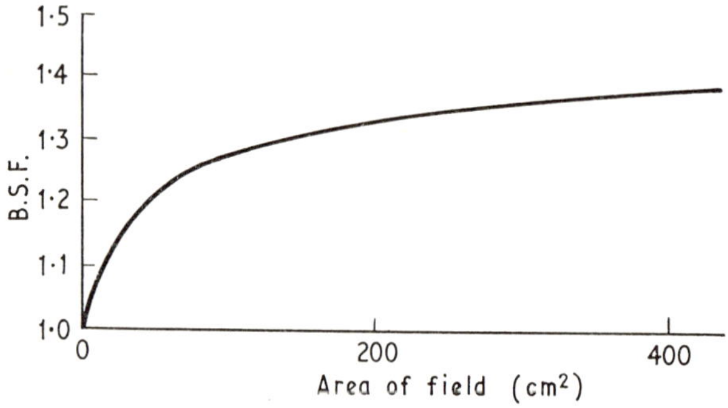

FIGURE 6.2 *Variation of BSF with Field Area for X-rays of about 2 mm Cu HVL*

If the absorbing material is homogeneous and the energy of the radiation within the material does not vary outside certain limits, the rads per röntgen will remain constant and percentage depth exposure may be equated with percentage depth dose. The percentage depth dose is defined [1] as the ratio, expressed as a percentage, of the absorbed dose D_x at any depth x to the absorbed dose D_0 at a fixed reference point on the central ray.

The primary dose decreases with depth, while the scatter dose contribution increases to a maximum and then decreases. At low X-ray energies, the decrease in primary dose near the surface is greater than the increase in scatter, so that the maximum dose rate is on the surface. As the photon energy increases, attenuation of the primary beam decreases while the range of the secondary electrons increases and they become directed more in the forward direction. At about 250 kVp, the maximum dose rate is a few millimetres below a water or soft tissue surface and at 22 MV it is several cm below [5].

The amount of scatter will be a maximum at the field centre, and will decrease towards the periphery. It may be appreciable outside the primary radiation field for orthovoltage radiation (Figure 6.3).

A number of methods [6, 7, 8, 9] have been published for calculating the BSFs and percentage depth doses of elongated or irregular shaped fields

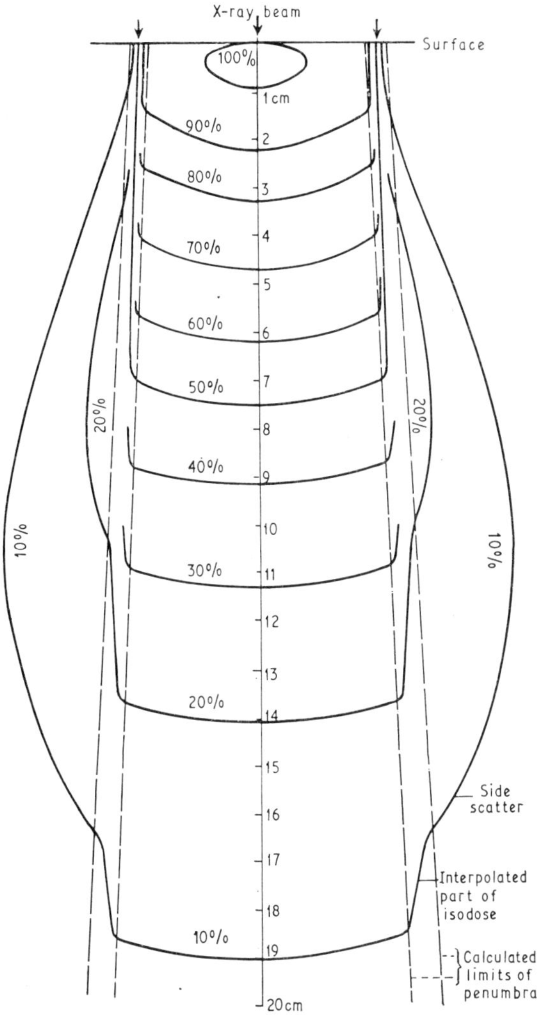

FIGURE 6.3 *Typical Isodose Curve for X-rays of about 2 mm Cu HVL, 500 mm FSD, and Field Area 100 mm × 100 mm*

from data applicable to square or circular ones. These calculations, usually tedious for orthovoltage radiation, become unnecessary for megavoltage radiation when field size and shape have little effect on dose distribution because the scatter is largely forward.

The energy spectrum of the radiation varies with the position of *B* (Figure 6.1*b*), being highest near the surface and the inside edge of the beam if side scatter is appreciable. Near the inside edge the scattered radiation is small in magnitude and the primary beam is a little harder than average, as it has passed obliquely through the tube filters. Energy measurements may be

TABLE 6.3 *Recommended Central Percentage Depth Dose Data*
(From ICRU Report 10d, *Clinical Dosimetry*, NBS Handbook 87, Washington D.C., NBS, 1963)

HVL or generating voltage	*Source of data*	*Beam definition*
0·01 to 1·0 mm Al	*Brit. J. Radiol.*, Supp. No. 10 (London, Brit. Inst. Radiol., 1961).	Open-ended applicators
1·0 to 8·0 mm Al	ibid.	Diaphragms
1·0 to 8·0 mm Al	Braestrup, C. B., "Depth dose measurements for 100, 120 and 135 kV röntgen rays," *Radiol.*, **42**, p. 258 (1944)	Diaphragms
1·0 to 8·0 mm Al	Wachsmann, F. and Dimotsis, A., "*Kurven und Tabellen fur die Strahlentherapie*," (Stuttgart, S. Hirzd, 1967)	Unspecified
0·5 to 3 mm Cu	*Brit. J. Radiol.*, Supp. No. 10 (as above)	Diaphragms
	Wachsmann, F. and Dimotsis, A. (as (above)	Unspecified
1·0 to 4 mm Cu	*Brit. J. Radiol.*, Supp. No. 10 (as above)	Closed-ended applicators
2 MV	ibid.	Diaphragms
4 MV	ibid.	Diaphragms
8 MV	Newbery, G. R. and Bewley, D. K., "The performance of the Medical Research Council 8 MeV linear accelerator", *Brit. J. Radiol.*, **28**, p. 241 (1955)	Diaphragms
15 MV	*Brit. J. Radiol.*, Supp. No. 10 (as above)	Diaphragms
20 to 25 MV	Shapiro, G., Ernst, W. S. and Ovadia, J., "Radiation dose distribution in water for 22·5 MeV peak röntgen rays", *Radiol.*, **66**, p. 429 (1956)	Diaphragms
	Laughlin, J. S. *et al.*, "Dose distribution measurements with the University of Illinois 25 MeV medical betatron," *Amer. J. Roent.*, **68**, p. 787 (1951)	Diaphragms
30 to 35 MV	Wideroe, R., "Integraldosen für 200 kV Röntgen und für Megavoltstrahlen," *Strahlenth.*, **110**, p. 1 (1959)	Diaphragms

made using two dosemeters with different energy responses, and comparing their readings at the same position [10].

Isodose Curves

In radiotherapy, control of dose throughout the irradiated tissue is of vital importance. It is convenient to prepare plans of dose distribution which apply to a particular X-ray energy, focal-skin distance and field size. These plans are usually in a form showing the dose as a percentage of the dose at a reference point on the central axis, equal dose contours being drawn for a plane. They are known as isodose curves and are relevant for uniform soft tissue (Figure 6.3).

Isodose curves may be constructed from measurements using, for instance, a monitor chamber and a search chamber, the latter movable in a large tank of water or other tissue-simulating material. Several methods [11, 12, 13, 14, 15] have been described for obtaining such measurements rapidly and accurately. Owing to the effects of applicators (p. 94) and to the difficulty in measuring surface exposure rates, curves should be normalized at 5 cm depth, or such other depth where this exceeds 5 cm as will ensure electronic build-up [1].

Isodose curves for medium-energy x-rays have been prepared [16] using the best available data. They are usually to be preferred to local measurements. About 250 charts cover values of HVL from 1·0 mm Cu to 3·0 mm Cu. For higher energies, there is a list [17] of sources of charts.

Central axis depth dose data are available for the whole range of energies from a beam of HVL 0·01 mm Al to 35 MV x-rays (Table 6.3).

PENUMBRA

If the focal spot is large or the beam defined far from the surface, the penumbra may be large. From calibration data (Chapter 5), its width can be calculated and may be drawn on the isodose curve. Isodose lines inside and outside the primary beam may then be joined by interpolation within the penumbra (Figure 6.3) without serious loss of accuracy. Alternatively, the dose distribution inside the penumbra may be measured with a small exploratory chamber [18, 19, 20].

Dosimetry for Moving-field Therapy

Isodose curves are inappropriate for moving-field therapy because, owing to the irregular shape of the body and the position of the centre of rotation, both the focal skin distance f and the area a of the entrance field at the body surface vary throughout the rotation (Figure 6.4). The dose, at a point in the body contributed by the field when it is in some particular position during its traverse, can be calculated from absorption and scatter data. The total dose at the point is the sum of the contributions from a sufficient number of field positions.

In order to calculate the dose in a form which is applicable whatever the exposure, a reference dose is required which does not change throughout the rotation. The air dose at the centre about which the field rotates is convenient for this reference.

This leads to the concept of the *tissue–air ratio*, or TAR [1], which is the ratio of the absorbed dose at P (Figure 6.4) to the air dose at the same point. The TAR value is nearly independent of the focal distance from the centre of rotation, but depends on the area of the field at this centre and the depth of overlying tissue. Tabulated values of TAR are available [2, 5, 16]. The principle can be extended to calculate the dose at any point in the body

[22, 23, 24, 25, 26, 27, 28]. All the methods are more or less tedious, and dose distributions for moving-field therapy are best carried out by computer. Calculations may be supplemented by measurements either in a phantom [29] or in some convenient body cavity.

Body Inhomogeneities

Depth dose data and isodose curves are applicable for homogeneous blocks of soft tissue, but these are seldom encountered in practice. The presence of body inhomogeneities such as fat, bone and lung may give rise to serious discrepancies between dose calculations and the dose actually received. The

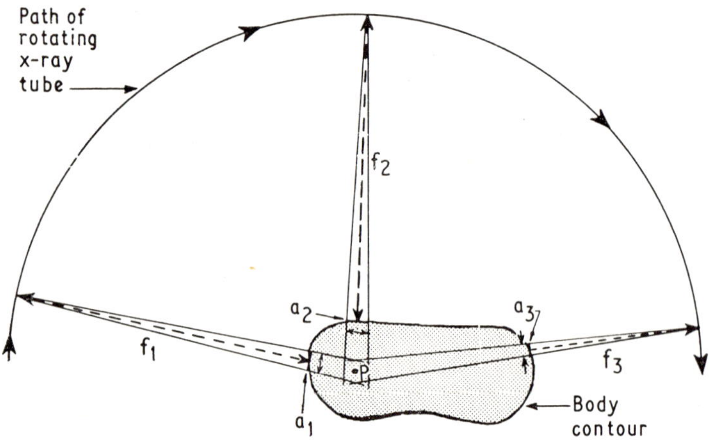

FIGURE 6.4 *Moving-field Irradiation*

f_1, f_2, f_3: effective focal skin distances; a_1, a_2, a_3: effective field sizes; P, centre of rotation arc.

problem is much less acute for megavoltage than for orthovoltage radiation, and it is a prime advantage of the former that dose distribution is largely unaffected by the presence of fat or bone, although correction for air in the lungs remains a problem.

Air in the lungs can modify the dose from megavoltage radiation by as much as 20 per cent, and therefore a correction should be attempted [30]. The presence of lung tissue and the air-filled cavity leads to a deficit of scatter and a surfeit of primary radiation, the latter predominating. But as the difference between the two varies with distance behind the lung, the practice of basing the correction on "transit" exposure measurements may lead to overcompensation [30]. Measurement of transit exposure is made at the side of the patient remote from the incident beam.

Fat has a comparatively low atomic number, so that for photon energies of say [31] less than 0·1 MeV, where photo-electron absorption is important, the absorption coefficient for fat is appreciably less than for unit density soft tissue. Consequently, if this soft tissue is irradiated through a layer of fat, the dose received will be higher than that calculated from standard depth dose data.

Bone contains minerals of relatively high atomic number, as well as having a high density, so that its absorption will be high, particularly for low-energy x-radiation. For a beam of x-rays of HVL 0·5 mm Cu, 10 mm of bone will reduce the dose [32] behind it by about 22 per cent, and 40 mm of bone by about 50–60 per cent. In clinical practice, the shadow effect behind bone is complicated by the irregular geometry. Boney structures contain soft tissue interstices [33], for instance, the Haversian canals (10–50 μ diameter). Now the ranges of secondary electrons may be of comparable dimensions [34]: 26·8 μ (50 keV), 1,032 μ (480 keV). Hence secondary electrons emitted from bone can materially increase the dose to soft tissue within bone, particularly when the photon energy is less than about 200 keV. At this energy, the rads per röntgen are much the same for all sizes of canal. Spiers [35, 36] has calculated the *permissible dose* to bone, that for soft tissue being taken as 100 per cent. On this basis, the permissible dose to bone is 90 per cent for x-rays of 0·06 Å, (500 to 1,000 tube kVp) and only 35 per cent for x-rays of 0·25 Å (100 tube kVp). But it rises to 48 per cent for x-rays of 0·5 Å (50 kVp tube), because then the absorption of scattered radiation in bone is high so that less reaches the soft tissues.

The overall thickness of tissue may be described in terms of the *equivalent body thickness*, which is that thickness of standard soft tissues having the same absorption as the tissue in question. The equivalent body thickness can then be used instead of the physical thickness in dose planning, either for stationary or moving fields [37, 38, 39]. Accuracy in determining the equivalent body thickness is limited by radiation scattered from the body, resulting in a false low equivalent body thickness. Johns [40] has devised a focusing system to counteract this.

When a block of tissue contains a cavity, it is possible to position a suitable metal compensator in line with the cavity so that the dose behind it and compensator is the same as behind a similar block of tissue without the cavity. The same principle can be used in compensating for variations in body thickness [41, 42]. The skin-sparing advantage of megavoltage radiation can be preserved by placing the compensator well above the body surface [43, 44].

Literature References

1 International Commission on Radiological Units and Measurements, *Clinical Dosimetry*, Recommendations of the ICRU, Report 10*d*. NBS Handbook 87 (Washington, D.C., Nat. Bur. Stand., 1963).

2 Hine, G. J. and Brownell, G. L., "*Radiation Dosimetry*" (New York, Academic Press, 1956).

3 Mabbs, D. V., "An experimental study of backscatter with x-ray beams of small area", *Brit. J. Radiol.*, **22**, p. 600 (1949).

4 Jones, D. E. A., "A note on backscatter and depth doses from elongated rectangular x-ray fields", *Brit. J. Radiol.*, **22**, p. 342 (1949).

5 "Depth dose tables for use in radiotherapy", *Brit. J. Radiol.*, Suppl. No. 10, 1961.

6 DAY, M. J., "A note on the calculation of dose in X-ray fields", *Brit. J. Radiol.*, **23**, p. 368 (1950).

7 EMMETT, M. L., "Surface backscatter for rectangular fields in the quality range from a half-value layer of 1 mm Al to cobalt-60 gamma radiation, *Brit. J. Radiol.*, **32**, p. 615 (1959).

8 EMMETT, M. L., "Treatment time allowances for leaded-down applicators", *Brit. J. Radiol.*, **32**, p. 702 (1959).

9 JONES, D. E. A., "Axial surface and depth doses for elongated rectangular fields", *Brit. J. Radiol.*, **32**, p. 352 (1959).

10 ALY, S. M. and WILSON, C. W., "Observations on the ionization produced by high-voltage radiation in moulded ionization chambers with walls of various atomic numbers", *Brit. J. Radiol.*, **22**, p. 243 (1949).

11 HAYBITTLE, J. L., SAUNDERS, R. D. and FABEN, R. V., "Plotting of isodose curves", *Brit. J. Radiol.*, **29**, p. 174 (1956).

12 ROSMAN, I. M. and ZIMMER, K. G., "An isodose plotter of simple design", *Brit. J. Radiol.*, **29**, p. 688 (1956).

13 HILL, M. J., "A dose-rate ratiometer", *Brit. J. Radiol.*, **33**, p. 331 (1960).

14 BEASLEY, W. R., MELVILLE, A. M. and KNIGHT, R. N., "An automatic isodose recorder", *Brit. J. Radiol.*, **34**, p. 648 (1961).

15 LARSSON, I., LIDEN, K. and STARFELT, N., "Automatic isodose recorder", *Acta Radiol.*, **1** (New Series), p. 29 (1963).

16 TSIEN, K. C. and COHEN, M., *Isodose Charts and Depth Dose Tables for Medium Energy X-Rays* (London, Butterworth, 1962).

17 INTERNATIONAL ATOMIC ENERGY AGENCY, "Single-field isodose charts for high-energy radiation. An international guide". *Technical Report Series*, No. 8 (Vienna, 1962).

18 JENNINGS, W. A., "Physical aspects of the roentgen radiation from a beryllium window tube operated over the range 2–50 kVp for clinical purposes", *Acta Radiol.*, **33**, p. 435 (1950).

19 JOHNS, H. E. and DARBY, E. K., "The distribution of radiation near the geometrical edge of an X-ray beam", *Brit. J. Radiol.*, **23**, p. 193 (1950).

20 GREENE, D., "Observations on the effect of chamber size on measurements at the edge of an X-ray beam", *Brit. J. Radiol.*, **35**, p. 856 (1962).

21 WHEATLEY, B. M., "An instrument for dosage estimation with fields of any size and shape", *Brit. J. Radiol.*, **24**, p. 383 (1951).

22 WORTHLEY, B. W. and WHEATLEY, B. M., "A generalized method for rapid dosage estimation with particular reference to 200 kV therapy", *Brit. J. Radiol.*, **25**, p. 491 (1952).

23 WHEATLEY, B. M., "A method for dose calculation with applications to moving-field therapy", *Brit. J. Radiol.*, **28**, p. 566 (1955).

24 STERN, B. E., "Dose calculation for moving-field therapy", *Brit. J. Radiol.*, **29**, p. 518 (1956).

25 JOHNS, H. E. *et al.*, "A system of dosimetry for rotation therapy with typical rotation distributions", *J. Can Ass. Radiol.*, **4**, p. 1 (1953).

26 JOHNS, H. E., "Physical aspects of rotation therapy", *Amer. J. Roent.*, **79**, p. 373 (1958).

27 TURNER, J. E., *A Fast Dosage System for Arc Therapy at* 200 kV (From the Author, Veterans Administration Research Hospital, Chicago, U.S.A.).

28 HARE, H. F. *et al.*, "Physical and clinical aspects of supervoltage rotational therapy", *Radiol.*, **57**, p. 157 (1951).

29 FOWLER, J. F. and FARMER, F. T., "Measured dose distributions in arc and rotation therapy. A critical comparison of moving- and fixed-field techniques", *Brit. J. Radiol.*, **30**, p. 653 (1957).

30 MASSEY, J. B., "Dose distribution problems in megavoltage therapy. 1. The problem of air spaces", *Brit. J. Radiol.*, **35,** p. 736 (1962).

31 WHYTE, G. N., "*Principles of Radiation Dosimetry*" (New York, Wiley, 1959).

32 HAAS, L. L. and SANDBERG, G. H., "Modification of the depth curves of various radiations by interposed bone", *Brit. J. Radiol.*, **30,** p. 19 (1957).

33 INTERNATIONAL COMMISSION ON RADIOLOGICAL UNITS AND MEASUREMENTS, *Physical Aspects of Irradiation*, Recommendations of the ICRU, Report 10*b*, 1962. NBS Handbook 85 (Washington, D.C., Nat. Bur. Stand., 1963).

34 WILSON, C. W., "Dosage of high-voltage radiation within bone and its possible significance for radiation therapy", *Brit. J. Radiol.*, **23,** p. 92 (1950).

35 SPIERS, F. W., "The influence of energy absorption and electron range on dosage in irradiated bone", *Brit. J. Radiol.*, **22,** p. 521 (1949).

36 SPIERS, F. W., "Radiotherapeutic physics II, Dosage in irradiated soft tissue and bone", *Brit. J. Radiol.*, **29,** p. 365 (1956).

37 O'CONNOR, J. E., "A transit dose technique for the determination of doses in inhomogeneous bodies", *Brit. J. Radiol.*, **29,** p. 663 (1956).

38 FEDERUK, S. O. and JOHNS, H. E., "Transmission measurement for cobalt-60 radiation with special reference to rotation therapy", *Brit. J. Radiol.*, **30,** p. 190 (1956).

39 JOHNS, H. E., "Physical aspects of radiation therapy", *Amer. J. Roent.*, **79,** p. 373 (1958).

40 JOHNS, H. E., *The Physics of Radiology* (Illinois, Chas. C. Thomas, 1964).

41 ELLIS, F., HALL, E. J. and OLIVER, R., "A compensator for variations in tissue thickness for high-energy beams", *Brit. J. Radiol.*, **32,** p. 421 (1959).

42 HALL, E. J. and OLIVER, R., "The use of standard isodose distributions with high-energy radiation beams—The accuracy of a compensator technique in correcting for body contours", *Brit. J. Radiol.*, **34,** p. 43 (1961).

43 ELLIS, F., "Accuracy in compensation for tissue heterogeneity in treatment by x-rays, supervoltage x-rays and electron beams", *Brit. J. Radiol.*, **33,** p. 404 (1960).

44 HALL, E. J. and OLIVER, R., "The use of metal compensators to correct for tissue heterogeneity in radiotherapy with high-energy radiation beams", *Brit. J. Radiol.*, **35,** p. 852 (1962).

7 | *X-ray Photography*

The action of X-rays on a photographic emulsion is similar to that of light, and film which is sensitive to light will respond also to X-rays. Photographic action depends on the energy absorbed in the emulsion layer, and as in an X-ray film this is only some fifteen microns thick, it will absorb little energy from a beam of penetrating X-rays. Several methods are available for increasing the efficiency of film to X-radiation. One of them is to use film coated

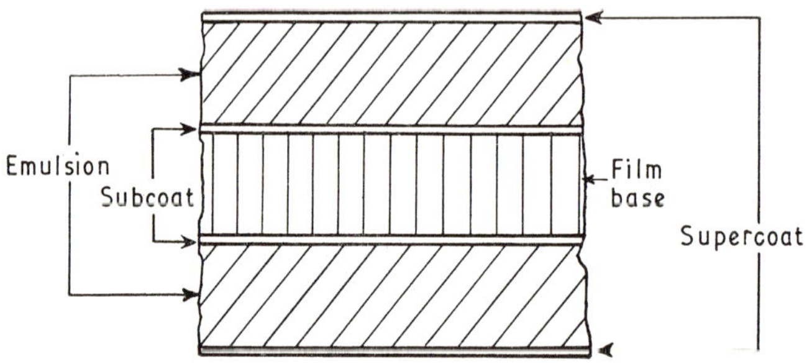

FIGURE 7.1 *X-ray Film*

with emulsion on both sides of the base (Figure 7.1). Another is to sandwich the film between metal intensifying screens, so that electrons released from the metal foil may add to the direct effect of the X-ray beam. A third method is to use fluorescent intensifying screens. The absorption of X-radiation in the screens produces a blue fluorescent light acting on the emulsion, and film intended for use with fluorescent intensifying screens must therefore be sensitive to blue light. Although sensitivity (p. 117) is high, contrast is usually lower than with non-screen film (Chapter 8).

The Emulsion

The photographic "emulsion" is a colloidal suspension of small crystals of a silver halide, usually bromide, in gelatin.

The suspension is produced by adding silver nitrate to a solution of potassium bromide in gelatin, when the gelatin retains the precipitated halide in suspension. The gelatin has several functions: it protects the unexposed grains from reduction by a developer, it keeps the grains well dispersed and it contains chemicals which increase sensitivity. It has the disadvantage that it is rather variable in its composition and properties, as it is of organic origin.

The emulsion layer coated on the base is protected by a thin layer of gelatin about one micron thick.

The Film Base

X-ray film base is usually about 200 microns thick. The ideal film base is flexible, strong, physically stable, optically perfect and chemically inert. Cellulose nitrate was used for many years, but it created a fire hazard and had indifferent keeping properties. It has been replaced by cellulose triacetate, or acetate butyrate, although these are inferior to the nitrate in some of their physical properties [1].

The emulsion will not adhere to the base, because gelatin is insoluble in organic solvents but soluble in water, and the reverse is true for cellulose triacetate. A sub-coat is therefore necessary.

The film base is usually dyed blue. This assists viewing because of the spectral response of the eye.

The Photographic Image

When an X-ray photon is absorbed in a crystal of silver bromide, electrons are freed from the bromide ions. The electrons tend to accumulate in regions of the crystal known as *electron traps* or *sensitivity specks.* The excess negative charge in the traps neutralizes adjacent silver ions, converting them to uncharged silver atoms and forming a latent image.

Thus, the latent image consists of metallic silver particles. In the exceptional case of very high exposure, enough silver is produced by this process to give a visible image; this is the *print-out effect.* Usually, however, the amount of silver present has to be increased many millions of times by *development.* This process increases the size of each speck until the silver bromide grain is entirely converted to metallic silver. It does not increase the number of specks. As the developer tends to reduce unexposed silver bromide to give further silver particles which appear as a background *fog*, there is an optimum time of development to give the best image.

Two mechanisms have been proposed to account for development; both may occur simultaneously.

In the first, the emulsion of silver bromide acquires an absorbed layer of excess bromide during preparation, and this layer gives the surface a negative charge, the distribution of which is distorted over a latent image. As the developer also carries a negative charge, it acts selectively on the areas of latent image.

The second mechanism requires that the silver particles of the latent image act as catalysts; they cause silver ions from silver bromide, which is slightly soluble in the developer, to move towards latent image nuclei where they are reduced to silver.

DEVELOPERS

A developer is essentially a reducing agent, but the composition of a particular developer is chosen to have additional desirable qualities.

Developing Agents

Until recently, developers containing metol and hydroquinone were the most common. Metol acting alone is a rapid developer giving a "soft" negative with low maximum density; hydroquinone is slower and gives excessive contrast. A mixture of the two gives a better result than either separately. However, metol-hydroquinone developers tend to stain, they have indifferent keeping properties and they can cause dermatitis. When phenidone replaces metol, these three drawbacks are mitigated, although phenidone closely resembles metol in developing characteristics. Another advantage is that as phenidone is used in low concentrations, it can be supplied in a small bulk for further dilution. Marketing of highly concentrated liquid developers has thus become practicable and popular, although powdered developers are still available.

Accelerators

Developers usually have a pH of about eleven, or sometimes higher for rapid developers. However, acid is released during the development process, so unless an *accelerator* is used to maintain the pH, this will fall, resulting in a decrease of speed of development. The accelerator may be sodium or potassium carbonate or hydroxide.

Preservative

Exhaustion of the developer occurs also because the developing agents are oxidized by the silver bromide and by atmospheric oxygen. A coloured oxidation product is formed which tends to stain the gelatin. A *preservative*, usually sodium sulphite, when added to the developer results in the formation of a harmless and colourless sulphonate. The preservative also lowers the rate of atmospheric oxidation. This can be reduced further by a floating lid, which reduces the area of developer exposed to air when the solution is not in use [2].

Restrainers

Fog can be diminished by the addition of *restrainers*, such as potassium bromide or benzotriazole.

Solvents and Diluents

Other ingredients sometimes added to developers are wetting agents, bactericides, and water softeners, or sequestering agents, to prevent scum formation in hard water.

Choice and Use of Developer

The choice of developer depends on the use to which it will be put. For instance, speed will be important in operating theatres; specially rapid developers are available when this is the case. If detail is required, a fine-grain developer should be used with a fine-grain film, otherwise prolonged development with a general purpose developer will allow the silver grains to coalesce, making the final image coarse-grained.

Developers supplied as concentrated solutions should be stored in a cool place. Care should be taken to ensure thorough mixing when a developer is prepared, particularly with liquid developers where no visible change takes place on dilution.

During use, in spite of containing both accelerator and preservative, a developer will gradually become exhausted; it will also decrease in volume, as every film carries some developer away with it. The life of the developer can be prolonged and its speed kept more nearly constant by the addition of a *replenisher*. This is more alkaline than the developer to counter the increase in acidity (Table 7.1). Relatively more hydroquinone than phenidone

TABLE 7.1 *A Typical Radiographic Developer and Replenisher Combination*
(From Brownhill, D., *X-ray Focus* (Ilford), *4*, p. 15, 1963)

	Developer	*Replenisher*
Sodium sulphite	50 g	50 g
Sodium carbonate	30 g	30 g
Sodium hydroxide	—	5 g
Hydroquinone	10 g	15 g
Phenidone	0·5 g	0·6 g
Potassium bromide	5·0 g	—
Benzotriazole	0·1 g	0·1 g
Water to	1·0 litre	1·0 litre

is used up during development. One reason for this is that phenidone can be regenerated from its oxidation product by the hydroquinone, this bieng oxidized in the process.

The replenisher may be used simply to keep the volume constant, but where precise control of developer speed is required over a long period, the *bleed system* is a more refined technique. In this, the natural loss of liquid is augmented and the bleed rate adjusted to maintain a constant bromide concentration in the developer. By this means the speed of the developer can be maintained for a much longer period [2].

DEVELOPMENT CONDITIONS

The three main factors affecting development are temperature [3], time and agitation. Most developers, except in automatic processers, are recommended for use within the range 18°C to 21°C. In temperate climates an immersion heater or thermostatically controlled jacket can maintain this temperature, but in the tropics, cooling may present difficulties. High temperatures cause the emulsion to swell and soften, but this is not serious below about 24°C. Special tropical developers have been prescribed for use at higher temperatures. Where the objective is speed, automatic processing units may operate developers at temperatures up to 30°C.

All the wet processing should be carried out at the same temperature, otherwise the emulsion may rupture locally, giving rise to an appearance resembling leather, known as *reticulation*.

Agitation is important in achieving uniform development. In an extreme case, air bubbles may prevent any contact between developer and emulsion over the bubble area, though this is unlikely if a wetting agent is used. Agitation should be continuous in dish development, where the film lies flat. In tanks, intermittent agitation is recommended, this being defined as a brief but thorough agitation at the start and at one-minute intervals throughout development. For a given contrast, development times may be about 25 per cent longer for intermittent than for continuous agitation.

The bromide ions produced during development have a retarding effect, and this is mainly in the areas of greater density; also, in the absence of agitation the developer may become locally exhausted in these regions. Consequently, what should be an area of uniform high density may have either a fringe or halo, or a central region of lower density. When two very small dense images are close together, slow development in their neighbouring areas may make them appear to be too far apart.

If an exhausted developer is used, it may produce streaking as it streams over the film.

FIXING

After development, the film has a visible silver image, but it is not in a state fit for viewing [1]. First, if exposed to light, the picture will *fade* because the silver halide grains are still light-sensitive, except those affected by exposure and consequently reduced during development. Secondly, as the unexposed grains remain in the emulsion, they spoil the transparency of unexposed parts of the film; thirdly, some developer will be retained in the emulsion and will cause over-development unless its action is stopped. Thus, it is necessary to "fix" a developed film.

Between developing and fixing, excess developer should be removed by washing in running water. In X-ray film processing, a tank with sprays is usually provided. Rinsing slows the developing action, but it is preferable to stop it by including an *acid stop bath*, effective because developers do not act in acid solution. A weak acid is suitable, such as $2\frac{1}{2}$ per cent potassium metabisulphite or one per cent acetic acid.

The fixer converts the undeveloped silver halide crystals into more soluble compounds, the complex sodium argentothiosulphates, which are removed by washing. They are not very stable compounds, and if the final washing is incomplete, some will remain in the emulsion to break down later, forming silver sulphide which shows as a yellowish-brown stain and also results in fading. With an exhausted fixer, the sodium argentothiosulphates become absorbed in the emulsion and are difficult to remove by washing, so that this also is a cause of staining.

Sodium thiosulphate, or *hypo*, is commonly used as a fixer. Acid is usually added to neutralize remaining traces of developer and to prevent staining. Thus the life of the fixer can be prolonged.

A hardening agent such as chrome alum or potassium alum is often added to the fixer. This is valuable particularly in warm weather and in the tropics, as if the gelatin is softened by heat, it may become partially detached from the acetate base. The pH of a hardening fixing bath [1] should be between 4 and 6·5.

Sodium sulphite may be added as a preservative; it prevents the formation of sulphur from the sodium thiosulphate.

When a film is placed in the fixer, it can be seen to clear gradually as the halide is dissolved. It should be fixed for a period twice as long as the time taken to clear. The speed of fixing can be increased by adding an ammonium salt, usually the chloride, to make the fixer a mixture of sodium and ammonium thiosulphates.

Fixing becomes slower as the bath becomes exhausted, when there will be a high concentration of dissolved silver and a low concentration of hypo as well as a rise in pH. Above a pH of about 6·5, a sludge of aluminium salts forms; its formation can be delayed by the addition of boric or tartaric acid.

The time taken to clear the negative is a useful indication of the state of the fixing bath. An economical and efficient method of control is to fill two tanks. The films are placed in one until clear. They are then removed to the second tank for an equal time. When the time to clear has doubled, the first fixing bath is discarded and the second takes its place. A fresh bath is then used as the second fixer.

The speed of a fresh fixer depends on several factors. Thin emulsions fix more rapidly than thick ones, and fine-grained than coarse. In addition, the speed of fixing increases with concentration of hypo, with temperature and with amount of agitation, although there is an upper limit to the fixing speed. In urgent cases, a film may be viewed almost immediately after it is placed in the fixer.

In a large department, it may be economically worth while to recover silver, at the same time regenerating the fixer. This can be done electrochemically.[8] Alternatively the silver may be deposited in a sludge by chemical means, or the spent fixer sold for silver recovery, but in neither of these cases is the fixer reused.

WASHING

The purpose of washing is to remove any soluble silver sodium complexes and fixer which remain. The washing process is essentially one of diffusion, so that although the amount remaining decreases with time, it is never zero. If the films are to be kept, washing must be thorough (p. 112). The emulsion may absorb up to ten times its own weight of water.

With water circulating at the ambient temperature, half-an-hour's to one hour's washing is adequate, when the rate of flow of fresh water is rapid enough to replace the water in the vessel about ten times per hour. Times can be shorter if agitation is increased, or if good keeping properties of the film are not required. If the water is warm, the risk of reticulation (p. 112) must be borne in mind.

Occasionally the washing water may have to be filtered, as solid particles if present can lodge in the emulsion, causing artefacts.

DRYING

The film may be hung up and left to dry either naturally or in the draught from a fan. Often, however, a heated (about 40–50 °C) drying cabinet is used; it should be provided with a fan. A wetting agent will prevent uneven drying. Where speed is more important than quality, a film may be washed in alcohol before drying.

The Darkroom

The darkroom must have a low radiation background and it should be conveniently situated. Where these two conditions are incompatible, radiation-absorbing building materials (p. 275) or lead shielding on walls, and, if necessary, on floor and ceiling, can reduce the exposure to acceptable levels.

Double doors, on the pattern of double French windows, make access possible without introducing light at the same time, but a light trap is more convenient. This consists of a short length of corridor with at least two right-angled bends. Its walls should be painted matt black, with a light-coloured guide-line. When processing units, work benches or other bulky articles have to be moved in or out of the dark room, and a light trap makes access difficult, it may be convenient to install a collapsible trap.

The dark-room should be big enough for comfortable working conditions and to allow for likely expansion. In a large hospital, it may well be about the size of one of the radiographic rooms [4].

Provision for passing films before or after exposure in and out of the dark-room can be by a wall-hatch, arranged with a light-proof sliding cover or interlock. X-ray protection is also incorporated in some of these hatches.[9]

Often, there are arrangements for passing out partially processed films from the darkroom,[10,11] sometimes in a "pass-through" tank of water into a drying room [4].

ILLUMINATION

General illumination is best provided in the darkroom using a central light with a spectrum suitable for the particular film or paper being processed. The illumination is reflected from walls and ceiling and may be supplemented by local lighting of the same colour. The reflection is enhanced by using white or pale glossy paint. The floor also may be light-coloured and of rubber, PVC, linoleum or other waterproof material.

Low-wattage bulbs in a special housing with ventilation, a diffusing screen, and a slide for a coloured filter, provide the working illumination. The coloured filter is made of dyed gelatin on glass and therefore will not withstand high temperatures, so the recommended bulb wattage should not be exceeded. These *safe-lights* are available for ceiling, wall or bench mounting.

The maximum sensitivity of the unadapted human eye is in the yellow-green region at about 5,500 Å. On adaptation, there is a shift to shorter wavelengths, and the fully dark-adapted eye has maximum sensitivity in the blue-green region [1] at approximately 5,150 Å .For this reason, when the photographic material is sensitive to all wavelengths of the visible spectrum, a safe-light in the green region is used, with the level of illumination kept low. An olive-green or an orange safe-light is suitable for X-ray films. Some films and papers are less sensitive to certain regions of the spectrum, usually red or orange, and then brighter safe-lights of these colours are the best choice. All sensitive materials will eventually fog if they are exposed for too long, even with the most suitable light filters. Unnecessarily difficult working conditions are often endured because the illumination is lower than it need be. In routine work it is worth while to experiment with the lighting using unexposed film, finally fitting lights which give no detectable fogging after, say, half-an-hour's exposure in the darkroom.

White light must be available when required, and during the day, this is most pleasantly supplied through a window. Shutters or a blind running in grooves fixed to the wall will provide a black-out. Whether or not there is a window, forced ventilation is desirable and must be provided through light-tight ventilators. For high-quality work, good ventilation is particularly important, as a humid dark-room can become very tiring.

A dark-room clock with a second-hand and an alarm can usually be sited so that it is illuminated without danger of fogging the films.

A viewing box[10,11] should be placed near the washing tank with facilities for catching the drips and for preventing contact between the viewing face of the box and the wet film.

DRY AND WET BENCHES AND PROCESSING UNITS

As dirty marks and other artefacts can be misleading, particularly in X-ray work, it is essential to adopt the highest possible standards of cleanliness, and to follow good practice by separating the dry and wet processes. Good working conditions are also important: the temperature should be high enough for comfort, but not above that of the wet processes (p. 111); benches should be at a convenient height and the manipulations arranged in a logical sequence round the darkroom [5]. Accidents will occur if obstructions lie about either on the floor or on the benches.

Teak and Formica are suitable materials for the dry bench tops. The film-hatch should open on to a dry bench and the film hangers, arranged in order of size, should be within easy reach. In a medical radiography department, it is of paramount importance that each film should be marked with the correct patient's name and any coding in use.

A selection of unexposed films can be kept in a film-hopper under the bench, but often the dark-room is too warm and damp for optimum storage conditions. A cool dry store room should be provided for bulk supplies of films and processing solutions. Films should be stored so that they are standing on edge, and adequate ventilation should be provided between and under the boxes. It is wise to mark all photographic materials with the date of delivery, as they deteriorate in time.

The wet processes may be carried out in dishes on the wet bench, but X-ray films are usually processed in special deep tanks,[9,10,11,12] with the films held in hangers vertically and the temperature controlled (p. 111). A lid over the tank will keep out light and dust.

Mobile processing units[9] are available for use in operating theatres. Speed is of paramount importance here, and if conventional methods are too slow, there is a fast Polaroid system[13] which will produce a finished radiograph in ten seconds.

The initial capital costs for automatic processing are likely to be acceptable in large photographic departments. Amongst these are a number of hospitals in the National Health Service, which spends altogether several million pounds per year on X-ray film [6].

The main advantages of automatic processing are a quicker reporting service and greater uniformity in processing and drying. Processing and drying are usually carried out at elevated temperatures and a dry film will be available in about five minutes. The developer and fixer are usually automatically replenished. Only the film loading tray need be in the dark. Automatic processing units[10,11,14,15] can handle two or three hundred films per hour.

Protective clothing, such as rubber or plastic aprons, should be worn in the dark room. Plastic tongs and rubber gloves help to prevent contamination of the skin.

Film Density

After processing, areas of the film which have been exposed to X-radiation appear darker than unexposed regions. As the *blackening* is more intense where the exposure was higher, the film is known as a *negative*, and in X-ray photography there is usually little purpose in proceeding to a positive print. The film blackening is due to metallic silver grains, and the consequent loss in film transparency is used as a measure of photographic density D.

Density is defined in terms of the film's ability to transmit light:

$$D = \log_{10} \frac{\text{Intensity of incident light}}{\text{Intensity of transmitted light}} \tag{7.1}$$

Within limits, the density is proportional to the amount of silver per unit area present. Thus, if two parts of the film have densities of two and one respectively, in the former approximately twice as much silver is deposited as in the latter; the transmitted light intensity will be 1/100th and 1/10th respectively of that of the incident light.

The silver particles scatter part of the incident light, so that the transmitted light emerges partly parallel to the incident light and partly in random directions. Thus the apparent density of the film will depend on whether the measuring instrument collects all the transmitted light or only the parallel beam, and the term density has to be qualified.

When the incident light is a parallel beam and only the transmitted parallel beam is measured, the density is known as the *specular* density. When the density is measured either with an incident parallel beam and by collecting all the transmitted light or with diffuse incident light and by collecting only the transmitted light emerging normally, the density is known as *diffuse* or *totally diffuse*.

A third quantity *double diffuse* density is measured when the incident light is diffuse and the total transmitted light is measured [1].

The difference between values of diffuse density and double diffuse density depends on the conditions of measurement and is usually small, but specular density is greater than either. The ratio known as the *Callier coefficient* or *Q-factor* is given by

$$\text{Callier coefficient} = \frac{\text{Specular density}}{\text{Diffuse density}} \tag{7.2}$$

Although numerically greater than one, it is usually less than two. The

Callier coefficient increases with grain size. It increases also with density to a maximum value at about $D = 0{\cdot}3$ and then decreases slowly.

THE DENSITOMETER

In crystallography, film badge dosimetry and some other applications, the film density is used as a measure of the incident x-ray exposure. Film density may be measured visually or by using a densitometer. In general, diffuse density is used in practical densitometry. Therefore, most densitometers incorporate an opal scattering screen to give complete and constant diffusion.

A densitometer consists essentially of a beam of light arranged to shine through the film and then on to a photocell. The amount of light falling on the photocell determines the current flowing through it and this can be measured by a galvanometer or microammeter. In practice, refinements[2,16] are necessary to give reproducible and accurate results [7].

The incident light is usually focused on to a small aperture the size of that portion of film to be examined. In crystallography, the aperture may have an area only a few microns in width and this will resolve the film grains. Such instruments are known as microdensitometers.

As the light source may fluctuate in intensity and the photocell may age, it is usual to split the incident beam, one part passing through the film and the other through a grey wedge.[17] The wedge is adjusted in position to give a photographic density such that two parts of the beam are of equal intensities. The absorption in the wedge is then equal to that in the portion of film being measured. A range of wedges may make it possible to read densities up to about $D = 6$. Modern wedges are made of grey glass.

The output may be linked to a pen for continuous recording of densities. Automatic densitometers have been developed for film badge dosimetry, where large numbers of films have to be measured. Different areas of each film are consecutively measured and their relative values used to estimate radiation quality (p. 55).

THE CHARACTERISTIC CURVE

The curve showing the relationship between the logarithm of the exposure and the film density is known as the characteristic curve (Figure 7.2).

Even without exposure, there is always some blackening known as fog. Beside that arising during development (p. 109) other unavoidable causes are cosmic ray bombardment and background radioactivity. Additional fogging can arise if the film is old, or if it has been faultily processed, or exposed during storage to certain gases, such as ammonia or hydrogen sulphide. The density F is due to fog.

The speed or sensitivity of an x-ray film at a given density, D, is governed by the exposure required to obtain this density under standard processing conditions. Thus

$$\text{Speed} = \frac{D - F}{X} \tag{7.3}$$

An allied parameter defines how much the blackening increases for a small increment in log exposure. This is contrast, where

$$\text{Contrast} = \frac{\mathrm{d}D}{\mathrm{d}\log X} \tag{7.4}$$

Contrast will be constant over the straight part of the characteristic curve, where it is equal to tan θ (Figure 7.2); it is also known as the gamma value.

Very low exposures result in no detectable blackening; gross over-exposure gives maximum blackening except for a possible small amount of reversal or *solarization* (Figure 7.2). The useful exposure range is normally well between these limits, and is known as the latitude.

Film contrast may be effectively enhanced by scanning the film with a light beam which, after transmission, is allowed to fall on a photomultiplier

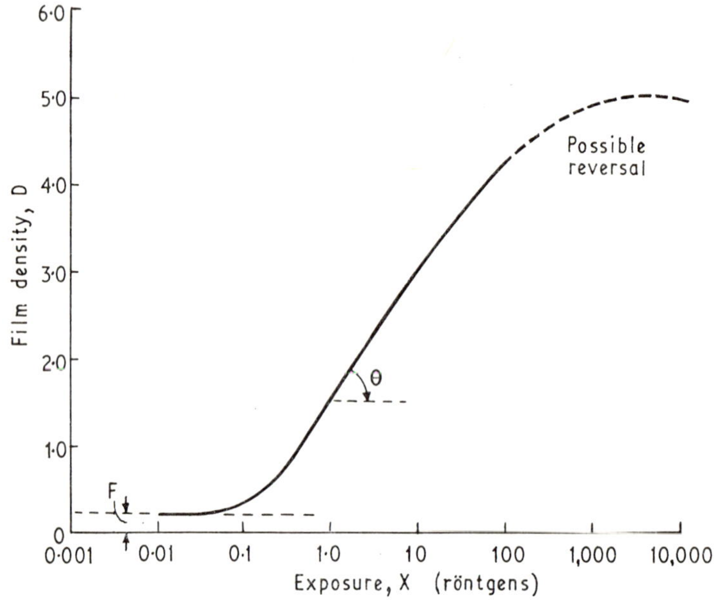

FIGURE 7.2 *A Characteristic Curve*

and the output signal is further amplified by a logarithmic amplifier.[17] The final voltage is then directly proportional to film density.

RECIPROCITY RULE

The reciprocity rule implies that a certain exposure X results in a certain density D whatever the exposure rate.

A number of electrons must be present to form a sensitivity speck (p. 109); however, each free electron has a short life within the trap. Consequently, unless the time interval between successive releases of electrons within a grain is below about 10^{-5} sec, the film density will depend on the exposure rate dX/dt, and there is failure of the reciprocity rule [8].

The reciprocity rule is valid for direct X-ray exposures, but fails for light exposures when salt screens are used.

CHOICE OF FILM

Film should be chosen with characteristics [9, 10] to suit the purpose in hand.

In an extreme case, when only two classes of objects have to be distinguished, a high contrast makes it easily possible for one object to appear black and the other "white". This is the situation in photographing print.

In medical work, many tissues differ little in absorbing properties, but need to be distinguished. In industry small flaws may lead to only slightly different thicknesses of absorbing material. For both situations half-tones are necessary and a high contrast film is required to give a visible change in density for a small change in film exposure (Chapter 8).

To achieve a desired effect it is generally preferable to vary the quality of radiation rather than the film characteristics. For instance, the contrast between two objects of different effective atomic number may be increased

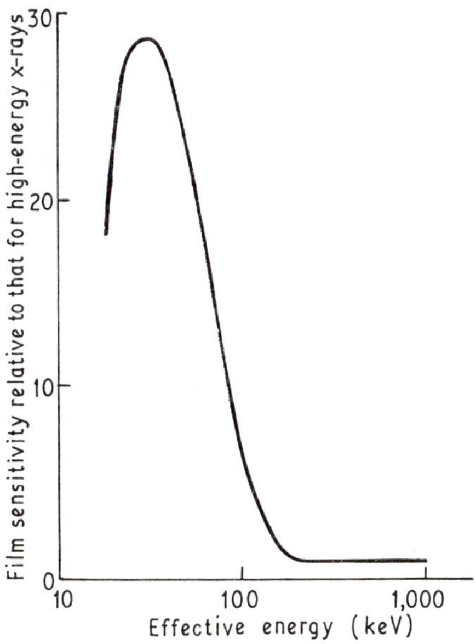

FIGURE 7.3 *Variation in Film Sensitivity with X-ray Energy*

by using x-rays of lower energy, and conversely, provided that photoelectric absorption is an important factor.

In exposure measurements, a wide range of densities may have to be measured accurately. As the useful range of film density is limited, the requirements of both high contrast and a large latitude are mutually exclusive. With low-energy x-radiation the effective latitude may be increased by substituting a film pack for a single film, the absorption in each film layer being appreciable; this method is used in crystallography [11]. In dosimetry, the effective latitude may be increased by using a film with emulsions of different speeds on either side of the base (p. 108).

High speeds are associated with large silver halide grains, and therefore a compromise has to be made between speed and resolution.

ENERGY DEPENDENCE OF BLACKENING

Suppose different parts of a film receive the same exposure but from x-radiation of different energies. Provided that the energy absorption per röntgen (p. 55) (the *F*-factor of the film) is approximately constant over the range of X-ray energies considered, the film density will also be constant. This is the case from about 200 keV to several MeV.

At lower energies, the *F*-factor increases to a maximum and then decreases as the X-ray energy decreases. The maximum is influenced by the absorption edges of silver and bromine [11]. The *F*-factor determines the film sensitivity (Figure 7.3).

In practice, it is more informative and more convenient to describe the energy dependence of the film's response in terms of exposures required for a range of film densities (Figure 7.4) rather than by density–energy curves.

The energy dependence of blackening is important in quantitative work.

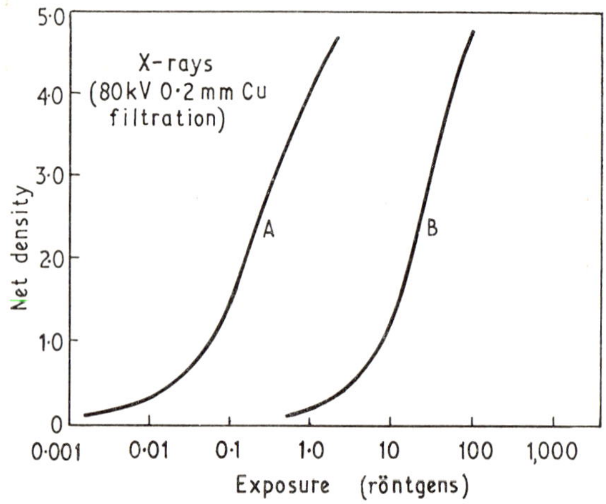

FIGURE 7.4 *Characteristic Curves for Monitoring Film*
A, both emulsions; B, slow emulsion only.
(Kodak Data Sheet FM 63. Kodak Ltd., London)

Other things being equal, exposure times and consequently radiation hazards in radiography will be minimized by using X-ray energies for which the emulsion has maximum sensitivity. In practice, other factors have to be considered; in particular the heavy absorption of soft radiation.

When film is being used as a dosemeter, there are different considerations. The sensitivity dependence on energy may be used to estimate the effective radiation energy, or the response may be flattened to be more uniform. In practice both these methods are used.

In the Laue method in X-ray crystallography (p. 195), "white" radiation is split into component wavelengths by the diffracting crystal; the film density of the individual reflections is affected by the relation between the absorption edge effect and the wavelength giving rise to the particular spot.

THE FILM AS PERSONAL DOSEMETER

The high sensitivity of a photographic emulsion to X-radiation makes film suitable for personal monitoring (p. 261), where exposures down to a few milliröntgens must be measured. However, when an individual receives a high exposure, a sensitive emulsion may give film densities so high that they are unmeasurable, and this is the condition when an estimate of the exposure is particularly necessary. This objection has been overcome[10] by coating a monitoring film with a fast emulsion on one side and a slow one on the other.

If the exposure completely blackens the fast emulsion, this can be removed. The exposure can then be estimated from the density of the slow emulsion. Within the usual range of exposures encountered in X- and γ-ray protection work, the slow emulsion does not respond on development, and the fast emulsion is used as the dosemeter. Using this double-coated film, exposures from about 0·3 mR to about 65 R (using the slow emulsion only) are measureable for X-rays of 80 kVp (Figure 7.4), filtered by 0·4 mm Cu. Hence a ratio of exposures of about 1 to 2 $\times$ 10^5 can be recorded.

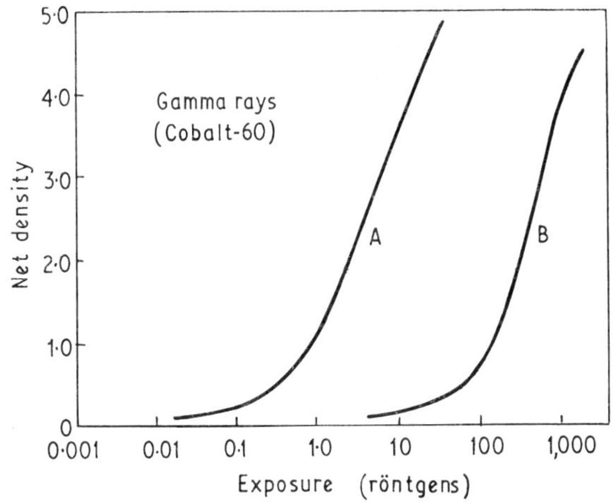

FIGURE 7.5 *Characteristic Curves for Monitoring Film*
A, both emulsions; B, slow emulsion only.
(Kodak Data Sheet FM 63. Kodak Ltd., London)

A batch of monitoring films is calibrated by exposing a sample and plotting a calibration curve showing the relationship between film density and exposure (Figure 7.5) or rads in air. The calibrating and monitoring films must be processed together. It is usual to calibrate with a gamma-ray emitter such as radium or cobalt-60, where the exposure rate is precisely known and where the gamma-radiation is of an energy high enough for the speed of the emulsion to be constant over a range of energies.

If the energy of the X-radiation to which the individual has been exposed is unknown, it is impossible to estimate directly the magnitude of the exposure. This is because the speed of the film may vary by as much as thirty times between the lowest and highest radiation energies (Figure 7.3). Consequently, an exposure of 10 mR from 30 keV X-ray photons would result in the same film density as an exposure of 300 mR from about 300 keV X-ray photons. This problem is solved by placing filters over parts of the film. If high-energy photons are incident, the filters will absorb little radiation, and the whole film will be of uniform density. On the other hand, low-energy photons will be absorbed by filters to give regions of low film density. The ratio of film densities with and without a filter is used to estimate the effective energy of the photons. In order to cover a range of photon energies, a number of different filters are used. After the x-ray energy has been estimated by this method, the relevant speed of the emulsion may be read from the calibration curve (Figure 7.3) and hence the exposure estimated.

A special film badge holder[18] [12] has been designed to house filters and the monitoring film (Figure 7.6). The aim is to make the exposure estimate as simple as possible. With its use, dose from β-rays and neutrons can be

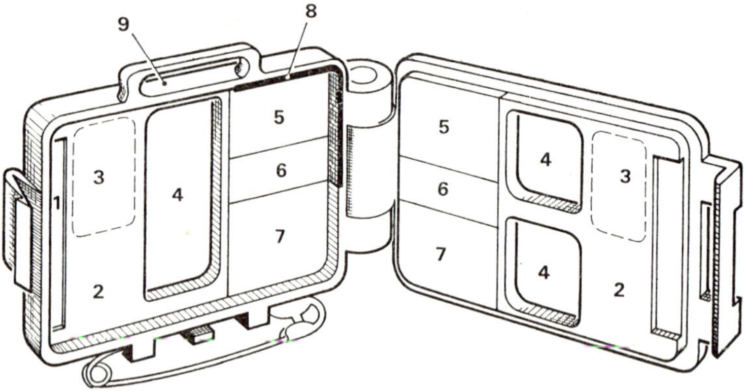

FIGURE 7.6 *RPS/AERE Film Badge Holder*

1. Recess for indium foil (neutron monitoring)
2. 300 mg/cm² plastic
3. 50 mg/cm² plastic
4. Open window
5. 0·012 in. lead + 0·028 in. tin
6. 0·012 in. lead + 0·028 in. cadmium
7. 0·040 in. Duralumin
8. Edge shield, 0·012 in. lead
9. Slot for wrist- or head-strap

(Jones, B. E., "New RPS/AERE film holder," *Film Badge Symposium, Harwell,* 12th Oct. 1961)

measured as well as that from x- and γ-rays. Only the last group will be considered here.

One filter is made of lead 0·012 in. thick plus tin 0·028 in. thick. It is so designed that for a range of energies, as the radiation energy decreases the loss in density due to absorption in the filter is counterbalanced by the gain in emulsion speed. Consequently under this filter, the response is independent of energy (to $\pm$10 per cent) from about 80 to 1,000 keV [12]. Within this energy range, the film may be calibrated using radium γ-rays.

For photon energies from about 50 keV to 80 keV, the apparent dose beneath the tin/lead filter is compared with that beneath a filter of Duralumin 0·040 in. thick and a correction factor applied (Figure 7.7).

Below about 50 keV, the correction factor becomes unacceptably large. Therefore, the emulsion (unfiltered) is calibrated with 30 keV x-radiation, and this calibration can be used between 25 keV and 60 keV with an error of $\pm$10 per cent. Below this range the apparent doses beneath unfiltered and Duralamin filtered regions are compared, and a correction factor applied (Figure 7.8). For regions of 15 keV and less, it should be possible to use the filtering effect of the two thicknesses, 300 and 50 mg/cm^2 of the plastic holder, for estimating exposure in a similar way.

Use of the holder making correction factors unnecessary over a wide range of X-ray energies has been achieved by matching filters to emulsion

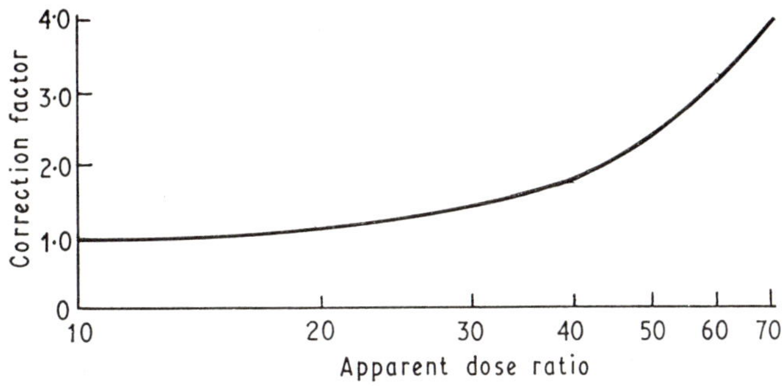

FIGURE 7.7 *Correction Factors*

$$\text{Apparent dose ratio} = \frac{\text{Dural filter}}{\text{Tin/lead filter}}$$

True dose = Apparent tin/lead dose × correction factor.

characteristics. Gross errors may be introduced if the badge holder and calibrations are used with different emulsions. A simpler film badge holder will suffice when type and energy of radiation are limited.

If an emulsion were available with a sensitivity independent of X-ray energy, exposure could be measured without the use of filters. However, a knowledge of the energy of the radiation is essential for a proper evaluation

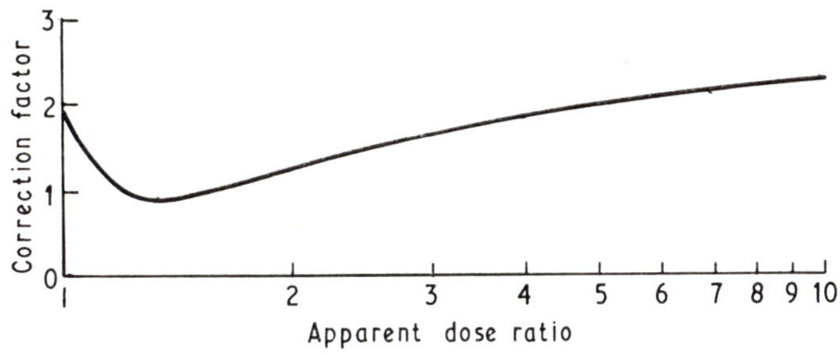

FIGURE 7.8 *Correction Factors*

$$\text{True Dose} = \frac{\text{Apparent dose on unfiltered film}}{\text{Apparent dose on Dural-filtered film}}$$

(Jones, B. E., "New RPS/AERE Film Holder," *Film Badge Symposium, Harwell,* 12th Oct. 1961)

of the hazard. Dealler [13] has pointed out that filters can be used for energy determination even for such an emulsion of constant sensitivity.

In the interests of standardization and with legal aspects in mind, personal monitoring in industry must be carried out by approved laboratories such as those of the Radiological Protection Service, to be reorganized as the National Radiation Protection Board.

Literature References

1 Holder, A., *The Ilford Manual of Photography* (Ilford, Ilford Ltd., 1958).

2 Brownhill, D., "The theory and practice of radiographic processing: Pt. I, Development", *X-Ray Focus* (Ilford), **4,** p. 15, (1963).

3 Towers, S. W., in *A Further Handbook of Industrial Radiology*, Ed. Wiltshire, W. J., Chap. 7 (London, Edward Arnold Ltd., 1957).

4 Ross, J. A. and Galloway, R. W., *A Handbook of Radiology* (London, H. K. Lewis, 1963).

5 McCombe, K. M., *Commercial Photography* (London, Technical Press, 1952).

6 Godber, G., "Radiology in the National Health Service", *Brit. J. Radiol.*, **36,** p. 545 (1963).

7 Rose, K. M. and Jeffery, J. W., "Errors arising from the photographic recording of x-ray intensities", *Acta. Cryst.*, **17,** p. 21 (1964).

8 Herz, R. H., in *X-ray Diffraction by Polycrystalline Materials*, Ed. Peiser, H. S., Rooksby, H. P. and Wilson, A. J. C., Chap. 32 (London, Chapman and Hall, 1960), (For Institute of Physics).

9 Morimoto, H. and Uyede, R., "A second comparison of various commercially available x-ray films", *Acta Cryst.*, **16,** p. 1107 (1963).

10 Klug, H. P. and Alexander, L. E., *X-Ray Diffraction Procedures* (New York, Wiley, 1954).

11 MacGillavry, C. H. and Rieck, G. D. (Eds.), *International Tables for X-Ray Crystallography*, Vol. III, (Birmingham, Kynoch Press, 1962).

12 Jones, B. E. in "*Proceedings of Symposium on Film Badge Dosimetry, Harwell, Oct.* 1961", Ed. Langmead, W. A. and Adams, N. (Harwell, U.K.A.E.A., 1962).

13 Dealler, J. F. B. in "*Proceedings of Symposium on Film Badge Dosimetry, Harwell, Oct.* 1961", Ed. Langmead, W. A. and Adams, N. (Harwell, U.K.A.E.A., 1962).

8 | *Radiography*

A radiograph of Frau Röntgen's hand was shown at a meeting of the Manchester Literary and Philosophical Society [1] in December 1896, one year after Röntgen's original communication on X-rays to the Physico-Medical Society of Wurzburg [2]. Medical men were quick to realize the value of the new rays in the detection of foreign bodies and the diagnosis of fractures, but technological applications were limited by the lack of penetration of the X-radiation available at that time. Industrial radiology developed rapidly later, particularly for the inspection of welds and castings during the second world war. There are now diverse uses of radiography, for example, the examination of paintings for alterations and the microradiography of alloys to show the distribution of the constituents.

Principles of Radiography

In radiography a beam of X-rays is passed through some part of interest of an object, and allowed to fall on a photographic film or other X-ray-sensitive material. The analogy with an optical shadow is close, but optical shadows usually distinguish only the completely opaque from the transparent,whereas the radiograph distinguishes varying degrees of transparency; it is a two-dimensional projection of the attenuation properties of the object to a particular radiation.

The interpretation of the radiograph demands a knowledge of the object and of X-radiation, and a radiologist is trained in his own field, be it medicine or metallurgy, before specializing in radiology.

QUALITY

The quality of a radiograph is not an absolute attribute. Quality is the extent to which the radiograph supplies the required information, although aesthetic considerations may also be taken into account.

The visualization of, for instance, a foreign body is less exacting than the examination of fine structure. Sometimes a plain radiograph will show the necessary features. At other times special techniques can extend the usefulness of a radiograph, by increasing artificially the contrast between the region of interest and its surroundings.

The details observable in a radiograph cannot be divorced from its viewing conditions. The uniformity and intensity of illumination affect the apparent film densities.

SUBJECT CONTRAST

The observable difference in density between two parts of a radiograph is known as the subject contrast [3]. The term should be distinguished from the photographic film property known as contrast (Chapter 7).

Suppose that a thickness d of foreign material must be detected within a block of thickness t (Figure 8.1).

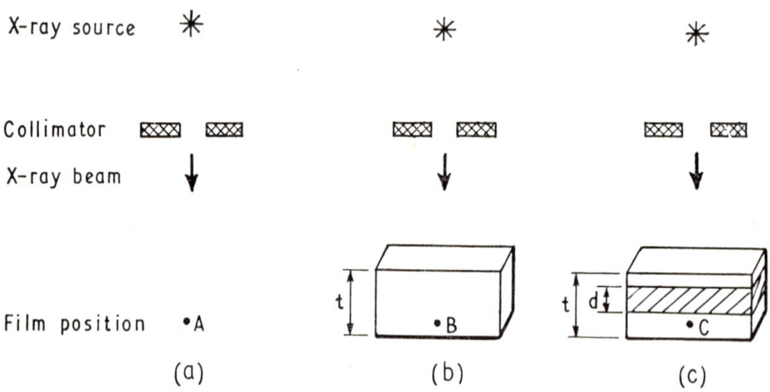

FIGURE 8.1 *Effect of a Thickness of Foreign Material on the Exposure at the Film*
(*a*) No absorber; (*b*) block of material of thickness t; (*c*) block of material of total thickness t, containing a thickness d of foreign material.

If μ_d = linear absorption coefficient of the foreign material } for the
and μ_t = linear absorption coefficient of the material of the block } x-radiation in use

also I_0 = intensity of radiation at the film at A, in the absence of any material
I_t = intensity of radiation at B
I_d = intensity of radiation at C

Then neglecting scatter

$$I_t = I_0 \exp(-\mu_t t) \tag{8.1}$$

$$I_d = I_0 \exp[-\mu_t(t - d)] \exp(-\mu_d d) \tag{8.2}$$

Hence

$$I_t/I_d = \exp[-(\mu_t - \mu_d)d] \tag{8.3}$$

and this ratio is independent of t.

For good subject contrast, blackening must be on the steepest part of the characteristic curve of the film (Chapter 7), and for this to be so, the value of I_t must be considered when selecting exposure conditions.

If the two materials differ enough in atomic number or density, a judicious choice of X-ray tube voltage may result in good subject contrast, with $(\mu_t - \mu_a)d$ differing sufficiently from zero. In medical radiography, μ_t and μ_a are often nearly equal, whatever the X-ray energy. In such a situation when, for example, the blood vessels must be examined, a *contrast medium* with a different effective atomic number and density may be introduced into the vessels (p. 150). At the other end of the scale, the variation in absorption of parts of a metallic specimen may be so great that it is impossible to arrange for all photographic densities to be within the latitude of the film. In this case the range of densities may be lessened by adding absorbing material to certain parts of the specimen (p. 134).

Choice of Kilovoltage

Two main considerations govern the choice of kilovoltage. The first is the quality of the radiograph and the second, which may be overriding in medical work, is the radiation hazard.

The quality of a radiograph is largely dependent on subject contrast, and therefore requires high value of $(\mu_t - \mu_a)$. If the two regions of equal thickness which are to be distinguished, have different atomic numbers, the tube kilovoltage should be selected so that photoelectric absorption in the material is numerically the important process. If, however, the two regions have similar atomic numbers but different electron densities, a higher X-ray energy resulting in relatively more Compton absorption will be advantageous, since the Compton linear absorption coefficient varies with the electron density (p. 36).

The quality of the radiograph may be adversely affected by unsharpness (p. 130), and the tube kilovoltage should be selected with this in mind.

For a tonal range, exposure at the film must be within the latitude of the characteristic curve [4]. Thus, although the range of exposure at the film should be confined to fairly narrow limits, the average exposure at the film depends on the energy of the X-rays, because film sensitivity is energy-dependent. However, neither the dose to the object nor the X-ray tube output required will be minimized by using radiation of maximum photographic sensitivity, if the attenuation of the X-rays by the object is excessive.

The choice of kilovoltage can affect the apparent width of an object such as a wire, which decreases in thickness towards the edge. Because of the variation in attenuation across the wire the range of densities may exceed the film latitudes. In this situation, the apparent wire width may depend on the film latitude rather than on the real width.

For the same reason, the choice of kilovoltage may affect also the definition and contrast.

SHADOW FORMATION

If the size of the focal spot is "small", an object will cast a shadow which is always larger than itself. If the size is "appreciable", a partial shadow, or penumbra will be associated with the full shadow, or umbra (Fig. 8.2). "Small" and "appreciable" in this context are relative to the size of the object and to the focal distances to object and image.

The umbra is the region behind the object in complete shadow, so that a straight line joining any point in the umbra to any point in the focal spot passes through the object. From a penumbra, only some such lines pass through the object.

If the object is smaller than the focal spot, the areas of penumbra will converge, to overlap below *P* (Figure 8.2), where a pseudo-image will appear. The magnification *DE*/*AB* of the pesudo-image will be different from the magnification *CF*/*AB* of the real image; indeed, *DE*/*AB* may be less than unity. Such diminution is known as *undercutting*. There is no region within

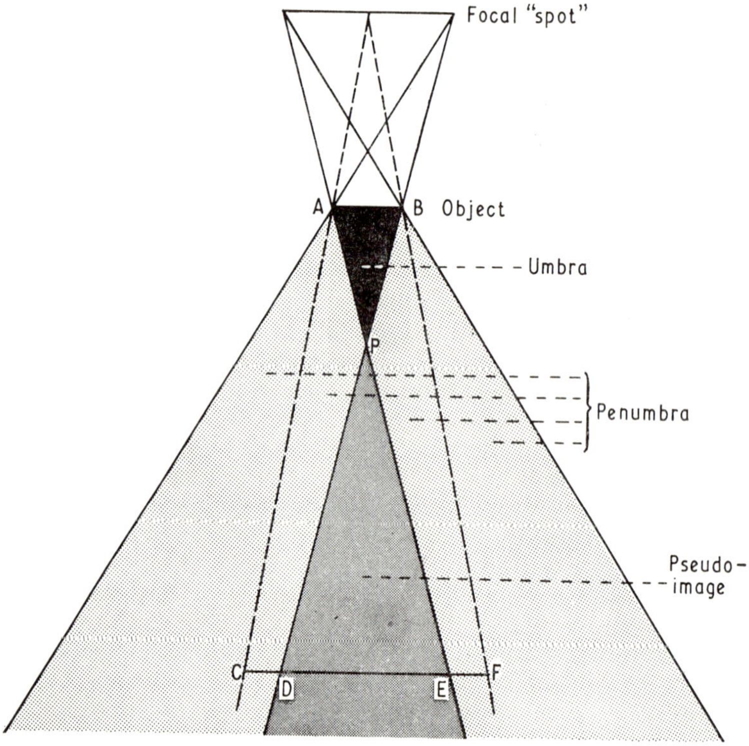

FIGURE 8.2 *Umbra and Penumbra*

the pseudo-image which is in complete shadow; this phenomenon may give the false impression that the object is radio-transparent. Undercutting is of practical importance in the study of fine structure, as in hair cracks in welds.

The film density varies across the penumbra. The variation is seldom smooth because the focal spot is often irregular in shape and in radiation emission.

Although for good definition a small focal spot is desirable, this limits the tube output available without overheating the anode. With a low output, long exposure times may be necessary. Movement blur will result if vibration can be transmitted to X-ray tube, to film or to the object. In medical work, long exposure times are particularly undesirable because breathing, heart beats and small involuntary movements take place even when the patient does not fidget. These drawbacks are present in an exaggerated form when dealing with infants and in veterinary radiography (p. 152).

The rival claims of small focal spot and high output can be partially reconciled by using a rotating anode tube or by a fine-focus tube (p. 221); it is usual for X-ray diagnostic tubes to have effective focal spots of from 0·2 to 2 mm^2 and to have ratings of several hundred watts.

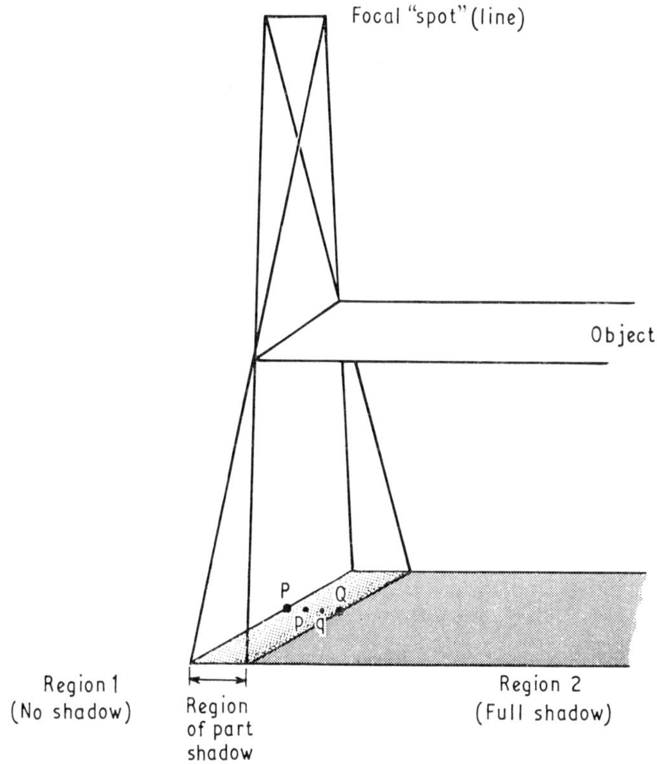

FIGURE 8.3 *Contrast and Definition*

Even with small focal spots, definition may be poor if the object is far from the film. When it is impossible to bring it closer, an increase in focus–film distance will improve the definition at the expense of increasing the exposure time.

Definition

Consider the radiography of an object with a sharp edge (Figure 8.3). Region 1 is outside the umbra and region 2 is inside. At P and Q respectively the film density begins to change from that of one region or of the other. The subject contrast [3] is

$$B_p - B_q$$

where B denotes density and pq is the smallest distance of observable difference in density and pq is within the region PQ. The *definition* is the rate of change of contrast with distance; thus, the definition is

$$(B_p - B_q)/pq$$

The *unsharpness* describes the converse of definition, and unsharpness may be measured by the distance pq. Unsharpness is, however, a general term and may measure the distance from a point just completely outside to one just completely inside the shadow. Geometrical unsharpness due to the size of the focal spot and the geometry of the apparatus is best analyzed mathematically by Fourier methods [5]. The width of the penumbra is only one of the causes of unsharpness; another cause is the grain of the film (Chapter 7), which creates fine structure discontinuities, and furthermore there is a tendency for grains to clump under the action of high-energy radiation [3].

Secondary radiation is another cause of unsharpness. Scattered x-radiation from the object or the emission of fluorescent light from a fluorescent screen both result in a diffuse rather than a sharp image.

A fourth source of unsharpness is movement blur.

If these separate components of unsharpness are denoted U_p, U_f, U_s, U_m, then the total unsharpness U is given by [3]

$$U^2 = U_p^2 + U_f^2 + U_s^2 + U_m^2 \tag{8.4}$$

Because of the square law relationship, the largest component of U is often the only one of consequence.

RESOLUTION

Two lines which are just distinguishable one from the other are said to be resolved. In radiography, resolution is usually expressed as the number of such resolved lines per mm.

EFFECT OF RADIATION ENERGY

In general, heteroenergetic radiation gives more contrast than monoenergetic. Consider a certain thickness of absorber and incident radiation, (*a*) of two energies and (*b*) of one energy, both systems arranged to give equal exposures in equal times at the film.

Now let the absorber thickness be decreased. The lowest-energy component will be proportionately increased the most at the film and as this is usually radiation with the highest photographic sensitivity, the contrast between the two thicknesses tends to be higher for the mixed energies than for monoenergetic radiation. If the thickness of absorption is increased, the effect is, in general, less marked than for a similar decrease in thickness. Constant-potential x-ray sets (and even more so gamma radiation from radionucleides) may therefore have disappointing radiographic properties.

PENETRAMETERS

The quality of a radiograph depends upon many factors such as focal spot size and shape, the geometry of the layout, and the film and its processing. It is important to have a method of assessing the quality of a radiograph in order to see if it is possible to reveal fine detail under particular conditions. An image quality indicator is called a penetrameter. It usually consists of either a step wedge, often with holes drilled in the steps, or a series of wires of graded diameters. If possible, the material of the penetrameter should be similar to that of the specimen to be radiographed. Penetrameters are mounted on the side of the specimen facing the radiation, and in the least advantageous position for good sensitivity. The quality of the image is judged from the smallest wire or step resolvable [6].

The Penetrameter sensitivity (Figure 8.4, *left*) is given by

(Thickness of the thinnest step or wire visible in radiograph × 100)/
(Total specimen thickness).

Sensitivities as low as 1 per cent can reasonably be expected for weld examination. This sensitivity relates only to the quality of the radiographic

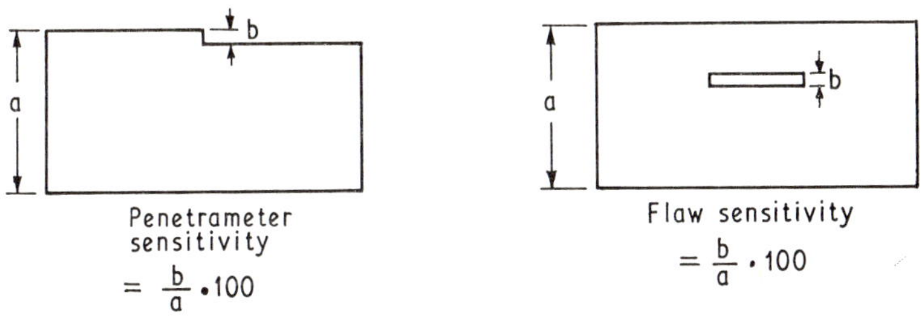

FIGURE 8.4 *Penetrameter and Flaw Sensitivity*
a, thickness of specimen; *b*, thickness of step or flaw, such that the thickness $(a - b)$ is just distinguishable from thickness *a*.

image and does not measure the size of any defect that might be revealed. Furthermore, the sensitivity of the two types of penetrameter is not the same under the same conditions, a wire type may give only a 2 per cent sensitivity when a step type is giving a 1 per cent sensitivity. This is because the wire image is smaller and more difficult to see (*vide ultra*). By analogy, the sensitivity of naturally occurring defects may be even greater numerically, as these are often even more difficult to detect (Figure 8.4, *right*). This sensitivity is known as the *flaw sensitivity*.

EXPOSURE

The optimum X-ray condition, in terms of kilovoltage and milliampere seconds, is governed by the X-ray exposure incident on the film, which generally must be within the latitude of the film characteristic curve. Thus, the X-ray tube output required depends on anode–film distance, the attenuation of the object and the efficiency of the photographic or other recording material.

Where there are few variables in more or less standard conditions, as in medical radiography, kilovoltage and milliampere seconds may be obtained from tables or a slide rule.[15] Exposure timing can be more effectively controlled by a radiation detector such as an ion chamber, placed near the film,[9] or by an automatic exposure device (Figure 8.5).

When there is a wide range of thickness, particularly of heavy metals, there may be no single optimum time. This is often the case when radiographing castings. In such cases, it is advisable to radiograph each part separately, but where this is impracticable, careful angling may reduce the difference in path lengths of the beam. Alternatively, suitable blocking material of absorption roughly equal to that of the castings can be used. In these applications, the wide latitude of a xerographic plate (p. 140) has advantages.

Intensification

The energy absorbed in a photographic emulsion is a small percentage of the energy in the x-ray beam, often less than one per cent. The overall efficiency of energy conversion may be increased by using an intensifying screen.

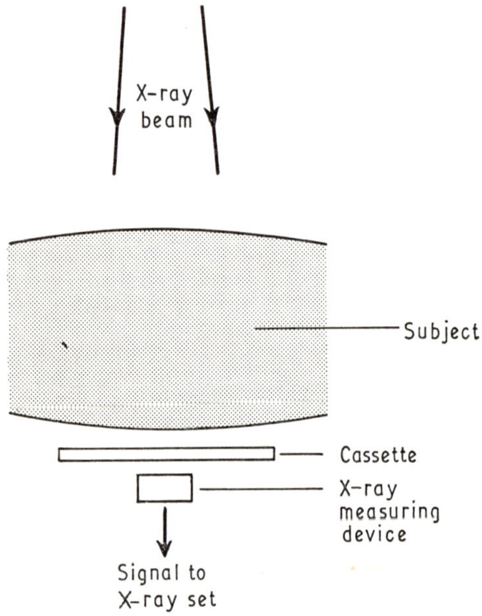

FIGURE 8.5 *Automatic Exposure Device*
After a predetermined total dose is reached, a signal is fed back to terminate exposure.

This increase in efficiency can lead to shorter exposure times and in industrial radiography enables more economical use to be made of existing x-ray equipment. The benefits are more substantial in medical work because the patient dose is reduced as the efficiency is increased.

Intensifying Screens

The principal of an intensifying screen [7] is that it absorbs radiation from the x-ray beam and generates secondary radiation which is made to blacken the film. Intensifying screens give a gain of efficiency at the expense of some loss in definition [8]. Two screens are usual, with the film sandwiched between them and held in close contact. Good contact can be checked by exposing with a wire grid on top of the cassette. The image of the wire should be sharp over the whole film.

Thick screens of coarse grain absorb X-radiation well, but result in low definition. Where detail is important, both back and front screens should be thin, but for general purposes it is more usual to use a thin screen in the front and a thick one at the back.

There are two types of intensifying screen.[9,10,11]

The most common type is a fluorescent screen, consisting of a paste of fluorescent salt, spread on a cardboard base. Fluorescent screens contain high atomic number elements and hence have high photoelectric absorption coefficients. The secondary fluorescent radiation is responsible for most of the blackening of the film, only a small contribution arising from the direct absorption of X-radiation in the film emulsion. The fluorescent radiation is in or near the ultra-violet and so heavily absorbed by any dirt present. A dirty screen may therefore give rise to artefacts. Screen and emulsion should be optically matched so that the maximum film sensitivity occurs at the wavelength of the fluorescent light. Emulsions suitable for use with screens are inefficient and unsuitable for radiography without screens [9].

Fluorescent screens should incorporate a salt with a high absorption at the incident X-ray energy. Zinc sulphide screens are efficient at 30 kVp, calcium sulphate at 90 to 130 kVp [10]. Fast tungstate screens, usually of calcium tungstate, are suitable for the range 40 to 120 kVp and are particularly efficient for the lower part of this range.

At higher kilovoltages, a second type of intensifying screen is more efficient. This is a lead screen, and secondary X-radiation, not fluorescent light, blackens the film.

The gain in efficiency from using a screen may be as high as one hundred-fold. It depends on the relative absorption of X-rays in screen and photographic emulsion. As this varies with X-ray energy, intensifying screens often appear to be faster when the beam is more highly filtered or when it is generated at a higher kilovoltage.

A refinement to improve definition is to incorporate a dye in the screen which absorbs the fluorescent light. Consequently light rays having a long path within the screen are preferentially absorbed. However, here once more, detail is improved at the expense of speed.

The optimum compromise between the conflicting claims of efficiency and definition depends on circumstance. For instance, in radiographing limb bones or in industrial work, speed may not be as important as definition, but where dose must be kept to a minimum as in obstetric radiography, and definition may be less critical, high-speed screens should be used. In veterinary work (p. 152) animals under constriction often tremble and the best definition is usually obtained by using the fastest screen available.

The Institute of Welding recommend [6] that preference be given to fine-grain films used either without screens or with lead intensifying screens, as the graininess of salt screens renders crack detection difficult.

When one exposure is arranged to take several radiographs as in multiple or multi-section radiography, the exposure at the different films may vary. This can be corrected by using compensating screens of different speeds, and graded sets are available for this purpose.

SCATTERED RADIATION

In addition to the primary beam, some radiation will reach the film after being scattered.

Unless secondary electrons arise in the neighbourhood of the film, their range will not permit them to reach the emulsion. Most scatter therefore arises from Compton scattered photons, subsequently absorbed in or near the film. The importance of the blackening due to scatter depends on this absorption relative to that of the primary photons absorbed in the emulsion. It thus depends on the energy of the X-ray beam, the atomic numbers of the components of the object and its surroundings and the various thicknesses. Scattered photons, being softer than the primary photons, are more readily absorbed in the emulsion and thus the need to reduce scatter accentuates the importance of using radiation of an energy high enough not to be excessively attenuated within the object.

As the scattered radiation does not arise from the focal spot, it detracts from good definition. Scatter arising from the object being radiographed is termed unsharpness (p. 130). For tube kilovoltages above about 500 kVp, scatter is predominantly forward, so that unsharpness from this cause is not very marked. Scatter arising from neighbouring objects results in a patchy blackening over the film.

There are a number of ways of reducing scatter.

Firstly, the kilovoltage must be carefully chosen, as there is a general reduction of scatter at higher voltages. The proportion of scattered to primary radiation can be reduced by filtering off the softer components of X-radiation as the beam leaves the tube. Filtration will also reduce scatter from specimen edges. However, excessive filtration can remove that part of the beam for which the emulsion is most sensitive and the part giving a differential effect for the less absorbent parts of the object.

Secondly, a filter between specimen and film will absorb the secondary radiation preferentially for two reasons: because the secondary radiation is softer than the primary and because filtration is more oblique. In medical radiography, this filter may take the form of an air gap between patient and film [11].

Thirdly, a lead backing to the film holder will absorb back-scattered radiation.

Fourthly, scattered radiation will be reduced by limiting the area of the beam. This can be done either by the use of an applicator or by lead cut-outs placed on the object.

Fifthly, blocking material may be effective when there are large and discontinuous variations in attenuation by the object. This is because the optimum energy of the beam depends on the attenuation by the specimen and so there cannot be a unique optimum voltage for different thicknesses. In addition, scatter from unblocked discontinuities may reach the film, but, if blocked, be absorbed in the blocking.

Finally, grids may be used to reduce scatter.

The Grid

Grids[9,13,19] are strips of lead built into radio-transparent material, the lead being of a height such that most of the scattered radiation is stopped by the strips, while the primary beam passes between them. The ratio of the height, *h*, of the lead strips to the distance, *d*, between them is known as the *grid ratio*. Common values for this ratio are between five and twenty; the higher the grid ratio, the better will be the discrimination against scattered radiation and the greater the contrast [12, 13, 14].

A stationary parallel grid (Figure 8.6) will cast shadows under the lead strips. When the field is wide compared with the focal distance, primary radiation at the periphery will arrive at an angle to the lead strips, so that their shadows will be excessive. Under these conditions, it is necessary to

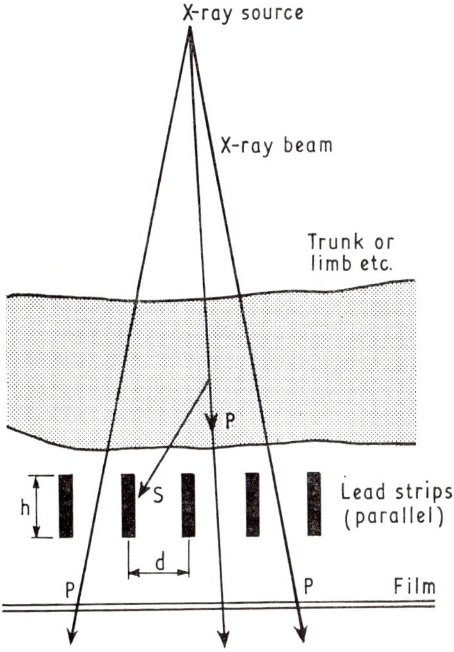

FIGURE 8.6 *The Parallel Grid*
P, primary radiation; *S*, secondary radiation.

have an *aligned* or *focused* grid (Figure 8.7), designed for a particular focal-grid distance, although this distance is not very critical.

A single grid is ineffective in absorbing scattered radiation arriving in a direction parallel to that of the lead strips, but performance can be improved by substituting two grids at right angles to each other for a single grid. If each of these *crossed* grids has a grid ratio of *G*, the effective grid ratio of the whole will be 2*G*. Such a crossed grid is more effective in eliminating scatter than a single one of grid ratio 2*G*. Crossed grids are arranged one above the other; they are effective particularly for tube kilovoltages below 100 kVp.

As all stationary grids prevent strips of primary radiation from reaching the film, their shadows appear as undesirable *grid lines* on the radiograph. Consequently, the grid may be made to move during exposure, in a direction perpendicular to that of the lead strips. In this way, the primary beam reaches all regions of the film for part of the exposure, and the grid lines are nearly eliminated. This is known as a Potter–Bucky grid or diaphragm [14]. A grid reduces the intensity of radiation falling on the film. To compensate for this, the tube current may have to be increased several times. Alternatively when the tube voltage is 100 kVp or so, this may be increased by 10 or 20 kVp, the precise increase depending on the grid ratio [15]. In medical radiography, this increase in exposure has to be weighed against the improvement in quality resulting from the use of a grid, and very high grid ratios are generally unjustified.

Grids are not often employed in industrial radiography, although they might be considered for thick specimens of light metals at voltages below

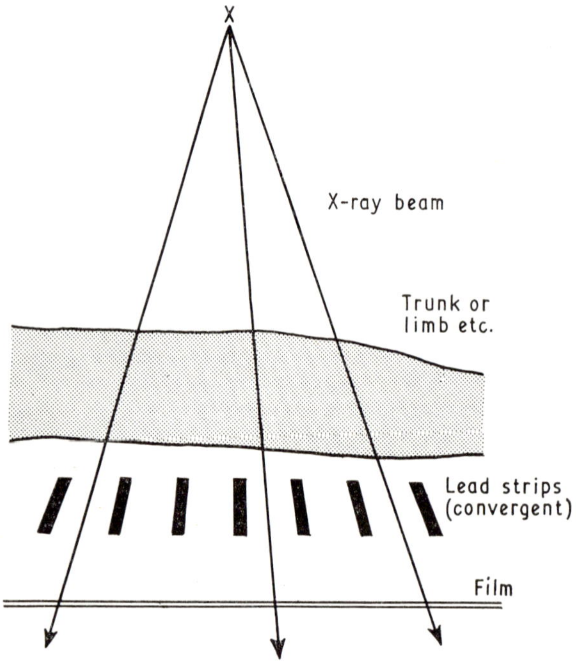

FIGURE 8.7 *The "Aligned" or "Focused" Grid*

about 150 kVp. Much industrial radiography is, however, carried out at higher kilovoltages, when the radio-opaque strips of a grid would be cumbersome and when scatter is less troublesome.

Image Registration

Although it has been reported that an exposure as low as 1 mR produces a sensation of a light flash to the human eye, direct viewing of a radiation beam is quite impracticable as well as being highly dangerous. A variety of materials absorb X-rays and produce a latent or visible image, and the choice of a suitable material depends on requirements: permanency of image, speed, detail and so on.

PHOTOGRAPHIC IMAGE

X-rays, like light, produce a latent image in photographic emulsions. Photography (Chapter 7) was the earliest and is still the most common method of recording the image.

Radiographs are normally examined directly as negatives. For exhibitions, if the original size is unsuitable, it may be altered by making a positive intermediate which is then photographed with suitable scaling up or down. Wording on paper or cards does not show up well in the dim light of a radiographic exhibition. A transparency of the wording, printed or stencilled in black ink on white paper should be prepared photographically. High-standard prints of radiographs must be supplied for reproduction in books and journals [16, 17].

Light can be excluded during radiography, either by using a film packed individually in a light-tight paper back or by placing an unwrapped film in a special holder, the cassette.

The Cassette

A cassette [10] is a holder for films during exposure. It may have a thin lead sheet on the side of the film remote from the tube to absorb backscattered

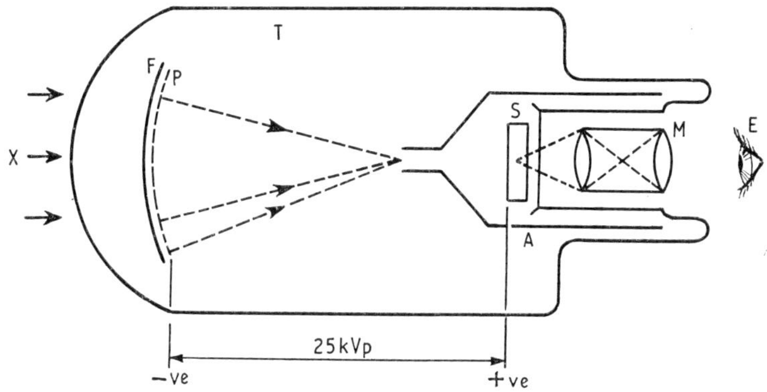

FIGURE 8.8 *Image Amplifier*
A, perforated anode; *E*, eye; *F*, fluorescent screen; *M*, microscope; *S*, small fluorescent screen; *X*, x-ray beam; *P*, photocathode.

(After Young, M. E. J., *Radiological Physics*, London, H. K. Lewis, 1967. *Courtesy of Philips Industries Ltd.*)

radiation. Cassettes made for use with intensifying screens (Chapter 7) are designed to give good and uniform contact between screens and film, by using felt pads or a self-adhesive backing material.

Both flat and curved rigid cassettes are available, some fitted with press-button fasteners so that a series of films may be joined together to photograph a large area; others are of flexible plastic material.

Cassettes are made in sizes to fit the standard films (Table 8.1), including small ones for dental work.

TABLE 8.1(*a*) *Standard Films for Industrial Radiography* (Ilford Ltd.)

Sheets (inches)	*Sheets* (mm) (for crystallography)
2 × 15	29 × 280
$3\frac{1}{4} \times 4\frac{1}{4}$	35 × 280
4 × 10	35 × 255
4 × 15	125 × 175
$4\frac{3}{4} \times 6\frac{1}{2}$	135 × 173
6 × 12	
6 × 15	*Rolls*
$6\frac{1}{2} \times 8\frac{1}{2}$	7 in. × 100 ft
7 × 17	70 mm × 100 ft
8 × 10	
10 × 12	
12 × 15	
14 × 17	

TABLE 8.1(*b*) *Standard Films for Medical Radiography* (Kodak Ltd.)

inches	*cm equivalent*	*inches*	*cm equivalent*
$4\frac{3}{4} \times 6\frac{1}{2}$	12 × 16·5	8 × 10	20·3 × 25·4
5 × 7	12·7 × 17·8	10 × 12	25·4 × 30·5
6 × 12	15·2 × 30·5	11 × 14	27·9 × 25·6
6 × 15	15·2 × 38·1	12 × 15	30·5 × 38·1
$6\frac{1}{2} \times 8\frac{1}{2}$	16·5 × 21·6	14 × 14	35·6 × 35·6
7 × 17	17·8 × 43·2	14 × 17	35·6 × 43·2

centimetres	*inches equivalent*	*centimetres*	*inches equivalent*
9 × 12	$3\frac{1}{2} \times 4\frac{3}{4}$	20 × 40	$7\frac{7}{8} \times 15\frac{3}{4}$
13 × 18	$5\frac{1}{8} \times 7\frac{1}{8}$	24 × 30	$9\frac{1}{2} \times 11\frac{3}{4}$
15 × 30	$5\frac{7}{8} \times 11\frac{3}{4}$	30 × 30	$11\frac{3}{4} \times 11\frac{3}{4}$
15 × 40	$5\frac{7}{8} \times 15\frac{3}{4}$	30 × 40	$11\frac{3}{4} \times 15\frac{3}{4}$
18 × 24	$7\frac{1}{8} \times 9\frac{1}{2}$		

The location of cassettes may present difficulties in industrial radiography and it is usually better to use a slightly oversize film to be sure of including the region of interest rather than risk repeating the radiograph. Magnets are useful to hold the cassettes in position with ferrous metals.

In a busy department, it may be expedient to install an automatic cassette conveyor to ensure an adequate supply of loaded cassettes and to remove these after exposure.

IMAGE AMPLIFIER

The brightness of the visual image may be much improved by an image amplifier.[4,15,19,22] Consequently the exposure or exposure rate is much less than for fluoroscopy and hazards are correspondingly reduced. The exposure of the radiologist is further diminished because he is more distant from the beam.

In an image amplifier, the X-rays are allowed to fall on a curved fluorescent screen *F* within an evacuated tube (Figure 8.8). The screen is backed by a photo-cathode *P* from which the ejected electrons are accelerated towards a perforated anode *A*. The electrons pass through the perforation on to a small fluorescent screen *S* which is viewed by a microscope *M*. The image may be viewed directly or by a television camera. The gain in brightness is due to the acceleration of electrons within the tube and to their concentration within a small area. The gain may be as much as a thousand times [18].

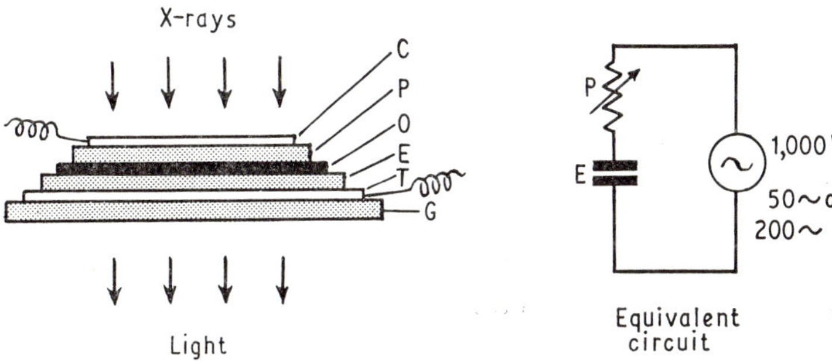

FIGURE 8.9 *X-Ray Image Convertor*
C, thin conductor; *E*, electroluminescent layer; *O*, optically opaque layer; *P*, photoconductor; *T*, transparent conductor; *G*, glass.
(*Thorn Electrical Industries Ltd.*)

Hay [19] has studied the extraction of the maximum possible information from the X-ray beam as it emerges from a patient, when television techniques are used.

Image amplification has made it practicable to take cine film without exposing the patient to an excessive dose. A camera [20] may photograph either the viewing or the television screen.

SOLID STATE RADIOGRAPHY

Conduction detectors[23] may be used to produce a picture of an X-ray field. A sheet of cadmium sulphide crystals can act in this way as an X-ray image convertor to give twenty to fifty times the brightness of a fluorescent screen (Figure 8.9).

In the absence of X-rays, the cadmium sulphide layer (photoconductor) has a high resistance and hence the applied voltage appears across it, rather than across the electroluminescent layer. X-radiation induces conductivity in the cadmium sulphide so that the applied voltage then appears across the electroluminescent layer, causing it to emit light.

There is no theoretical limit to the area. For diagnostic X-rays, the cadmium sulphide layer should be about 0·5 mm thick. At higher X-ray energies, the layer may be increased to improve the absorption properties, but with slight loss in definition. Maximum gain is obtained at a frequency of 50 Hz at low dose rates and about 200 Hz for higher dose rates. The current consumption at 50 Hz is approximately 0·1 mA per square inch of plate area. The speed of response is slow, so that the method is unsuitable for fast-moving images. On the other hand, the image may be stored on the screen for up to about one minute after the incident X-rays are switched off, thus reducing exposure times.

FLUOROSCOPY

Instantaneous visualization can be obtained by using a fluorescent screen. The dose to a patient is, however, higher than in photography and if the fluorescent screen is viewed directly, precautions must be taken to make sure that the radiologist does not receive an excessive dose (Chapter 14). In industrial work only the latter hazard has to be considered.

In contrast to a photographic emulsion, a fluorescent screen does not integrate the incident radiation, and the screen brightness depends not on total dose but on dose rate; this must not be excessive so that exposure times can be long enough to give useful information without over-exposure of personnel. For safety, the fluorescent screen should be viewed in a mirror. When possible the apparatus should be totally enclosed and viewed through lead glass.

The screen material is usually of zinc calcium sulphide which fluoresces at about 530 nm, and this is in the region of maximum sensitivity of the eye. A dark-adapted eye will respond to a threshold some four thousand times less than one not so adapted. Full adaptation takes about half-an-hour, but the relative eye sensitivity starts to rise steeply after about a quarter of an hour [21]. During the waiting time, other tasks can be carried out if red goggles[9] are worn.

Fluoroscopy is useful to give instantaneous visualization and to show movement. It can be cheaper than photography in the routine examination of articles such as packaged goods. Disadvantages are that unsharpness is high and contrast is low.

Fluoroscopy now is used only when it is particularly appropriate, because of associated radiation hazards and because alternative techniques are now available.

XEROGRAPHY

Xerography is a dry process (Greek "xeros" = dry) employing a photoconductive layer deposited on a conducting base plate. If this layer is charged and subsequently exposed to X-radiation, the induced photoconductivity will enable the surface charge of an exposed element to leak away to the base plate, to a degree depending on the radiation dose absorbed by that element. Consequently the charge pattern will reproduce the radiation exposure variation over the surface.

The best photoconductor is amorphous selenium, deposited as a uniform film 50 to 100 μ thick, although anthracene or sulphur may be used. The surface is uniformly charged by exposing it in the dark for up to one minute to the corona discharge from a wire electrode at about 8 to 15 kVp [22]. The surface potential will then be a few hundred volts. The sensitized plate may be kept in an x-ray cassette for a considerable period.

The speed [23] of an xerographic plate is about half as fast as non-screen film and about one-tenth as fast as screen film. The characteristic curve has a slope much less steep than that for a photographic film and the latitude is large and the contrast low. This low contrast is mitigated by an outlining effect described below, although this exaggeration of outlines can lead to misinterpretation of the object.

During exposure, the air in the neighbourhood of the plate surface becomes ionized. This will preferentially discharge the areas of high charge, and therefore gives rise to a band of charge deficit at the edge of an image, and within this band image formation is impaired. This is known as *undercutting* (*see* also p. 128). The lateral transfer of charge can be reduced by applying a voltage, slightly greater than that applied to the plate during charging, between the selenium layer and the dark slide of the cassette. The resulting electrostatic field tends to confine the ion movement to a direction normal to the surface of the plate.

After exposure the charge pattern present is invisible and has to be *developed.* This is done by a charged powder which is deposited in proportion to the charge density of the plate. In the *powder cloud method* a powder is whirled into a cloud and some particles will be charged positively, some negatively. The powder is then made to impinge as a jet on a charged metal plate, thus removing all the particles of one sign; the remaining particles, all charged to the opposite sign, are allowed to fall on the xerographic plate where they are deposited according to the charge pattern on the plate, making these variations visible. White cosmetic powder is suitable for development. Amorphous selenium is grainless and the resolution depends on the size of the powder particles, which may be only 25 μ.

During development, the powder drifts towards areas of high charge. This builds up the charge at the edge of the image at the expense of the neighbouring area, giving rise to the *outlining effect.*

Undercutting and outlining tend to act in opposite directions, but the net effect is to exaggerate the edges.

When the xerograph is no longer required, the powder can be wiped off or transferred to a sheet of adhesive paper and the plate reused. However, the image will not be completely removed unless the plate is exposed to intense light. The life of a plate is limited only by mechanical damage to its surface. The only expendable material is a very small quantity of powder.

Charging and developing take from one to sixty seconds and the finished picture can be available about ten seconds after exposure. These advantages have to be weighed against the special equipment and low sensitivity, when comparing xerography[19] with radiography. The low sensitivity may well rule out xerography for ordinary medical purposes. It has, however, been used with success for treatment planning with the 4 MeV linear accelerator, where the extra dose is of no consequence for radiotherapy patients [24, 25] and where it is a great advantage to have a wide latitude and to see on the same picture bone and soft tissue detail, as well as air cavities, and the outline of the skin. The skin dose is a few rads.

In industrial radiography, the outlining effect combined with the low contrast enables detail to be made visible in specimens of widely varying thickness.

Radiographic Techniques

A simple radiograph gives a shadow picture of the whole region; it may not give an adequate image of the area of interest, perhaps because this is overshadowed by some other part, perhaps because it is insufficiently contrasted with the neighbouring material.

In examining welds, it may even be necessary to cut away part of the structure in order to insert the cassette at an advantageous position, and to reweld after radiography is completed.

In medical work, the thickness of tissue may sometimes be reduced by compression. Equation (8.3) shows that this will not increase the contrast between the object of interest and the soft tissue surrounding it, but it can be rewarding both because it will reduce the necessary exposure, and hence the absorbed dose, and because it will reduce scatter.

Many ingenious techniques have been developed to facilitate the extraction of information from a radiograph. Sometimes enhanced detail is required, sometimes the exact location of a part.

STEREORADIOGRAPHY

In looking at some object, each eye sees a slightly different view. The views differ more the nearer the object is to the eye. This phenomenon gives

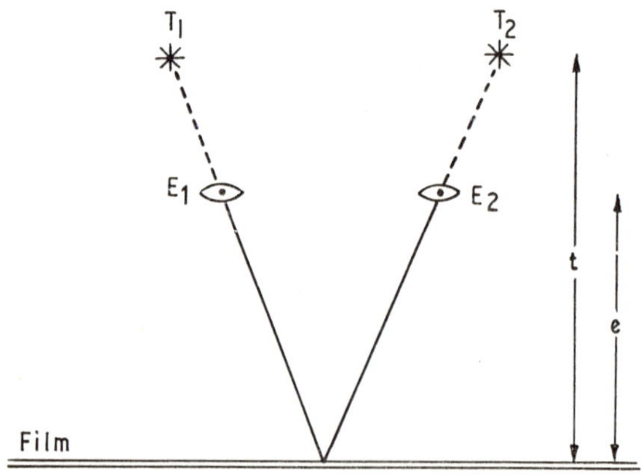

FIGURE 8.10 *Principle of Stereography*

the sensation of seeing things in depth. It is the purpose of stereoradiography to reproduce this three-dimensional effect and so to inform the viewer whether some object is in front of, or behind, another.

Two radiographs must be taken corresponding to the views from the two eyes. The tube position may be fixed and the object moved. More often, the object is fixed and the two radiographs taken with two different positions of the tube.

Suppose the positions of the focal spot are T_1 and T_2 (Figure 8.10). If the two films are subsequently projected on a "stereoscope"[24] so that each is viewed by one eye at E_1, E_2, then if

$$\frac{T_1T_2}{t} = \frac{E_1E_2}{e} \tag{8.5}$$

where E_1E_2 = distance between the eyes = 63 mm approx.
t = focus-to-film distance
e = eye-to-film distance

the eyes will subtend the same angle at the film as will the two focal spots, and the view will appear to be a three-dimensional one. Individuals with much better sight in one eye than in the other may, however, experience difficulty in obtaining a stereoscopic effect.

Stereoradiography is useful in locating foreign bodies, for instance when removing gunshot from a wound. The positions of neoplasms and other pathological changes can be found using stereoradiography and it also has applications in dentistry (p. 152) and in viewing structures visualized using contrast media [26].

In industrial radiography, stereography can be useful in elucidating the directions of complex cracks.

LOCALIZATION AND MEASUREMENT

An object must be viewed from at least two directions before its position in space can be determined. Various techniques are available: stereoradiography (*vide ultra*), or measurements can be made either from two mutually perpendicular radiographs or from one radiograph on which there are two views; the last is known as the tube shift method.

Consider the general problem of measuring the distance between two points A and B, both of which can be made to produce an X-ray image. The general solution of this problem can be applied to measurements of the lengths of cracks, foreign bodies or other objects, and to their distances from points of reference.

In the method when two mutually perpendicular radiographs are taken, the distance AB (Figure 8.11) is calculated from the resolution of the image along the three axes x, y, z. Correction for magnification can be made by means of a calibrating ring R.

The tube focus must be moved relative to the object from the position F'_{xy} for the first exposure to the position F''_{yz} for the second exposure. The two films are placed so that each is perpendicular to the X-ray beam, the first being in the plane xy (say) and the second in the plane yz. The images of AB are $A'B'$ and $A''B''$ respectively.

Compared with AB, the image $A'B'$ will in general be foreshortened because of the angle between AB and the plane xy. At the same time $A'B'$ will be a magnified ($\times m'$) image due to the divergence of the beam.

Resolve the image into the three perpendicular co-ordinate directions x, y, z. Let $A'C'$ be the projection of $A'B'$ in the x-direction and $C'B'$ in the

y-direction. Similar argument holds for $A''B''$, where the magnification is m'' and the projections $A''C''$ and $C''B''$ in the y- and z-directions respectively.

Now, by Pythagoras' theorem,

$$AB = \sqrt{\left[\left(\frac{A'C'}{m'}\right)^2 + \left(\frac{C'B'}{m'}\right)^2 + \left(\frac{C''B''}{m''}\right)^2\right]} \tag{8.6}$$

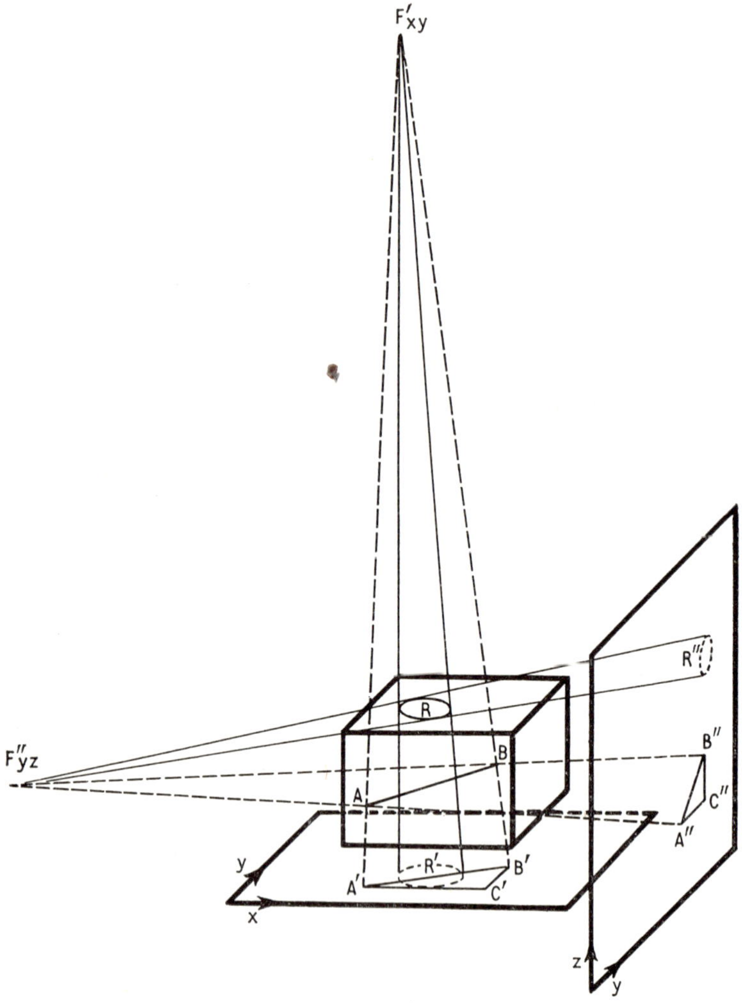

FIGURE 8.11 *Location by Two Radiographs at Right Angles*

so that AB can be calculated from measurements on the films of the projections, provided that m' and m'' are known.

The focus–film distance is normally known, and sometimes the distance between film or focus and AB is measurable. When this is so, m' (and m'') are readily calculated (focus–film distance divided by focus–AB distance).

When this is not the case, a radio-opaque ring, R, of known diameter can be placed in line (in one of the radiographs) with AB (Figure 8.11). The virtue in using a ring lies in the property that, even when the ring is not perpendicular to the X-ray beam, there is always one diameter which gives an unforeshortened image, from which the magnification m, *at the ring*, can be calculated.

On the yz plane, the magnification at the ring is the same as that for AB. Hence, in this case

$$m = m''$$

and both are known. From this, the real distance between ring and A (or B) in a plane parallel to the yz plane can be calculated. From this last dimension, the magnification m' can be calculated, since

$$m' = \frac{F'_{xy}\text{-to-film distance}}{F'R + (R \text{ to } AB)} \tag{8.7}$$

Now that m' and m'' are known, AB can be calculated from equation (8.6).

A useful check on the measurements can be performed by comparing $B'C'/m'$ with $A''C''/m''$ since both are the projection of AB in the y-direction.

THE TUBE-SHIFT METHOD

In the tube-shift method, two exposures are made on one film, the focal spot being moved from F' to F'' between the exposures, in a direction parallel to the plane of the radiograph and to one of its edges (Figure 8.12). The two images A' and A'' of A (and B' and B'' of B) will be separated by a distance depending on $F'F''$ and the height of A (and B) above the film.

By similar triangles

$$\frac{F'F''}{A'A''} = \frac{(F\text{-to-film distance}) - AA'}{AA'}$$

Hence, as $F'F''$ is known and $A'A''$ can be measured, AA' can be calculated, and so can m', the magnification at A. Similarly for B and the magnification m'' at B. In practice the unmagnified projection of AB on the film can be obtained by dividing $A'B'$ (or $A''B''$) by the mean of m' and m'', although more accurately, the reciprocal of the magnification of AB

$$= \frac{1}{m'} + \frac{1}{m''}$$

The third dimension of the 3rd projection of AB is the difference in height above the film of A and B. This is known; hence AB can be calculated. If the focus–film distance and the tube shift are kept constant, a ruler can be calibrated to read directly the height of a point above the film, from the image shift.

It is usually possible to assess the accuracy of the method by measuring a point at a known height.

There are variations in these methods, for instance markers at known positions may be used for calibration. One marker may be placed on top and one under the object. After shifting the tube, the movement of the point of interest on the radiograph relative to that of each marker will give its height relative to that of the markers.

Choice of method depends on several factors. The tube shift method is the most economical in film. It is particularly suitable when radiography is possible from only one direction as for a large metal disc. When the points of interest are near the centre of an object having high absorption in not more

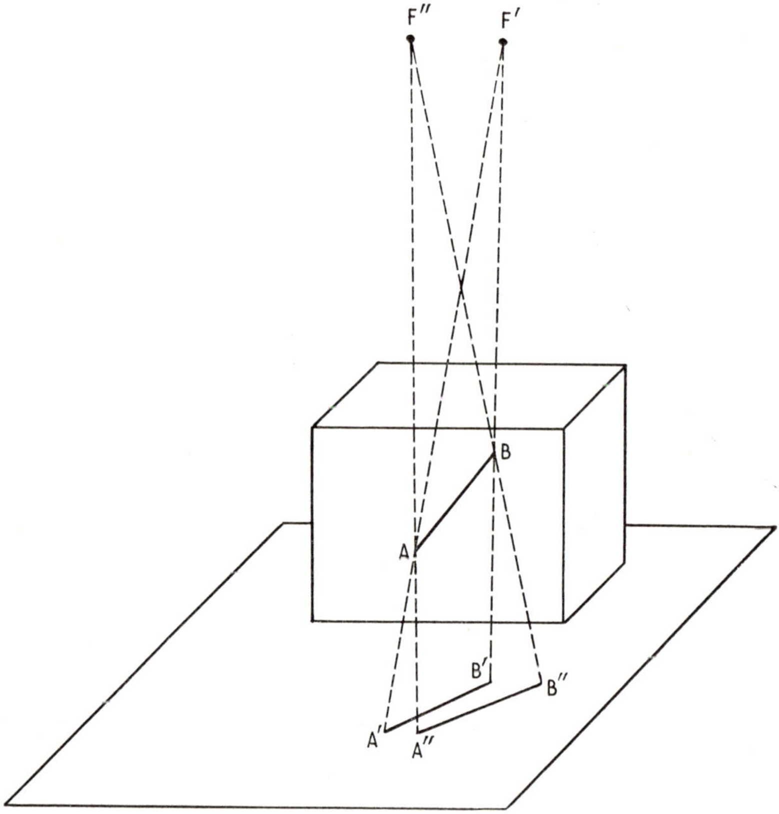

FIGURE 8.12 *Location by Tube-shift Method*

than one dimension, it may be more accurate to take two mutually perpendicular radiographs in the other two dimensions. This is often the case with the human body. It is essential that the patient does not move between the two exposures. In industry, markers with the tube shift method are most commonly used.

Precise measurements are required in medical work to determine the size of the heart or aorta, in pelvimetry or in a reconstruction of the positions of radioactive sources after a radio-therapeutic implant. In industrial work the position of a flaw in a weld may be required, before cutting it out and rewelding.

TOMOGRAPHY

Tomography is a method for reducing the unwanted background. It gives a sharp image of a section of the specimen while blurring the images of parts outside that section. Special equipment[13,19,21b,25] is necessary, and is available from most manufacturers of X-ray equipment.

During the exposure of up to a few seconds, the X-ray tube focal spot is made to move from position T_1 to position T_2 (Figure 8.13).

Consider an object at P. Its image moves from P_1 to P_2 during the exposure. Now the film is made to travel in a direction opposite to that of the tube and

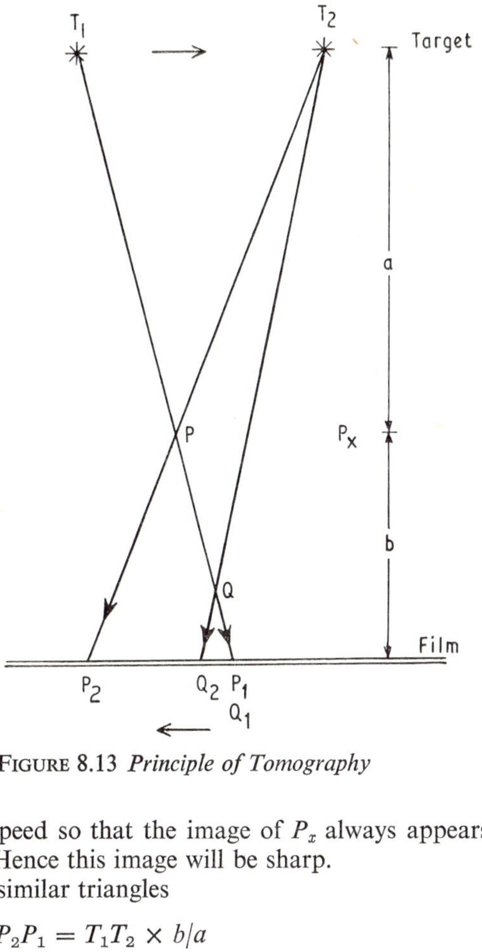

FIGURE 8.13 *Principle of Tomography*

at a speed so that the image of P_x always appears at the same spot on the film. Hence this image will be sharp.

By similar triangles

$$P_2P_1 = T_1T_2 \times b/a \qquad (8.8)$$

where a is the target-to-P distance

b is the film-to-P distance.

Therefore any other position P_x at the same height b above the film will also cast a stationary shadow on the film.

Now consider another point Q at a different height above the film. The image of this moves only a distance P_1Q_2 during the exposure, and this distance is less than the film moves. Hence the shadow of Q moves across the film during exposure and will appear as a blur, contributing only to the general blackening in that part of the film.

Similarly, a point further from the film than *P* will produce a blurred image, as this moves more quickly than the film.

When investigating a small lesion, a localizing film using a calibrated rule to read directly in depths of object, can be a great help in reducing the exposure and film wastage otherwise incurred in finding the correct level for the tomograph [27].

Tomography is useful particularly when there is much overlying bone as in the chest, and lung tomographs can be obtained which show rib shadows only at the lateral margins of the thorax [28]. Another application of tomography is to use it to produce a radiograph of that layer of the body required for dose planning in radiography [29].

It is evident from Figure 8.13, that a series of tomographs at different levels may be obtained for only one exposure, when a stack of cassettes is used [30]. An additional advantage of this *simultaneous tomography* is that any movement such as in respiration is identical for all films. To compensate for extra absorption, more sensitive films and screens[9] should be used at positions further from the tube.

As exposures for tomography tend to be long, body movement may cause objectionable blurring of the image. In tomography of the heart, this has been overcome by using an electrocardiograph to trigger consecutive intermittent short exposures so that all are taken with the heart at the same stage in its cycle [31].

The possibilities of tomography have been largely neglected in industrial radiography.

SUBTRACTION RADIOGRAPHY

Subtraction radiography is a means of making the part of interest stand out from the background.

Two radiographs of the same view and magnification are taken under different conditions. They are designed to change the radiation transmitted through the part of interest differentially compared with other parts. The two original exposures may be made to differ from each other in medical work by using a contrast medium for one radiograph [32]. In a wider field different x-ray energies may be used for the two radiographs. In this way, the contrast between elements of different atomic numbers may be enhanced. For example, one element may have predominantly Compton absorption for both negatives, but for another element, photoelectric absorption may be predominant at the lower energy. A positive transparency is made from one of the radiographs so that its exposure over the background area is complementary to the other radiograph. This positive is placed exactly over the other negative with both over an unexposed film. The sandwich of the three films is then exposed to light so that an image of the blackening of the two films together is produced on the unexposed film. As one of the two films is a positive and one is a negative, the final image is a subtraction radiograph of the two original exposures.

For optimum results, the relative intensity of the negative of one exposure and the positive of the other must be carefully controlled. Provided this is so, the unwanted background may be eliminated.

MICRORADIOGRAPHY

Microradiography shows fine structure by the radiography of small specimens or of thin sections of large specimens. Sometimes this technique is referred to as X-ray microscopy, but this term is liable to be misleading and to suggest an analogy with the electron microscope.

The term microradiography describes two systems: point projection microradiography and contact microradiography. In projection microradiography a very small "point" X-ray source is used and the magnification is produced by direct image enlargement. In contact microradiography, the object is radiographed in contact with the film emulsion and the image is enlarged optically (Chapter 1). Very fine grain emulsions must be used for the original radiograph [33, 34].

Microradiography has been applied to a range of problems in biology, medicine, metallurgy and geology. Resolutions of the order of 100 nm have been obtained [35, 36].

DYNAMIC STUDIES

Sometimes dynamic radiographic studies must be made. As X-rays cannot be focused, the picture on a radiograph is always larger than the object. This, in itself, makes plain radiographs unsuitable for cinematography, where 35 mm films are usual.

It is possible to reduce the final image in size by photographing the picture produced by X-rays on a fluorescent screen. However, as each fluorescent image needs a higher exposure than for a radiograph, the high dose required for multiple exposures precludes this as a practical method for making moving pictures in medical work. Image amplifiers (p. 139) have transformed the situation and make cine-radiography [37] a practical proposition. Both pathological and physiological processes have been investigated; an example of the latter is the opening and closing of the larynx during swallowing [38].

Alternatively, dynamic processes can be investigated by taking serial radiographs. Using flash radiography individual exposures can be as short as nano-seconds (p. 227). This method has possibilities for studying moving machinery and projectiles.

Applications Particular to Medicine

Examination of inaccessible parts of the human body has such obvious benefits, that it is not surprising that much skill and ingenuity has been exercised in increasing radiographic contrast, as this is naturally limited.

The interpretation of radiographs is impossible without a knowledge of medicine, and is the responsibility of medical radiologists.

MASS MINIATURE RADIOGRAPHY (PHOTO-FLUOROGRAPHY)

A film for a plain chest X-ray is typically 14 in. × 17 in., so that it would be expensive both in film and storage costs to survey large sections of the population by this technique.

In mass miniature radiography, the X-rays are made to fall on a fluorescent screen, forming a visible picture. This picture is reduced in size by an optical system of mirrors and lenses and is then thrown on to a photographic film held against a curved surface so that all parts are in focus. These miniature (35 mm, 70 mm, 100 mm or 4 in. × 5 in.) films show less detail than a full-sized radiograph, but patients showing significant abnormalities are recalled for further examination. The dose received depends to some extent on the patient's size; average figures can be as low as 0·23 rad on the skin and 0·05 mrad to the gonads for males [39].

In some countries mass survey examinations of the chest are carried out by fluoroscopy, with undesirable high doses of about one hundred times the individual gonad dose from photo-fluorography [40].

CONTRAST MEDIA

Some regions of the body may be more clearly visualized by the introduction of foreign materials which have low, or more commonly high absorption coefficients. Such *contrast media* should be non-irritant and readily excreted; new materials have to be rigorously tested with these attributes in mind. Sodium, iodine and barium salts are usual for this type of work, and various compounds[26,27,28,29,30,31] have been developed for different examinations.

Using contrast media, dynamic studies can be made by means of cine-radiography [37], fluoroscopy and serial radiographs. For instance, if a drink of barium sulphate in water is taken, the gastro-intestinal tract can be visualized and the time taken for the passage of the drink will indicate if digestion is proceeding at the normal rate. The passage of the barium may be made to trigger the exposure, thus economizing in the number of swallows and the radiation exposure [41]. Demonstration of gastric emptying with a contrast media may be used as an aid to post-operative management [42, 43].

Opaque media in cavities may elucidate adjacent spaces or structures as well as the cavity itself. Thus, they may reveal the existence and size of tumours [44].

It is usually a simple procedure to introduce contrast media into the gastro-intestinal tract, but more sophisticated methods have to be used for some other systems. For instance, in lymphangiography, or visualization of the lymph vessels, these are first injected with an aqueous solution of patent black, and then a suitable lymph trunk can be selected for injecting the radio-opaque medium. The lymph vessels remain opacified for a few minutes [45, 46].

In a similar way, a contrast medium may be injected into the blood stream for a variety of examinations. When it is injected into the right auricle of the heart through an arm vein, or more rarely by a long thin tube from the femoral artery, the heart and great vessels can be visualized on a radiograph. This is known as cardioangiography; it is performed under local anaesthesia. Contrast media injected into the blood stream [47, 48] may be used to demonstrate aortic aneurysms, neoplasms and other pathological changes, varicosities of both arteries and veins and blood supplies to malignant growths [49]. Intravenous pyelography (IVP) is a common method of assessing the morphology and function of the urinary system [50]; the contrast medium fills the kidneys and ureters after injection into a vein.

Although it is usual to employ high-density materials in examinations of the gastro-intestinal tract, the cardiovascular system and the genito-urinary system, in some other cases gases such as air, oxygen or carbon dioxide are introduced as low-density materials. These can be used by themselves as in pneumoencephalography for studies of the brain, or in perineal air insufflation in urinary system investigations.

They can be used also in conjunction with a high-density medium, for example double contrast cystography with barium sulphate and carbon dioxide has been used in the diagnosis and staging of bladder cancer [51]. Early gastric cancer has been diagnosed by a double contrast technique using air and a colloidal barium spray [52]. Double contrast examination of the colon has been advocated as a routine procedure [53].

BONE

Röntgen himself discovered that bone absorbs more radiation than soft tissue, so that it casts a deeper shadow than its surroundings. This is partly because the effective atomic number (14) of bone is higher than that of soft tissue (7·6), and so the photoelectric absorption coefficient is higher for bone, and partly because the difference in density affects the Compton absorption.

At 30 keV, the mass absorption coefficient of bone is about four times that of soft tissue, whereas at 60 keV the factor is less than two and at 150 keV about one. Above this voltage, therefore, differences in linear absorption coefficient are due to differences in density. However, although contrast between bone and soft tissue is high at low voltages, in selecting the optimum kilovoltage, other factors have to be considered (p. 127), in particular overdosage of the proximal skin. Adequate contrast for most purposes is obtained at kilovoltages of 80 to 120 kVp.

More sophisticated use has been made of differential absorption effects to find the concentration of inorganic substances in bone and teeth [54]. In this technique two radiographs are taken, one at 62 kVp and the other at 250 kVp, and the film densities compared photoelectrically. At the lower energy, the absorption will vary sharply with effective atomic number, but this is not so at the higher energy. Therefore differences in contrast may be used as a measure of mineral (high atomic number) content. The particular kilovoltages were chosen to give a simple relationship between density measurements and mineral content.

This method of measuring the mineral content of bone has been used to investigate whether osteoporosis is a predisposing cause of fracture. The mineral content varies with age and after a rise from the sixth to the eighteenth year, shows a parabolic fall.

SOFT TISSUE RADIOGRAPHY

Using intensifying screens and a low kilovoltage[4] (about 50 kVp), it is possible to distinguish fat (density = 0·92 g/cm^3) from other soft tissues such as muscle and connective tissue (density = 1 g/cm^3).

Soft tissue radiography has been developed particularly for breast examinations. Menstrual cycle and lactation changes can be demonstrated as well as tumours [55]. Lipomas and fibro-lipomas show up as low density areas and calcification of soft tissues as areas of high density.

Air and gas are easily visualized in soft tissue radiography, which may be used to demonstrate gas gangrene [43].

Applications Particular to Dentistry

The high mineral content and consequent high X-ray absorption of teeth make dental radiography a useful diagnostic tool. The subject is restricted, so that techniques can be standardized and it is usual for a dentist to be his own radiographer and radiologist.

As teeth are small, high resolution and an X-ray tube with a small focal spot are required. A Potter–Bucky diaphragm is unsuitable.

With the normal periapical film, $1\frac{1}{4}$ in. × $1\frac{5}{8}$ in., it is usual to radiograph two teeth on one film, and the whole mouth with 32 teeth can be radiographed on fourteen films.

Several techniques have been developed for radiographing a number of teeth at one exposure. The bite-wing method uses a special film pack with a wing on which the patient bites to hold the film in position. The coronal two-thirds of maxillary on mandibular teeth may be recorded on one film, and five exposures will cover all the teeth. A second method is *panoramic tomography* (p. 146) and is used for mass dental radiography of children and of adults in the services. The X-ray tube is positioned at the back of the patient's head and the film placed outside the mouth; both tube and film rotate. In an alternative arrangement, the X-ray tube is fixed and both patient and film rotate [56]. In a third method, the focal spot of a special X-ray tube is introduced into the mouth and a film wrapped round the face. All the teeth of a jaw may be radiographed on one film [57].

Other special dental films ($2\frac{1}{4}$ in. × 3 in.) are available for radiographing the dental arches. These may be used with intensifying screens in an intra-oral cassette. They are valuable in a general survey of the mouth and in particular for locating unerupted or impacted teeth.

Dental caries is ordinarily recognizable by other means, but radiographs can demonstrate cavities too small to be otherwise detected and sometimes they are instrumental in discovering unsuspected extensive cavities [58].

In locating malformations or pathological changes, it is sometimes helpful to take stereoscopic views [59, 60].

The differential absorption effect of high atomic number elements compared with those of lower atomic number has been exploited to determine the mineral content of teeth in the same way as of bone (p. 151). As in bone, the mineral content rises from the sixth to the eighteenth year and then falls parabolically. Carious teeth have a lower mineral content than healthy teeth of the same age [54].

Careful direction of the beam and the use of suitable cones and lead aprons make dental radiography a safe procedure. Without these precautions, even the gonads can be unnecessarily irradiated.

Veterinary Applications

Special limitations govern the use of X-rays in veterinary work. Animal sizes vary so much that a very wide range of equipment is required. Economic considerations weigh much more heavily than in medicine and dentistry, and may rule out the radiography of large animals where high-voltage equipment is necessary. Sophisticated techniques developed for medicine are usually precluded both on the grounds of expence and because they are difficult to carry out without the co-operation of the "patient".

The great variation of thickness encountered makes selection of the optimum exposure conditions difficult, unless an exposure meter is used. As a rough guide, if 80 kVp is optimum for some view of a medium sized animal, 8 kVp should be added (or subtracted) for each inch thickness more (or less) than that of the standard view.

The same 10 per cent criterion holds when 50 kVp is optimum [61].

It is usual to anaesthetize small animals and to radiograph them supported with plastic foam cushions on a table. Wild animals often tremble when their movement is restricted, but dogs and cats may be hobbled for x-ray examinations [62]. Large animals should be arranged to stand comfortably and the film placed in a self-supporting cassette holder [61]. If movement blur is to be small it is essential to keep exposure times short. This prerequisite may rule out radiography [63] altogether, when suitable x-ray machines are unavailable. Flash radiography (p. 227) has a place here.

X-rays are useful to detect foreign bodies and fractures, and sometimes tumours, renal calculi, osteofibrosis and rickets in animals [64].

Applications Particular to Industry

Industrial radiography is concerned mainly with the casting and fabrication of metals. All metals, from the lightest, beryllium, $Z = 4$, to the very heavy such as uranium, may be of interest, but those commonly requiring radiography may be grouped into light and heavy. The former are mainly aluminium, $Z = 13$, and magnesium, $Z = 12$; the latter are ferrous metals, $Z = 26$. Other common metals such as copper, zinc or lead are not often used in highly stressed applications, and thus seldom require radiography.

In the light group the linear photoelectric absorption coefficient is greater than the linear Compton absorption coefficient at photon energies below about 60 keV. As the photoelectric coefficient is proportional to the fifth power of the atomic number while the Compton absorption coefficient is linearly proportional to Z, differences in absorption coefficient will depend very markedly on their atomic numbers when x-rays of low energies are used. At higher energies when Compton absorption is predominant, differences between the various metals are less marked.

When the photon energy is greater than 1·02 MeV, absorption by pair production can take place, although it is not predominant unless the energy is much higher, for example, 15 MeV for aluminium [65]. Pair production attenuation is proportional to Z^2, so that at these high photon energies the higher atomic number metals have the higher linear absorption coefficients.

GENERAL FABRICATION AND ASSEMBLY

Radiography of concealed components is often used as a control in assembly work, for example in radio valve manufacture or for rivet inspection in aircraft structure fabrication. Although in most cases radiographic examination is unlikely to supersede conventional inspection methods, owing to cost and inconvenience, it can be used with advantage in difficult assembly problems to supplement the visual inspection. Manufacturing standards may improve with the knowledge that hidden assemblies are liable to be radiographed.

CASTING

Most metals in common use melt at temperatures which can be achieved on an industrial scale. Thus, aluminium melts at 660°C, iron at 1,535°C, copper at 1,083°C and zinc at 420°C. Casting is therefore a convenient method of shaping a metal, either into its final form or, for instance, into an ingot on which further work such as rolling may be performed. Any defects in castings may have repercussions in subsequent fabrication processes. Radiography is a useful tool in detecting and identifying such defects, but a knowledge of the casting process is required for the successful interpretation of these radiographs.

A pure metal normally solidifies at a constant temperature, its freezing point. It will remain at this temperature until freezing is complete. The latent heat of freezing represents the difference in energy in the liquid and solid states of the metal. When no foreign particles or nuclei are present, the liquid metal may cool to many degrees below the freezing point before solidification begins; then the latent heat causes the melt temperature to rise to the freezing point. This is known as super-cooling.

In alloys the freezing process is usually far more complicated. Continual changes in composition of the liquid and solid phases occur during freezing, and further phase changes may occur even after solidification. Such phase changes are usually accompanied by heat production [66, 67].

There are two main types of casting [68]: one using sand or temporary moulds and the other using metallic or other permanent moulds. In both cases, the moulds are made as nearly as possible to the final dimensions, after allowing for shrinkage and finishing.

Sand-casting is particularly suitable for very large objects. The sand is usually clay-bonded, and is rammed when moist round the pattern. The pattern is removed later, to leave the mould. A casting with hollow parts may have the cores for these made from specially bonded and baked sand. The metal is poured into the mould through *runners* or *gates*. When the mould is filled, excess metal flows into *vents* and *risers* which also allow gas to escape during the casting. After the casting is complete, metal remaining in the runners and risers is removed and re-used.

Metal *chills* may be included in the moulds to assist cooling in local regions. Core supports, known as chaplets or studs, are other metallic components of the mould. All should be of metal similar to the casting.

The *lost wax* process is a modification of sand moulding, and is used for producing precision castings. In this method a wax pattern is made in an accurately machined master mould. A special sand mould is made round the wax mould, which is then melted and removed to leave a hard, accurate sand mould.

A metallic mould is known as a die. Its surface is smoother than that of a sand mould and it can be made to finer tolerances. The liquid metal may enter the die under gravity or be injected at high pressure; sometimes centrifugal casting is used. Repetition die casting presents important design problems; the casting section should be kept uniform and thin; the position

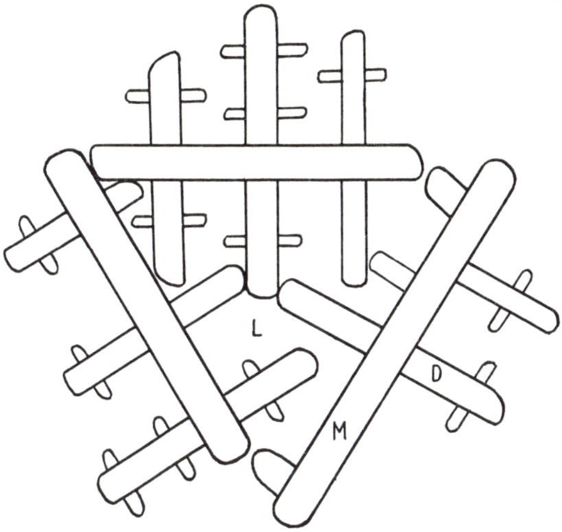

FIGURE 8.14 *Dendritic Segregation (known as Coring)*
L, liquid metal; *M*, metal crystal; *D*, dendrite arm.
When the dendrites meet, fine pores may develop parallel to the dendrite arms.

of the parting line of the dies should be chosen carefully to avoid leaving an unsightly mark where the overfill or *flash* has been removed.

Casting defects arise from imperfect filling of the mould. There may be voids, or foreign material may be included in the casting.

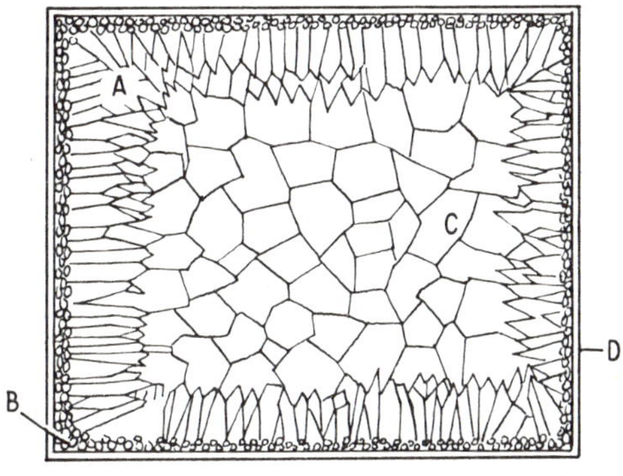

FIGURE 8.15 *Microstructure of Ingot to Show Grain Structure*
A, columnar grains; *B*, chill grains; *C*, equiaxial grains; *D*, mould.

The solidified metal often grows into pine-tree-shaped grains known as dendrites (Figure 8.14), which extend and thicken so that impurities may be trapped between their branches. Shrinkage may occur between dendrites. The number of dendrites formed depends on the number of nuclei present.

Castings usually solidify first from the side; when this process is rapid, columnar grains are formed (Figure 8.15). As the metal solidifies it contracts so that the level of liquid metal is being continuously lowered towards the centre (Figure 8.16). This may eventually lead to a "pipe" or other form of core defect.

Gases may be evolved during solidification; they produce porosity or even large blowholes in the casting. They can originate from dissolved gases in the liquid, or in the wall of the mould, or from chemical reactions. For instance, dissolved copper oxide may react with hydrogen dissolved in the copper to produce steam.

Casting Radiography

The purpose of casting radiography is to identify and localize any flaws [3] which may be important in the function to which the casting may be put.

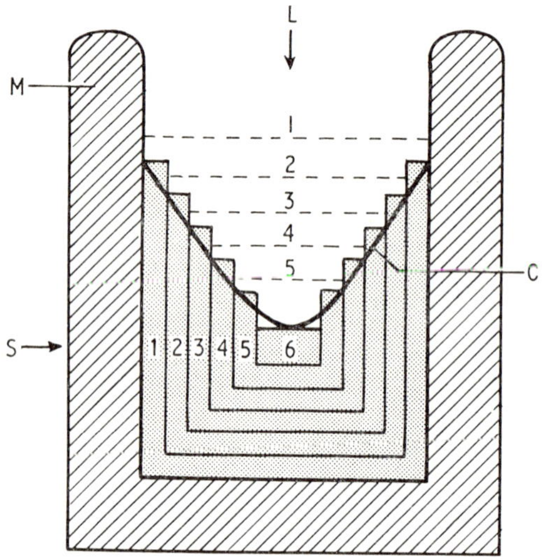

FIGURE 8.16 *Freezing of a Casting to Produce a Concave Surface or "Pipe"* The freezing is assumed to take place in a succession of stages (1–6). The liquid levels are also shown. *L*, successive liquid levels; *S*, successive solid stages; *C*, concave casting surface; *M*, mould.

Besides knowledge of casting defects and their radiographic appearance, there must be good liaison and mutual confidence between radiographer and foundryman.

The X-ray tube should be run at a voltage appropriate to the atomic number, and to a lesser extent, the thickness of the casting. It may be desirable to use a grid to obtain the best results (p. 134). Where contrast is excessive, this can be reduced by increasing the hardness of the radiation, or by limiting the range of absorption paths (p. 134).

It may be necessary to detect even small flaws in what may be a thick casting of irregular shape [69, 70, 71] and which may be made of one of a variety of metals. The experienced worker can sometimes anticipate likely faults and make it unnecessary to radiograph the whole casting, thus improving the quality of the picture. For example, abrupt changes of section are subject to stress cracking; large flat surfaces are likely to hold slag.

Each metal has its characteristic defects: light metals are in general chemically active and oxidize readily; aluminium dissolves hydrogen.

Before commencing the radiograph, the heads and runners should be cut off. Dirt and sand should then be removed by sand-blasting. However, it may be advantageous to delay final fettling, and thus to avoid removing traces of heads and runners which may be the cause of flaws.

Orientation is very important in flaw identification and it is usually necessary to radiograph the flaw from two directions. Otherwise, for example, a fine long void viewed along its axis can be confused with a gas bubble.

Some important casting defects such as thermal strain cannot be demonstrated radiologically. However, most common defects: voids, cracks and tears, cold shuts, segregation and inclusions are susceptible to detection and identification by X-radiography.

Certain defects can appear on the radiograph as areas of either high or low photographic density. One of these is a concentration of any of the constituents of an alloy: *segregation* (Figure 8.14). It depends on the cooling rate of the alloy and thus may differ in degree within regions of the same casting. Another is due to the *inclusion* of foreign matter. These inclusions can arise from the mould or from oxidation to form slag. Inclusions, such as slag, which have been molten at some stage, will be rounded in shape.

Other defects give rise to areas of high photographic density on the negative. Large gas bubbles appear as discrete rounded shapes; multiple small bubbles constitute *porosity*. Other gas-filled spaces may arise from shrinkage. All these are known as *voids*. Cracks may occur in the casting at temperatures either near the freezing point, when they are called *hot tears* and have a jagged outline tending to taper, or at lower temperatures when they are called *stress cracks* and are approximately straight and uniform in section. A *cold shut* is a discontinuity caused by the failure of the molten metal to unite either with another molten stream or with solid metal. It will appear as a dark line on the negative.

Diffraction mottling is a rather unusual phenomenon which can be important in radiography. If a casting has large perfect grains, diffraction effects may be visible provided that the X-radiation has wave-length components to satisfy Bragg's equation. Diffuse patches may appear even in regions remote from the primary beam, their position altering with the tilt of the casting. Diffraction mottling may be a limiting factor for resolution in microradiography, particularly in copper-based alloys [3].

WELDING

In recent years welding methods have become increasingly important and intensively developed. This is because modern industrial design often requires strong light structures with smooth lines, particularly for modern aircraft and rockets, and for bridges.

The process of welding is the bonding of metals, usually by applying heat. There should be complete fusion or union at the weld and to achieve this additional metal is sometimes added (Figure 8.17).

The term "welding" does not include soldering or brazing because in these cases the bonding is by the diffusion of a molten alloy with the two pieces to be joined.

As clean surfaces are essential for a good weld, a flux may be used to remove surface oxides, or as a protection against oxidation during the welding process. Sometimes the last is so troublesome that the weld has to be made in a vacuum, or in an inert gas such as argon.

When the two surfaces to be joined are very clean and in intimate contact, pressure alone may produce welding. The traditional blacksmith's practice

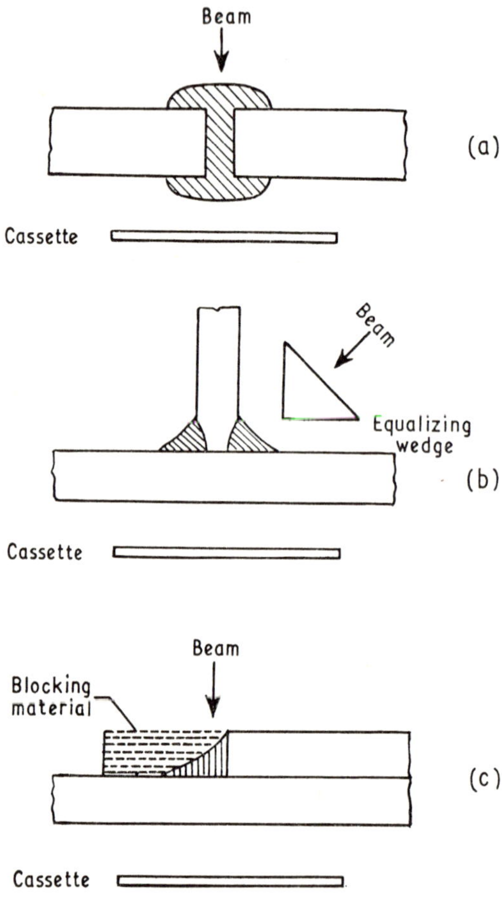

FIGURE 8.17 *Radiographing Typical Welds*
(*a*) Butt weld; (*b*) fillet weld; (*c*) lap weld.

is to heat, but not melt, the metal, when pressure welding is achieved by hammering.

More often the metal is locally melted, and this is known as fusion welding. Heating may be by a tungsten or carbon electric arc, or by an oxy-acetylene blowlamp. Another method is to heat the metals by passing a heavy current, although this is unsuitable for large specimens made of metals of high electrical conductivity, such as copper. High thermal conductivity also makes it difficult to obtain local melting.

Welding Radiography

Welds have to withstand considerable mechanical stress with absolute reliability. Non-destructive methods of weld testing are of paramount importance and weld radiography is possibly the most important application of industrial radiography [3].

Acceptance standards vary between different countries and between different applications in each country. In general, defects of globular form are less serious than angular defects; the same defect may be more serious if the stresses are to be dynamic rather than static; standards can be somewhat relaxed if the materials can undergo plastic deformation. Although any form of crack or region of incomplete fusion usually renders a weld unacceptable, exceptions can be made by a radiographer having a knowledge of both welding and the purpose to which the weld will be put. Without such exceptions, rejection may prove to be unnecessarily and prohibitively expensive. In contrast, for some structures such as pressure vessels in ships, no flaws are admissible.

Some weld defects are not visible radiographically. Examples are the presence of cooling stresses: the embrittlement due to grain growth and phase changes due to overheating. These defects can be revealed by destructive tests such as metallographic examination and torsion tests.

For fine structures to be demonstrated radiographically the film needs a fine grain and high contrast. Photographic density should be at least 2·0 when using lead intensifying screens and between 1·3 and 2·3 for fluorescent screens. Adequate brightness is essential for viewing [6].

Before radiography, the weld should be examined for surface markings which may indicate hidden defects. If the surface is very irregular the weld should be ground to eliminate resulting radiographic artefacts.

The cassette should be placed against that part of the weld which is of most interest, such as the root or the fusion face. In an extreme case, it may be necessary to cut away part of the structure so that the cassette can be inserted in the most advantageous position.

When possible, the beam should be normal to the weld with the cassette immediately behind it (Figure 8.17*a* and *c*). In other cases, suitable beam direction and the positioning of the cassette may require considerable judgment. For instance, in a fillet weld (Figure 8.17*b*), the beam direction often cannot be perpendicular to the cassette and equalizing wedges or blocking material may be required [3]. Radiography is of particular use in examining butt welds (Figure 8.17*a*).

The nature of the radiographic image is diagnostic. For example, lines can appear as several types. When straight with sharp edges, they arise from lack of fusion, which may be due to dirty surfaces or inadequate fluxing. Incomplete penetration can give a similar appearance or an intermittent or wavy shadow.

Fine lines wandering over the weld are due to cracks. Cold cracks show as more continuous and straighter lines than those for hot tears. *Crater cracks* are fine cracks radiating from a depression where the arc or flame was removed. Large specimens suffer from high thermal stresses and are more likely to crack than small ones.

As in casting, air or other gases may be trapped, or liberated, to give, for example, porosity voids.

Less well-defined shadows may be due to inclusions. These may arise from dirt originally present or from oxidation during the welding process in faulty techniques.

Grain growth may be revealed by diffraction mottling [3].

Collections of radiographs illustrating selected weld defects are available [72, 73]. Interpretation should be verified by sectioning and metallographic examination when the sample is rejected.

Applications in Art

Radiography is increasingly used to examine paintings [74]. As the object is thin and flat good contact may be made with the cassette over the whole area and there is no scatter. Shadows may arise from wooden panels or supporting members.

Energies between 15 and 30 keV may be used for oil paintings with dense pigments and between 5 and 15 keV for drawings and water colours, although consideration must be given to absorption edge effects. Where the paint thickness varies considerably over the canvas, as for example in some modern techniques, or where dense highlights occur, it may be advisable to take radiographs at different voltages in the various regions. Paint thickness is usually of about the order of 50 microns.

A wide variety of metals are used in pigment bases. In general, older painters used more dense pigments based on lead; more modern pigments usually incorporate lighter metals such as titanium. Where metallic gold occurs it is mostly in the form of leaf of negligible absorption. In many paintings the radiographic image gives the distribution of the white lead in the original, so that the less dense areas in the radiograph correspond with the whiter areas of the original. Thick lead priming may cause considerable loss of radiographic detail.

Radiography may elucidate the artist's technique with regard to brushwork or delineation, and also the stages of development of the painting with regard to alterations and overpainting. Discrimination is possible between old and more modern parts or restoration in a painting. The condition of the canvas and wooden supports with regard, for example, to decay or worm holes, may also be revealed.

Radiography has been applied to the study of porcelain, and the method of coring and any repairs using metal supports may be revealed [4]. The original decoration of corroded pots and other objects has sometimes been shown by radiography.

Literature References

1 SCHUSTER, N. H., "Early days of roentgen photography in Britain", *Brit. Med. J.*, p. 1164 (1962).
2 KAYE, G. W. C., *X-Rays* (London, Longmans, 1929).

3 WILTSHIRE, W. J. (Ed.), *A Further Handbook of Industrial Radiology* (London, Edward Arnold, 1957).
4 CROWTHER, J. A., (Ed.), *Handbook of Industrial Radiology* (London, Edward Arnold, 1949).
5 LUBBERTS, G. and ROSSMAN, K., "Modulation transfer function associated with geometrical unsharpness in medical radiography", *Phys. Med. Biol.*, **12,** p. 65 (1967).
6 INTERNATIONAL INSTITUTE OF WELDING, *Handbook on Radiographic Apparatus and Techniques* (London, Institute of Welding, 1961).
7 VAN DER PLAATS, G. J., *Medical X-Ray Techniques* (London, Cleaverhume, 1960).
8 UVORIN, P., "Measurements of the parallactic unsharpness caused by the film and the intensifying screens in radiography", *Brit. J. Radiol.*, **32,** p. 418 (1959).
9 OSBORN, S. B., "The implications of the reports of the Committee on Radiological Hazards to Patients (Adrian Committee) 1. Variations in the radiation dose received by the patient in diagnostic radiology", *Brit. J. Radiol.*, **36,** p. 230 (1963).
10 JOHNS, H. E., *The Physics of Radiology* (Springfield, Chas. C. Thomas, 1961).
11 ARDRAN, G. M. and CROOKS, H. E., "The reduction of scatter fog in chest Radiography", *Brit. J. Radiol.*, **37,** p. 477 (1964).
12 JACOBI, C. T. and HAGEN, D. E., *X-ray Technology* (St. Louis, C. V. Mosby, 1960).
13 HILLS, T. H., "Comparative performance of grids in relation to their stated ratio", *Brit. J. Radiol.*, **32,** p. 102 (1959).
14 SCHALL, W. E., *X-rays* (Bristol, J. Wright, 1961).
15 BOLDINGH, W. H., "Quality and choice of Potter-Bucky grids", Part VI, *Acta Radiol.*, **56,** p. 202 (1961).
16 GIBSON, G. L., *Copying and Duplicating Medical Subjects and Radiographs* (Springfield, Chas. C. Thomas, 1953).
17 KODAK LTD., *Kodak Data Sheet* G.N.2.
18 NEMET, A. and COX, W. F., *Proc. Instn. Elect. Engrs.*, **103** (B), p. 345 (1956).
19 HAY, G. A., "Quantitative aspects of television techniques in diagnostic radiology", *Brit. J. Radiol.*, **31,** p. 611 (1958).
20 SAMUEL, E. and SUMERLING, M. D., "An assessment of the use of the 70 mm camera in radiological practice", *Brit. J. Radiol.*, **37,** p. 620 (1964).
21 CHAMBERLAIN, W. E., "Fluoroscopes and fluoroscopy", *Radiol.*, **38,** p. 383 (1942).
22 OLIPHANT, W. D., "Xerography, a symposium. I. Apparatus and method of use", *Brit. J. Radiol.*, **28,** p. 543 (1955).
23 HILLS, T. H. and STANFORD, R. W., "Xerography, a symposium. II. The present medical application", *Brit. J. Radiol.*, **28,** p. 545 (1955).
24 FARMER, F. T., "Use of xero-radiography for megavoltage treatment planning", *Bull. Hosp. Phys. Ass.*, **8,** p. 7 (1961).
25 FARMER, F. T., FOWLER, J. F. and HAGGITH, J. W., "Megavoltage treatment planning and the use of xero-radiography", *Brit. J. Radiol.*, **36,** p. 426 (1963).
26 ODEN, B., BELLMAN, S., and FRIES, B., "Stereo-microlymphangiography", *Brit. J. Radiol.*, **31,** p. 70 (1958).
27 SIMON, M., "Precision tomography with simple apparatus", *Brit. J. Radiol.*, **27,** p. 266 (1954).
28 MCDOUGALL, J. B., *Tomography* (London, H. K. Lewis, 1940).
29 HARTLEY, J. B., "Localization by transverse tomography", *Brit. J. Radiol.*, **34,** p. 550 (1961).
30 LODGE, T., *Recent Advances in Radiology* (London, Churchill, 1955).

31 Russell, J. C. B., "Stroboscopic tomography of the heart", *Brit. J. Radiol.*, **37,** p. 440 (1964).
32 Levick, R. K. and Mitchell, J., "A simplified method of subtraction radiography and its application to renal arteriography", *Brit. J. Radiol.*, **35,** p. 843 (1962).
33 Cosslett, V. E., Engstrom, A. and Pattee, H. H. (Eds.), *X-ray Microscopy and Microradiography* (New York, Academic Press, 1957).
34 Clark, G. L., *Applied X-rays* (London, McGraw-Hill, 1955).
35 Cosslett, V. E., *X-ray Microscopy* (Cambridge, University Press, 1960).
36 Cosslett, V. E., "X-ray microscopy retrospect and prospect", *Brit. J. Radiol.*, **34,** p. 1 (1961).
37 Ardran, G. M., "The value of cineradiography in medicine", *Brit. J. Radiol.*, **37,** p. 819 (1964).
38 Ardran, G. M. and Kemp, F. H., "Opening and closing of the larynx during swallowing", *Brit. J. Radiol.*, **29,** p. 205 (1956).
39 *Radiological Hazards to Patients, Interim Report of the (Adrian) Committee* (London, H.M.S.O., 1959).
40 United Nations, *Report of the United Nations Scientific Committee on the Effects of Atomic Radiation*, Suppl. No. 16 (A/5216) (New York, United Nations, 1962).
41 Hughes, D. and Wood, M., "X-ray set triggering for barium swallow radiography", *Brit. J. Radiol.*, **32,** p. 699 (1959).
42 Matheson, N. A. and Dudley, H. A. F., "Contrast radiography: an aid to post-operative management", *Lancet*, 1963, i, p. 914.
43 Samuel, E., "Contrast radiography", *Lancet*, 1963, i, p. 1052.
44 Crispin, A. R. and Fry, I. K., "The perianal space shown by barium enema", *Brit. J. Radiol.*, **36,** p. 319 (1963).
45 McLaren, J. W., *Modern Trends in Diagnostic Radiography* (London, Butterworth, 1960).
46 Kinmouth, J. B., "Lymphangiography in man. A method of outlining lymphatic trunks at operation", *Clin. Sci.*, **11,** p. 13 (1952).
47 Grainger, R. G., "New cardiovascular contrast media: sodium iothalamate (Angio Conray) and methylglucamine iothalamate (Conray), *Brit. J. Radiol.*, **37,** p. 568 (1964).
48 Kendall, B., "Cerebral angiography using Conray", *Brit. J. Radiol.*, **37,** p. 581 (1964).
49 Strickland, B., "The place of arteriography in tumour localization", *Brit. J. Radiol.*, **34,** p. 555 (1961).
50 Powell, T. *et al.*, "Intravenous pyelography—a comparative trial of ten methods in patients with good renal function", *Brit. J. Radiol.*, **40,** p. 30 (1967).
51 Doyle, F. H., "Bladder cancer double-contrast cystography and a bladder analogue", *Brit. J. Radiol.*, **36,** p. 306 (1963).
52 Foti, M., "Early diagnosis of gastric cancer by double-contrast spray method", *Brit. J. Radiol.*, **36,** p. 407 (1963).
53 Scott-Harden, W. G., Philip, L. D. and Moule, B., "Radiology of the colon", *Brit. J. Radiol.*, **40,** p. 15 (1967).
54 Oeser, H. and Krokowski, E., "Quantitative analysis of inorganic substances in the body", *Brit. J. Radiol.*, **36,** 274 (1963).
55 Egan, R. L., "Mammography: report of 2000 studies", *Surgery*, **53,** p. 291 (1963).
56 Blackman, S., "Rotational tomography of the face", *Brit. J. Radiol.*, **33,** p. 408 (1960).
57 Blackman, S., "Recent advances in dental and oral radiographic techniques", *Brit. Dental. J.*, **109,** p. 399 (1960).

58 WORTH, H. M., in *A Text-Book of X-ray Diagnosis*, Ed. SHANKS, S. C. and KERLEY, P. Vol. III, Part II, *The Teeth and Jaws* (London, H. K. Lewis, 1957).
59 SILVA, C. A., "Stereoroentgenograms in dentistry", *Oral Surg.*, **9,** p. 757 (1956).
60 BENKOW, H. H., "Appliance for identical radiographs and stereoradiography", *Dent. Radiog.*, **30,** p. 21 (1921).
61 SMITH, R. N., *Radiography for the Veterinary Surgeon* (Bristol, John Wright, 1960).
62 *Holiday's Surgical Diseases of the Dog and Cat.* Ed. McCunn, J. (London, Baillière, 1947).
63 WOOLDRIDGE, W. R., *Farm Animals in Health and Disease* (London, Crosby Lockwood, 1954).
64 HUTRYA, F., MAREK, J. and MANNINGER, R., *Special Pathology and Therapeutics of the Diseases of Domestic Animals* (London, Baillière, 1946).
65 PRICE, W. J., *Nuclear Radiation Detection* (New York, McGraw-Hill, 1958).
66 REED-HILL, R. E., *Physical Metallurgy Principles* (New York, Van Nostrand, 1964).
67 BAILEY, A. R., *A Textbook of Metallurgy* (London, Macmillan, 1960).
68 SEYBET, A. V. and BURKE, J. E., *Procedures in Experimental Metallurgy* (New York, Chapman and Hall, 1953).
69 *Proceedings of the 4th International Conference on Non-destructive Testing* (London, Butterworth, 1964).
70 ROCKLEY, J. C., *An Introduction to Industrial Radiology* (London, Butterworth, 1964).
71 HALMSHAW R., (Ed.), *Physics of Industrial Radiology* (London, Heywood, 1966).
72 HOULDCROFT, P. T., YOUNG, J. G. and SMITH, A. A., *Classified Radiographs for Defects in Aluminium Fusion Welds* (London, British Welding Research Assn., 1958).
73 INTERNATIONAL INSTITUTE OF WELDING, *Radiographs of Welds*, Commission V, 1961 (Internat. Inst. Welding).
74 REES JONES, S., "Physics and painting", *Bull. Inst. Phys.*, **11,** p. 157 (1960).

9 | *Radiotherapy*

In the early days of radiology, X-rays were generated at low kilovoltages. For this reason, it was not possible to radiograph thick or bony regions without giving a high dose to the skin. Three weeks after an attempt to demonstrate the presence of a foreign body in the head of a patient, he lost some of his hair. It was obvious that the area of irradiation and of depilation coincided and the conclusion that X-radiation had a biological effect was unexpected but inescapable [1].

A period of indiscriminate exploitation of X-radiation followed. Dose measurement was rudimentary and the mode of action of radiation unknown, so it is not surprising that unsuitable conditions were treated by this method, nor that diseases which have since proved amenable to radiotherapy were given doses that proved to be too high or too low. There is still much to learn, and radiotherapy is only recently beginning to rest on a comprehensive scientific basis [2]. However, starting with the trial and error methods of the early workers, experience has already led to the establishment of a sound practice of radiotherapy.

A number of conditions and diseases can be successfully treated by radiotherapy: cancer is the most important, but not the only one of these. In the treatment of cancer, the value of radiotherapy should be assessed not only by counting those who would not otherwise have survived, but also by remembering that radiotherapy may be less mutilating than surgery and that palliative treatment may relieve distress even when it cannot cure.

However, treatment by radiation carries with it an element of risk, because effective doses are those which, in general, can cause serious injury to normal tissue as well as a small and long term risk [3, 4, 5] of carcinogenesis (Chapter 14).

A genetic hazard arising from radiation reaching the gonads has also to be considered for those of, or under, childbearing age.

For these reasons, radiotherapy should be limited to diseases which untreated would be fatal, and to painful or incapacitating conditions for which there is no other effective treatment. Some skin diseases may be cured by comparatively low doses, but even here radiotherapy is unjustified unless there are special considerations [6]. Ankylosing spondylitis is one of the few non-malignant diseases for which radiotherapy is a form of treatment accepted by many clinicians. In this treatment, the whole length of the spine and the sacro-iliac area may be irradiated [7]. The great majority of these patients improve after radiotherapy and are able to return to work.

Although malignant diseases form the largest group suitable for radiotherapy, only certain types respond satisfactorily. Radiation acts as a destructive agent, killing the malignant cells or rendering them incapable of sustained division, or perhaps by causing them to differentiate [1, 8, 9, 10, 11, 12].

Radiotherapy, like surgery, cannot be successful when the cancer has metastasised to areas outside those treated. Even when this has not occurred, it may be impracticable to give a dose high enough to destroy the primary tumour, should this be radio-resistant Furthermore, the radio-sensitivity of a particular cell, normal or malignant, depends on its state in the mitotic cycle [12], as well as on the presence of any chemical protectors or radio-sensitizers. In particular, it varies with the oxygen tension, so that an otherwise radio-sensitive cell becomes comparatively radio-resistant when its oxygen tension is low [13, 14].

This phenomenon is thought to account for recurrencies in some irradiated tumours in the following manner: a tumour may have areas deficient in blood vessels and consequently poorly supplied with nutrients and oxygen; such area tend to become necrotic. In the border area between an area of necrosis and one with a good blood supply, there may be cells which are viable but in which the oxygen tension is low. Such cells will be radio-resistant and therefore may survive doses sufficient to sterilize well-oxygenated cells, and these survivors may subsequently give rise to a recurrence of the tumour.

The oxygen tension may vary widely within a single tumour [15, 16]. This is particularly true for large tumours which tend to have necrotic centres. But even small tumours may have anoxic cells, as congested blood vessels can lead to a low oxygen tension in nearby tissue, and hence to small areas of necrosis [17].

The effect of oxygen on radio-sensitivity is large for X-rays and other low LET radiations. If the radio-sensitivity of anoxic cells is taken as unity, it rises rapidly to two at about 10 mmHg partial pressure, and then more gradually to 2·7 at about 100 mmHg partial pressure. The pressure in well vascularized tissue is considered to range from 40 to 100 mmHg.

The oxygen tension in a tumour may be raised by giving the patient oxygen to breathe. This makes little difference to the radio-sensitivity of cells previously well oxygenated, but it increases that of previously poorly oxygenated cells. Unfortunately, this differential increase in radio-sensitivity is usually considered to be worth while only with oxygen at several atmospheres pressure, so that the patient has to be treated in a pressure chamber [18]. Du Sault [19] has questioned the desirability of giving high pressure oxygen. When treating mice with spontaneous tumours, she obtained the greatest cure rate with 95 per cent oxygen and 5 per cent carbon dioxide at atmospheric pressure. The vasodilation and respiratory stimulation of carbon dioxide resulted in more oxygen being carried to the tumour than with pure oxygen, which depresses the respiratory rate.

As the differential radio-sensitivity of anoxic and oxygenated cells is small for radiations of high LET, the possibilities of neutron therapy [20, 21] have been re-examined with a view to eliminating the need for artificial oxygenation.

A tumour may recur after many years. So-called latent periods between treatment and recurrence can be at least partly accounted for by the time necessary for a few malignant cells to multiply to a number large enough to constitute a palpable tumour. A growing tumour has, say, $(1 + x)$ times as many cells in one generation as in its predecessor. The growth rate of some tumours can be measured, either by radiography or, if they are accessible, directly. Hence, if the duration of the cell cycle can be estimated, it is possible to calculate x.

Now, the volume of a single cell is about 10^{-6} mm^3. Therefore the number of cells in a certain tumour can be estimated, and the number of cell generations necessary for a few cells to give rise to the number present in the tumour. Calculations on these lines show that tumours are undetectably small for a very long time, and this conclusion is not invalidated if multiplication rates for cells are not constant throughout the growth of the tumour [22, 23, 24].

Paterson [10] has defined the tumour lethal dose as "that dose of radiation which produces complete and permanent regression of a tumour *in vivo* in the zone irradiated". In this sense cure cannot be pronounced until death and when post-mortem examination reveals no sign of the disease. Practical considerations demand a shorter term evaluation when, for instance, comparing results of different methods of treatment. Five-year survival figures have been adopted generally as a useful compromise between the disadvantage of waiting for the deaths of all patients and misleading short-term evaluations. However, five years is an unnecessarily long period for rapidly lethal tumours such as those of the lung, and too short a period for accurate prognosis for slowly growing tumours such as some carcinomata of the breast [25].

A distinction must be made between radio-sensitivity and radio-curability: if the former is defined as the rate of regression of a tumour and the latter as the permanency of regression after radiotherapy, the two are not necessarily correlated [26].

In general, it is possible to give only one radical course of treatment because heavily irradiated normal tissue will not tolerate a second and because tumour cells may acquire radio-resistance [10], possibly because of impaired blood supply and consequent anoxia. However, it is not necessarily good practice to prescribe the maximum dose that will be tolerated; such a dose may increase the incidence of radiation damage to normal tissue, with no, or a marginal, increase in effect on the tumour. For these reasons successful radiotherapy depends on clinical judgement in dose prescription, as well as on accurate planning and treatment.

Radiotherapy may give rise to systemic effects, particularly if the *integral dose* is high. Integral dose [27, 28] is defined as the energy imparted to matter by ionizing particles in the region of interest, usually the whole body, and is measured in gramme-rads [29]. It is the integrated product of each element of mass δm of the region, multiplied by the absorbed dose, D in rads in that element; it may be the limiting factor for tolerance when large volumes are irradiated. An approximate calculation of the integral dose delivered by X-rays may be made from a knowledge of body contour, field area, central axis depth dose and skin dose [7].

Integral dose is a concept useful for comparing different techniques for the same patient [30], although even here it has its limitations, as the site of irradiation may radically affect the reaction. Furthermore, there is no reason to suppose that, for example, a mass m absorbing a dose $2D$ is biologically equivalent to a mass $2m$ absorbing a dose D, although the integral dose is $2mD$ in each case [31, 32].

There are limitations also when integral dose concepts are used for evaluating systemic effects on different patients, as it has been shown that, given the same integral dose, thinner patients are likely to suffer greater effects than are fatter ones [33, 34, 35].

With limited success, attempts have been made [36, 37, 38] to reduce the side effects of radiotherapy, by administering radio-protective drugs.

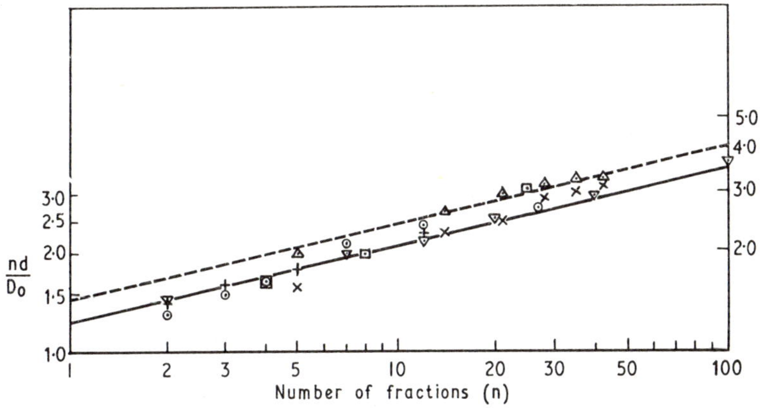

FIGURE 9.1 *Relationship between Total Dose (nd) divided by* D_0 *(Equivalent Single Dose), and the Number of Fractions n*
Eight sets of observations of human skin and tumour reactions.
(Fowler, J. F. and Stern, B. E., *Brit. J. Radiol.*, **33**, p. 389, 1960)

Severe local and systemic effects make it impossible to deliver a single x-ray dose high enough to cure a deep-seated tumour, but it has long been known that a higher dose can be tolerated if it is fractionated and delivered over a period of days or weeks [7]. For orthovoltage radiation, the dose for skin tolerance may be three times as high if it is delivered over five weeks, as if it is delivered in a single dose [39]. Fractionation appears to favour the differential effect on the tumour relative to side effects, although the *tumour-killing dose* is also higher when fractionated than when delivered as a single dose.

One reason why a single dose D has a greater biological effect than the same total dose given in n fractions, each of dose D/n, is that recovery takes place between the fractions. Another explanation is that a high dose is more than proportionately as effective as a lower one. Both these explanations may apply, but the latter seems to be of the greater importance in radiotherapy [40].

Fractionation regimes have been chosen by tradition, personal experience and clinical acumen, as well as by generalizations based on published results [41, 42, 43, 44, 45]. Collected data show some consistency (Figure 9.1), and they have been used to form the basis of rational explanations, particularly in terms of cell survival models [40, 46, 47, 48].

Studies of cell population kinetics suggest that new regimes of fractionation may reduce the differential effect of radiation on cells with different oxygen tensions [48]. Fractionation phenomena are bound up also with the LET of the radiation: the effect of fractionation on the relationship between total dose and biological effect is substantially different for neutrons from that of the lower LET X-radiations [49, 50].

It has been customary to fractionate deep therapy into something of the order of fifteen to twenty treatments spread over three weeks or a month, and to deliver a total tumour dose of about 4,000 to 6,000 rads, although a few clinicians have advocated much longer periods of three or four months and have consequently given higher doses [51, 52]. More recently, three treatments per week have been found to give results not worse than five treatments per week in the same overall time, and the former arrangement is obviously more convenient for patient and treatment centre. The search for optimum fractionation regimes is being pursued [53, 54], but progress cannot be rapid as each patient must be treated in a manner currently believed to be the best, and the long-term results for large numbers of patients must be assessed.

Choice of Method of Treatment

Only X-ray therapy will be considered here, although large sources of radioactive isotopes, particularly cobalt-60, are often used as alternatives.

A principle of treatment is that radiation shall be confined as far as possible to the tumour, including where appropriate, neighbouring regions of likely spread, but excluding other normal tissues. In general, superficial lesions are treated with low-energy X-radiation at a short focal distance, and deep-seated lesions by high-energy radiation at a long focal distance.

Supervoltage[4a,32,33] radiation [55, 56, 57, 58, 59] has a number of advantages over orthovoltage radiation for deep therapy, and has largely superseded it. The most obvious advantage is its greater penetrating power with high depth dosage and consequent simplicity of treatment plan [60]. Other, and perhaps more important advantages are the small differentials in the absorption of radiation by bone and fat compared with soft tissues [61, 62], and the sparing of the skin. In addition, there is little side- or back-scatter, so that treatment volumes can be sharply defined and there is little difference in depth dose between fields of different sizes. The chief disadvantage is the high cost and complexity of the machine, although the former is to some extent offset by high dose rates obtainable from, for instance, the linear accelerator, so that many patients may be treated in one day on one machine.

SUPERFICIAL LESIONS

These lesions are accessible and therefore easy to locate. The dose should fall off rapidly at depths below those to be treated. In practice, some decrease in dose at the base of the tumour is acceptable; for instance, Jennings [63] has used radiations with a half-value-depth approximately equal to the thickness of the lesion.

Very soft radiation for radiotherapy is available from a beryllium-window tube. When this is run at 2 kVp, and with no additional filtration of the beam, the HVL is only 0·1 mm of tissue. This type of tube is versatile as it can be run at voltages up to 50 kVp, and by appropriate filtration, the beam may be hardened up to a HVL of 15·8 mm tissue [64, 65]. Tubes with glass windows are also available for superficial (grenz ray) therapy. Operated at 15 kVp they have a HVL of 1 or 2 mm tissue [66]. Tube voltages up to about 140 kVp are used [67, 68] for superficial therapy.

It is usual to restrict the area of treatment by placing lead cut-outs on the skin. Exposure rates are usually high, and may be 1,000 R/min.

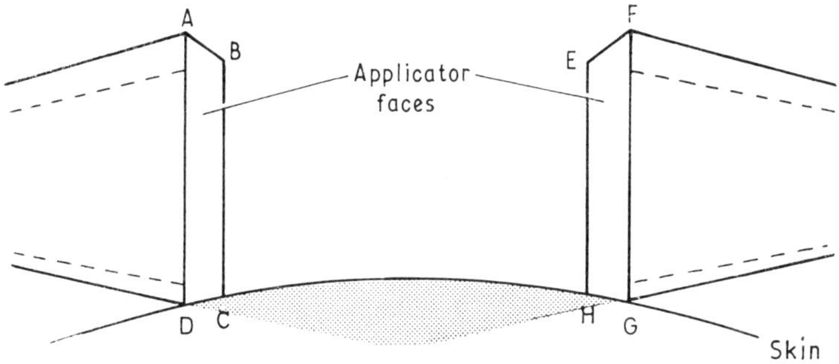

FIGURE 9.2 *The Glance-field technique*
The two applicator faces *ABCD*, *EFGH* are parallel, or sometimes are slightly inclined. Typical dimensions are 15 cm × 12 cm for each applicator face. A few cm thickness of tissue, shown shaded, is irradiated by the primary beams.

Benign haemangiomas of the skin have been treated by superficial x-radiation with a total dose of up to 1,500 rads. Malignant diseases such as basal and squamous cell carcinomata of the skin, rodent ulcers and tumours of the buccal cavity require a higher dose.

A different method of achieving a rapid fall-off in depth dose is by using the glance-field technique (Figure 9.2) [69, 70]. Comparatively large areas may be so treated, and the method has been developed for treating the breast and chest wall.

Another special technique is sieve radiotherapy, (p. 174). Although this is more commonly used for deep-seated tumours, Jolles [71, 72] has used it for treating squamous carcinoma of the skin, anus and vulva, particularly as a palliative measure in advanced cases.

DEEP THERAPY

As the adult human trunk is typically 200 to 250 mm thick, tumours may well be 100 mm below the body surface. The problem of treating such a tumour with an adequate dose without seriously damaging more superficial tissues is solved by using penetrating (high-energy) radiation and one of the many techniques available for concentrating high dose volumes where they are required. Sometimes simple methods are effective, for instance compression of the patient can materially reduce the depth below the body surface of an abdominal tumour.

When a large volume of the trunk must be irradiated to a uniform dose, a framework known as a trunk bridge [73] makes the treatment simple to plan and execute and it eliminates the possibility of local "hot-spots" from the overlapping of neighbouring fields.

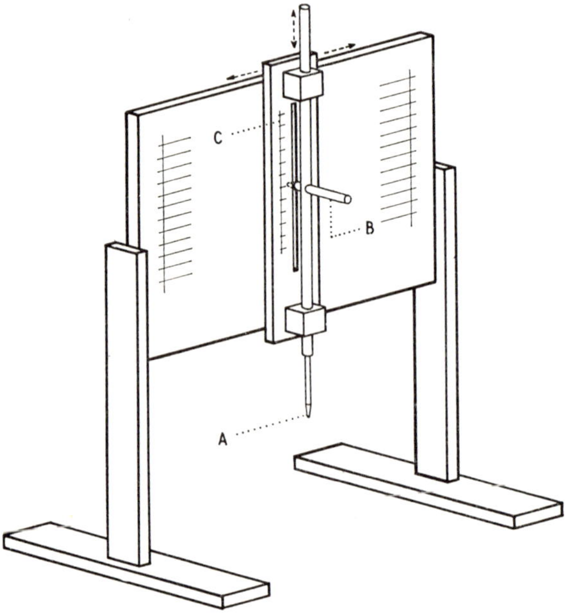

FIGURE 9.3 *Pantograph for Body Contours*
A pointer *A*, when moved over the body contour, causes pen *B* to trace the contour. *C* measures the height of bony landmarks, etc.

(Stern, B. E. and Hodges, G. B., *Brit. J. Radiol.*, **30**, p. 613, 1957)

Except for the simpler techniques, optimum dose distribution can be obtained only when a detailed treatment regime is planned individually, in the following main steps:

1. Location of volume to be treated with reference to the body contour and bony landmarks. Indication of special regions, such as the eye or the rectum, where the dose must be kept low.
2. Prescription of dose and its fractionation.
3. Selection of technique and treatment plan.
4. Setting-up and treatment of the patient.

The first two steps are the responsibility of the radiotherapist, the third of the physicist in cooperation with the radiotherapist, and the fourth of the radiographer, under the therapist's supervision.

Location of Volume to be Treated

The patient should be treated in a reproducible position, and unless there are reasons to the contrary, prone and supine positions have much to recommend them. When asked to lie down, few patients will lie precisely flat, so that steps must be taken to ensure that they are not lying a little to one side.

In appropriate techniques, a certain level, often at the centre of the X-ray fields, is selected for the body contour, on which the treatment is planned. Several methods are available for reproducing the body contour on paper [74, 75, 76, 77], (Figure 9.3).

Prescription of Dose

It is usual to plan the prescribed dose so that variations within the tumour do not exceed ± 10 per cent. The dose over a cross-section of a volume containing the tumour is obtained by using isodose curves (p. 101) to obtain

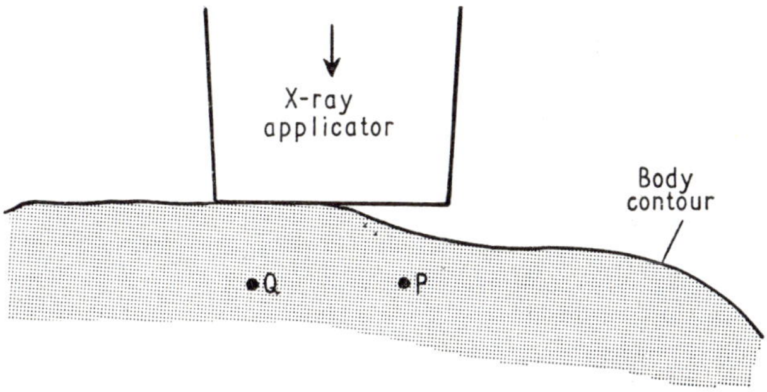

FIGURE 9.4 *Use of Bolus*
In the absence of bolus packing in the air space between X-ray applicator and body contour, at *P* both the primary and scatter dose, and at *Q* the scatter dose will, in general, be different from that of the isodose curve.

the contribution from each field, which are then summed. The selection of optimum dose distributions is aided by automation [78, 79]. Corrections may have to be made for body inhomogeneities. The dose distribution over different cross-sectional areas of tumour may decrease due to fall-off in dose towards the periphery of the field; it may be further modified by changes in the body contour along the length of the field (Figure 9.4), except where bolus [80] packing is used.

Selection of Technique and Treatment Plan

As a technique will be selected to give an adequate tumour dose while sparing normal tissue, certain techniques tend to become standard for particular anatomical sites.

The Single Field

As the dose for serious damage to normal tissue is much the same as that for treatment of the tumour, it is evident from central axis depth dose tables (Table 9.1), that the application of a single field cannot usually deliver a suitable dose to the tumour without overdosing normal tissue. Using supervoltage therapy [81] where the maximum dose is a centimetre or two below the skin, and in the exceptional case when this maximum dose area coincides with that to be treated, the dose behind the tumour falls off so gradually that, even here, a single field technique is not usually appropriate.

TABLE 9.1 *Central Axis Percentage Depth Dose Table for Two Field Sizes and X-rays of* (a) 250 kVp, 2 mm Cu HVL *and* (b) 8 MV
((*a*) From Clarkson, J. R., *Brit. J. Radiol.*, 1953, Supp. 5.); ((*b*) From Newbery, G. R. and Bewley, D. K., *Brit. J. Radiol.*, **28**, p. 241, 1955)

	(*a*) 250 kVp 500 mm FSD		(*b*) 8 MV 1,000 mm FSD	
Area of field (mm^2)	5,000	20,000	5,000	20,000
Depth in tissue (mm)				
0	100	100	13	21
20	87	93	100	100
50	63	72	89	90
100	30	39	70	72
150	15	21	53	57
200	7	11	41	45

The Multi-field Technique

In this technique [82], several fields are directed towards the area to be treated (Figure 9.5), so that they all contribute to the tumour dose, while

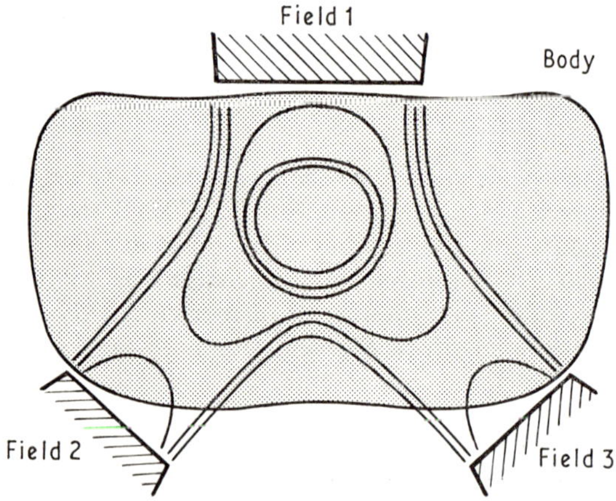

FIGURE 9.5 *Multi-field Technique showing Typical Isodose Contours*

any part of the skin is irradiated by the primary beam of not usually more than one field. With judicious choice of entry ports and directions, almost any dose distribution may be achieved at almost any depth, provided that the X-radiation has a HVL of at least about 2 mm Cu and a focal skin distance of at least about 500 mm. For X-rays generated above 2 MV, only a small number of fields is usually required, and two opposing fields will deliver a dose substantially constant over the interposing tissue, for almost any field separation likely to be met in clinical practice [82].

Moving-field Therapy

The logical development from multi-field techniques is for the entry port to vary continuously over the body surface; either the patient or the X-ray field

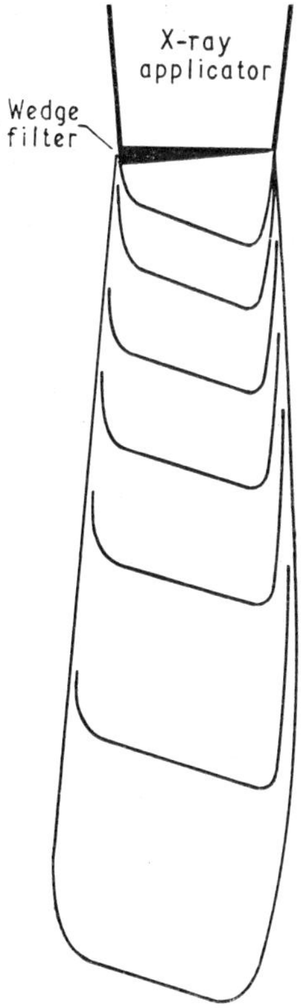

FIGURE 9.6 *Wedge Filter and Sketch of Resulting Isodose Curve*

may rotate [83, 84, 85, 86]. There is little difficulty in keeping the skin dose low even with orthovoltage radiation, but supervoltage radiation is to be preferred because of the smaller effect of body inhomogeneities.

Some X-ray machines are designed for complete rotation about the patient. This is a simple technique, but if the body contour differs much from a circle, 360° rotation involves treating through large thicknesses of normal tissue (Figure 6.4). This drawback may be overcome by limiting the arc through which the X-ray source moves: so-called pendulum therapy. Treatment may be from more than one arc [87, 88].

Rotation in the horizontal plane may be obtained when a patient lies on a rotating couch under a fixed inclined beam [89], or with a special X-ray set moving in a spiral above a stationary patient [90, 91].

Wedge Filters

Suppose that an X-ray beam has a wedge-shaped filter fitted (Figure 9.6). The intensity of radiation will decrease towards the thick edge of the wedge, with consequent distortion of the isodose curve [82, 92, 93, 94, 95, 96, 97, 98, 99]. A particular wedge can be designed to give isodose lines at a desired, say 45°, angle to the central axis. The angle tends to decrease with depth for orthovoltage, but not for megavoltage radiation [82].

A wedge used with megavoltage radiation must be placed at some distance from the skin, to prevent secondary electrons emitted from the wedge contaminating the X-ray beam.

If treatment is delivered through two fields at right angles to each other, each with a wedge and the thick edge nearer to the other field, a region of uniform dose at a few cm depth will result, with a rapid fall-off in dose behind the volume irradiated by both fields. This arrangement is useful for treating sites at only moderate depths and close to two surfaces such as the larynx and the antrum. Wedge fields have been used also for treating the cervix uteri [100].

Particularly for orthovoltage radiation, the dose distribution for a pair of line wedge filters deteriorates for large-area fields. If a large tumour is close to three surfaces, it can be treated by three mutually perpendicular wedge fields with a more uniform dose than from two such fields. This principle has been used in the *point-wedge*, where each of the three mutually perpendicular fields has a point wedge [101] stepped to give a "hot" point at one corner and a "cold" one at the diametrically opposite corner.

Sieve and Grid X-ray Therapy

Orthodox radiotherapy principles accept that the whole tumour must be treated by an adequate dose, and to this end elaborate steps are taken to ensure that dose variation is kept between narrow limits. And yet there is a school of thought which advocates deliberate variation of dose throughout the treated volume: fractionation in space as well as time [102, 103]. The effect is to produce alternate shifts of treated and spared tissue. The rationale of this technique is not clear. It may be that, in the low-dose regions, radiation damage stimulates normal cells more than tumour cells to proliferate, as it is known that normal cells are characterized by a low intrinsic growth rate, but a high responsive growth rate, while the reverse is true for malignant cells [104] or "diffusable substances" from the treated area may control growth elsewhere. Sieve and grid techniques are less damaging to the skin than comparable orthodox fields; these techniques are valuable for treating through skin already damaged by a first course of radiotherapy. They are most often used as palliative treatment with orthovoltage X-rays for recurrent growths.

The sieve technique involves irradiation through a chessboard (Figure 9.7*a*), consisting of alternate squares of lead or lead/rubber and radio-transparent areas, each of side 0·5 to 1·5 cm. After a substantial part of the total dose is delivered, the opaque and transparent areas are reversed.

In grid therapy (Figure 9.7*b*), the X-ray field is delivered through a piece of lead or lead/rubber with rows of circular holes of 1 cm diameter, the centre of each hole being 1·4 cm from its neighbour. The position of the holes relative to the skin is unchanged throughout treatment.

Multi-field techniques may be used where appropriate.

Maximum, minimum and "average" dose at places of interest are usually calculated.

X-RADIOTHERAPY COMBINED WITH OTHER FORMS OF TREATMENT

When the mode of action of two forms of treatment is different, it may be possible to derive benefit from both.

It has been argued that the treatment of choice for carcinoma of the breast is simple mastectomy followed by a radical course of radiotherapy to

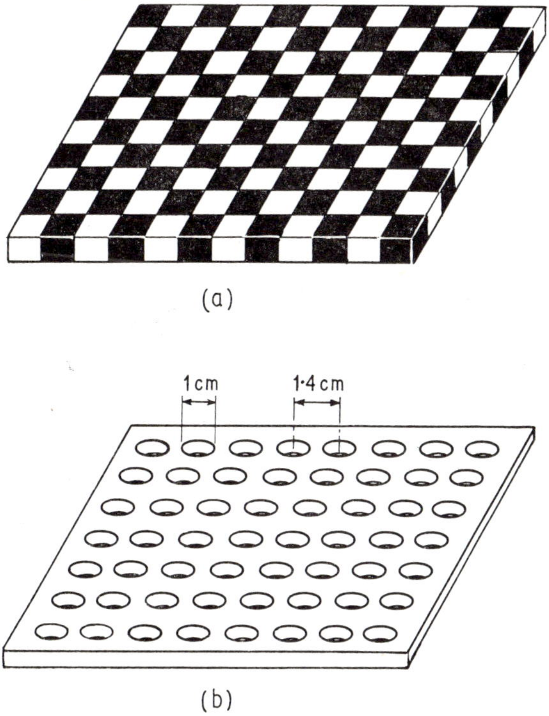

FIGURE 9.7 (*a*) *The sieve;* (*b*) *the grid*

the whole area [105, 106]. Other clinicians advocate radical mastectomy succeeded or preceded by radiotherapy, for certain stages of the disease [7, 107]. Pre-operative irradiation is thought to diminish the risk of dissemination of malignant cells at operation. On the other hand, it may be easier to deliver an adequate dose to any remaining tumour cells after the surgical removal of a bulky tumour. Consequently both sequences have their place in selected cases of various malignant diseases [108, 109, 110, 111]. Search has been made for agents which would increase the radio-sensitivity of malignant cells [112, 113], but with only limited success.

Some chemical substances, such as the nitrogen mustards, retard tumour growth. Their action is temporary and their usefulness is further limited by toxicity. Some tumours are hormone-dependent and they too can be controlled for a time by appropriate drugs, but eventually the dependence will be lost. As the mode of action in chemotherapy and radiotherapy is different, it is sometimes possible to give both, with greater palliative effect than if either were used singly [114, 115, 116, 117].

Sometimes treatment is with two types of radiation. For carcinoma of the cervix uteri, the local disease is often best treated by intracavity radioactive sources. The dose falls off rapidly in a lateral direction, owing to the short source–tumour distance [88, 118, 119]. As this is also a likely direction of spread of the disease, intracavity treatment is usually supplemented by X-radiation, except in early cases. Megavoltage X-rays are the most convenient [120]. With orthovoltage X-rays, two opposing fields will not deliver an adequate depth dose without overdosing the skin, unless the patient is thin. It is therefore customary to use additional lateral fields [10, 121] or arc therapy [87, 88]. Absorption of radiation in the bones of the pelvis may be a problem, unless megavoltage radiation is used. Radiographs of the intracavity applicators show these seldom to be central in the body, so that careful planning for each case is essential if areas of dangerously high or low dosages are to be avoided.

Setting-up the Patient

Care is necessary to ensure that treatment is carried out precisely as planned.

The entry port at the skin must be of the planned area and at the planned focal distance. In addition, it must be at a precise position on the patient with the central axis of the beam at a precise angle to the patient, accuracy being particularly important for small fields. Both the X-ray beam and the treatment couch should be capable of controlled and fine adjustments in a variety of directions. Special devices, such as the back pointer and the pin and arc [7, 120, 122], have been developed to help in the alignment of X-ray beam and patient.

For high-energy machines with short treatment times, it is economic to introduce special low-energy X-ray sets with the same dimensions and adjustments. These are used for the time-consuming plans and initial setting-up, thus freeing the high-energy machines for treatments.

Before treatment, the setting up should be checked by using the exit beam to expose a film [123].

The patient must not move during treatment; he must be resting comfortably, with suitable supports, or sometimes with more elaborate methods of fixation [124]. It is usual to keep him under observation by viewing through a radiation-opaque window or by more sophisticated methods [125].

Precision setting-up has to be repeated for each treatment; sometimes this is aided by the use of a plaster cast or similar device, made for each patient with seatings for the various fields. The cast may be made of tissue-simulating wax [126], or special depolymerized rubber may be moulded for the same purpose [127].

For X-rays generated above 1 MV, these methods tend to displace regions of maximum dose towards the skin.[82] A compensating filter [128] is to be preferred; this is placed at the exit of the beam from the machine and remote from (above) the skin, and is so shaped that the absorption in it compensates for lack of absorption due to irregularities in the body contour.

The dose delivered may be controlled by timing, provided that careful and frequent calibrations are carried out; alternatively, a dosemeter [129] may be installed, which terminates the treatment after a predetermined dose has been given.

Dose Specification

Dose should be specified in such detail that others may repeat or analyse the technique [82]. Much early statistical data cannot be used for lack of these details.

In preparing reports for publication, a check-list (Table 9.2), has been recommended [82], although the items on the list may have to be augmented in reports of complex or unusual techniques.

TABLE 9.2 *X-ray Therapy Check List*
(From ICRU Report 10d, *Clinical Dosimetry*, NBS Handbook 87, Washington D.C., NBS, 1963)

A. *The Target Volume*
  1. Shape and dimensions of the target volume
  2. Relation of target volume to:
    (*a*) External contour of the body in the treatment position
    (*b*) Centre of tumour if other than centre of target volume
    (*c*) Neighbouring structures of interest
    (*d*) External landmarks or reference points

B. *The Type and Energy of Radiation*
  1. The type of radiation (x-rays, etc.)
  2. Energy of radiation
    For x-rays (i) up to 2 MV; kV or MV and HVL
    (ii) above 2 MV; MV only
  3. Output data
    (*a*) For x-rays: exposure rate at a specified point
    (*b*) Instrument and chamber or other dosimeter type
    (*c*) Monitor exposure control on apparatus
  4. Auxiliary information:
    X-ray tube or accelerator beam current

C. *The Method of Irradiation*
  1. Fixed portals
    (*a*) Number of portals used and location of entry portals
    (*b*) For each portal:
      (i) Pertinent distances to point(s) of interest, source to surface, target to site of interest (usually centre of target volume), surface to site of interest
      (ii) Dimensions and shape of field
      (iii) Position and orientation of portal
      (iv) Auxiliary devices: special filters or compensators
      (v) Portal film demonstration of accuracy of beam direction
  2. Rotation therapy
    (*a*) Type of motion

TABLE 9.2—(*continued*)

(*b*) Pertinent distance
(i) Source to axis
(ii) Axis to point(s) of interest
(*c*) Angles and arcs
(*d*) Shape and size of portal (usually at axis)
(*e*) Auxiliary devices

D. *The Absorbed Dose and its Pattern*
1. Daily and cumulative doses in the target volume:
(*a*) Maximum
(*b*) Minimum
(*c*) Others when pertinent
2. Doses outside target volume:
(*a*) High dose regions outside target volume
(*b*) Doses in regions of radiosensitive tissues
3. Isodose distribution in one plane, through centre of target volume when feasible and pertinent
Extended to other planes, in three dimensions, when necessary
4. Derivation of values:
(*a*) Source of data used for derivation of absorbed doses on basis of homogeneous medium:
(i) Backscatter factors
(ii) Depth dose or tumour/air ratio factors
(iii) Other corrections (field elongation, finite patient thickness, etc.)
(iv) Exposure to absorbed dose conversion factors
(*b*) Type, magnitude and basis of corrections for bone, air or other heterogeneity. Direct measurements, transit measurements, calculated measurements, calculation corrections, etc.

E. *The Time Distribution*
1. Total number of sessions (a session is a treatment or group of treatments given during one patient visit)
2. Total overall time and the distribution of the sessions in this time when pertinent
3. Unusually high or low dose rates, unusual distributions of time factors, etc., when pertinent.

Literature References

1 SPEAR, F. G., *Radiations and Living Cells* (London, Chapman and Hall, 1953).

2 BRITISH JOURNAL OF RADIOLOGY, "Dose fractionation in radiotherapy", *Brit. J. Radiol.*, **36,** p. 305 (1963).

3 SIMPSON, C. L. and HEMPELMANN, L. H., "The association of tumours and Roentgen-ray treatment of thorax in infancy", *Cancer*, **10,** p. 42 (1957).

4 UNITED NATIONS SCIENTIFIC COMMITTEE, *Report on the Effects of Atomic Radiation*, Gen. Assembly, Official Records, 17th Session, Suppl. No. 16 (A/5216) (New York, United Nations, 1962).

5 BRINKLEY, D. and HAYBITTLE, J. L., "The late effects of artificial menopause by x-radiation", *Brit. J. Radiol.*, **42,** p. 519 (1969).

6 *Radiological Hazards to Patients, Second Report of the* (*Adrian*) *Committee* (London, H.M.S.O., 1960).

7 CARLING, E. R., WINDEYER, B. W. and SMITHERS, D. W. (Eds.), *British Practice in Radiotherapy* (London, Butterworth, 1955).

8 GLUCKSMANN, A., "Radiation histology II. The response of human tissues to radiation with special reference to radiotherapy", *Brit. J. Radiol.*, **25,** p. 38 (1952).

9 MITRA, S. and DEE, P. K., "Differentiation and radiation effects on cancer cells", *Brit. J. Cancer*, **8**, p. 107 (1954).
10 PATERSON, R., *The Treatment of Malignant Disease by Radium and X-rays* (London, Edward Arnold, 1948).
11 INTERNATIONAL ATOMIC ENERGY AGENCY, *Biological Effects of Ionizing Radiation at the Molecular Level*, Proc. Symp., Brno, 1962, Proc. Ser., 1962 (Vienna, I.A.E.A.).
12 BACQ, Z. M. and ALEXANDER, P., *Fundamentals of Radiobiology* (Oxford, Pergamon, 1961).
13 GRAY, L. H. *et al.*, "The concentration of oxygen dissolved in the tissues at the time of irradiation as a factor in radiotherapy", *Brit. J. Radiol.*, **26**, p. 628 (1953).
14 GRAY, L. H., "Oxygenation in radiotherapy. I. Radiobiological considerations", *Brit. J. Radiol.*, **30**, p. 403 (1957).
15 CATER, D. B. and SILVER, I. A., "Quantitative measurements of oxygen tension in normal tissues and in the tumours of patients before and after radiotherapy", *Acta. Radiol.*, **53**, p. 233 (1960).
16 EVANS, N. T. S. and NAYLOR, P. F. D., "The effects of oxygen breathing and radiotherapy upon the oxygen tension of some human tumours", *Brit. J. Radiol.*, **36**, p. 418 (1963).
17 THOMLINSON, R. H., "Pre-therapeutic experiments with the fast neutron beam from the Medical Research Council cyclotrotron. III. A comparison of fast neutrons and X-rays in relation to the 'oxygen effect' in experimental tumours in rats", *Brit. J. Radiol.*, **36**, p. 89 (1963).
18 Churchill-Davidson, I., Sanger, C. and Thomlinson, R. H., "Oxygenation in radiotherapy. II. Clinical applications" *Brit. J. Radiol.* **30**, p. 406 (1957).
19 DUSAULT, D., "The effect of oxygen on the response of spontaneous tumours in mice to radiotherapy", *Brit. J. Radiol.*, **36**, p. 749 (1963).
20 BERRY, R. J. and ANDREWS, J. R., "The effect of radiation ionization density (LET) upon the reproductive capacity of mammalian tumour cells irradiated *in vivo*", *Brit. J. Radiol.*, **36**, p. 49 (1963).
21 FOWLER, J. F., MORGAN, R. L. and WOOD, C. A. P., "The biological and physical advantages and problems of neutron therapy", *Brit. J. Radiol.*, **36**, p. 77 (1960).
22 BRUES, A. M., WEINER, A. E. and ANDERVONT, H. B., "Relations between latent period and growth rate in chemically induced tumours", *Proc. Soc. Exp. Biol., Med.*, **42**, p. 374 (1939).
23 COLLINS, V. P., "Wilms' tumour: its behaviour and prognosis", *J. Louisiana State Med. Soc.*, **107**, p. 474 (1955).
24 COLLINS, V. P., LOEFFLER, R. K. and TIVEY, H., "Observation of growth rates of human tumours", *Amer. J. Roent.*, **76**, p. 988 (1956).
25 COLLINS, V. P., "Breast cancer: The influence of treatment that fails to cure", *Cancer*, **9**, p. 1177 (1956).
26 GLUCKSMANN, A., *British Empire Cancer Campaign, Annual Report*, **26**, p. 190 (1948).
27 MAYNEORD, W. V. and CLARKSON, J. R., "Energy absorption II. Part I. Integral dose when the whole body is irradiated", *Brit. J. Radiol.*, **17**, p. 151 (1944).
28 MAYNEORD, W. V., "Energy absorption III. The mathematical theory of integral dose and its application in practice", *Brit. J. Radiol.*, **17**, p. 359 (1944).
29 INTERNATIONAL COMMISSION ON RADIOLOGICAL UNITS AND MEASUREMENTS, ICRU Report 10*a*, 1962. NBS Handbook 84 (Washington, D.C., Nat. Bur. Stand., 1962).

30 FOWLER, J. R. and FARMER, F. T., "Measured dose distributions in arc and rotation therapy: a critical comparison of moving- and fixed-field tech niques", *Brit. J. Radiol.*, **30,** p. 653 (1957).
31 FOWLER, J. F. and STERN, B. E., "Dose-rate factors in integral dose estimations", *Brit. J. Radiol.*, **31,** p. 316 (1958).
32 FOWLER, J. R. and STERN, B. E., "Dose-rate effects: some theoretical and practical considerations", *Brit. J. Radiol.*, **33,** p. 389 (1960).
33 GOLDSCHEIDER, G. and STERN, B. E., "Factors influencing the occurrence of reactions during the treatment of carcinoma of the cervix uteri by radium", *Brit. J. Radiol.*, **26,** p. 370 (1953).
34 COURT BROWN, W. M. and MAHLER, R. F., "Integral dose, body size and site of irradiation as factors determining the severity of radiation sickness", *J. Fac. Radiol.*, **5,** p. 408 (1954).
35 JOLLES, B., *British Empire Cancer Campaign Annual Report*, **32,** p. 408 (1954).
36 BALDINI, G. and FERRI, L., "Experimental and clinical research on the radio-protective action of cysteinamine and cystamine", *Brit. J. Radiol.*, **30,** p. 271 (1957).
37 COURT BROWN, W. M., "A clinical trial of cysteinamine". *Brit. J. Radiol.*, **28,** p. 325 (1955).
38 HEALY, J. B., "A trial of cystamine in radiation sickness", *Brit. J. Radiol.*, **33,** p. 512 (1960).
39 PATERSON, R., "*The Treatment of Malignant Disease by Radiotherapy*" (London, Edward Arnold, 1963).
40 FOWLER, J. F. and STERN, B. E., "Fractionation and dose-rate. II. Dose–time relationships in radiotherapy and the validity of cell survival curve models", *Brit. J. Radiol.*, **36,** p. 163 (1963).
41 STRANDQUIST, M., "Studien über die kumulative Wirkung der Roentgenstrahlen bei Fractionierung", *Acta. Radiol.*, Suppl. No. 55 (1954).
42 PATERSON, R., "Studies in optimum dosage", *Brit. J. Radiol.*, **25,** p. 505 (1952).
43 DU SAULT, L. A., "Influence of time factors on dose response curves", *Amer. J. Roent.*, **87,** p. 567 (1962).
44 ELLIS, F., "The dose–time relationship in radiotherapy", *Brit. J. Radiol.*, **36,** p. 153 (1963).
45 SAMBROOK, D. K., "Clinical aspects of fractionation and dose–time relationships", *Brit. J. Radiol.*, **36,** p. 174 (1963).
46 LAJTHA, L. G., OLIVER, R. and ELLIS, F., "Rationalisation of fractionation in radiotherapy", *Brit. J. Radiol.*, **33,** p. 634 (1960).
47 OLIVER, R., "Theoretical implications of cell survival data in relation to fractionated radiotherapy treatments", *Brit. J. Radiol.*, **36,** p. 178 (1963).
48 OLIVER, R. J. and LAJTHA, L. G., "Problems of radiosensitivity in mixed cell populations with particular reference to radiotherapy", *Brit. J. Radiol.*, **34,** p. 659 (1961).
49 BEWLEY, D. K. *et al.*, "Pre-therapeutic experiments with the fast neutron beam from the Medical Research Council cyclotron. VII. Experiments on the skin of pigs with fast neutrons and 8 MV X-rays, including some effects of dose fractionation", *Brit. J. Radiol.*, **36,** p. 107 (1963).
50 FOWLER, J. F. *et al.*, "Experiments with fractionated X-ray treatment of the skin of pigs", *Brit. J. Radiol.*, **36,** p. 188 (1963).
51 BACLESSE, F., "La roentgentherapie seule dans le traitement des cancers du sein operable et inoperables", *Assn. Française Chirurgie* 51e *Congrès Français de Chirurgie, Paris*, 1948.
52 BACLESSE, F., "Roentgen therapy as the sole method of treatment of cancer of the breast", *Amer. J. Roent.*, **62,** p. 311 (1949).

53 Botstein, C., "Periodic fractionation of treatment", in *Progress in Radiotherapy*. Vol. II, ed. Buschke, F. (New York, Greene and Stratton, 1962).
54 British Institute of Radiology, "Preliminary report of the working party of the British Institute of Radiology on the effects of dose fractionation in radiotherapy", *Brit. J. Radiol.*, **36,** p. 382 (1963).
55 Meredith, W. J., "Some aspects of supervoltage therapy", *Amer. J. Roent.*, **79,** p. 57 (1958).
56 Fletcher, G. H., "The role of supervoltage therapy", *Amer. Cancer Soc., Proc. Symp. on Radiother. Cancer*, 1960, p. 179 (1961).
57 Miller, C. W., "Recent developments in linear accelerators for therapy", *Brit. J. Radiol.*, **35,** p. 182 (1962).
58 Farmer, F. T., "Supervoltage therapy: A review of present-day facilities and techniques". *Phys. Med. Biol.*, **6,** p. 505 (1962).
59 Spiers, F. W. and Meredith, W. J., "Statement of dosage in megavoltage radiation therapy: recommendations of the Faculty of Radiologists", *Clin. Radiol.*, **13,** p. 163 (1962).
60 International Atomic Energy Agency, *Single-Field Isodose Charts for High Energy Radiation*, Tech. Report Ser. No. 8, 1962 (Vienna, I.A.E.A.).
61 Lamerton, L. F. and Winsborough, M., "An approach to the problem of scattering of high-voltage radiation in non-homogeneous and limited media", *Brit. J. Radiol.*, **23,** p. 236 (1950).
62 Haas, L. L. and Sandberg, G. H., "Modifications of depth dose curves of various radiations by interposed bone", *Brit. J. Radiol.*, **30,** p. 19 (1957).
63 Jennings, W. A., "The achievement of optimum depth dosage distributions from the physical standpoint", *Cathode Press*, **8,** p. 4 (1951).
64 Jennings, W. A., "Physical aspects of the roentgen radiation from a beryllium window tube operated over the range 2–50 kVp for clinical purposes", *Acta. Radiol.*, **33,** p. 435 (1950).
65 Jennings, W. A., "A theoretical study of radiation outputs and qualities from a beryllium window tube operated at low kilovoltages (10–50 kVp), *Brit. J. Radiol.*, **26,** p. 193 (1953).
66 Bucky, G. and Combes, F. C., *Grenz Ray Therapy* (New York, Springer, 1954).
67 Lamerton, L., "A physical investigation of the radiation from a low-voltage x-ray tube (cautery technique), *Brit. J. Radiol.*, **13,** p. 136 (1940).
68 Pack, G. T. and Ariel, I. M. (Eds.), *Treatment of Cancer and Allied Diseases. I. Principles of Treatment* (London, Pitman Medical, 1958).
69 Binks, C. and Sherriff, E. V., "The measurement of skin dose with particular reference to post-operative treatment of carcinoma of the breast", *Brit. J. Radiol.*, **27,** p. 283 (1954).
70 Evans, W. G. and Mabbs, D. V., "Treatment of superficial benign lesions with x-rays", *Brit. J. Radiol.*, **35,** p. 866 (1962).
71 Jolles, B., "The sieve in radiotherapy", *Brit. J. Radiol.*, **25,** p. 395 (1952).
72 Mitchell, R. G., "Physical aspects of the sieve method", *Brit. J. Radiol.*, **25,** p. 403 (1952).
73 Easson, E. C. *et al.*, "The Manchester trunk bridge for the treatment of radiosensitive tumours", *Brit. J. Radiol.*, **30,** p. 311 (1957).
74 Murphy, W. T., *Radiation Therapy* (Philadelphia, W. B. Saunders, 1959).
75 Treherne, J. D. and Greening, J. R., "A device for obtaining body contours", *Brit. J. Radiol.*, **25,** p. 664 (1952).
76 Spicer, B. A., "A body contour device", *Radiog.*, **20,** p. 41 (1954).
77 Stern, B. E. and Hodges, G. B., "A pantograph for body contours", *Brit. J. Radiol.*, **30,** p. 613 (1957).
78 Sterling, T. D., Perry, H. and Bohr, G. K., "A practical procedure for automating radiation treatment planning", *Brit. J. Radiol.*, **34,** p. 726 (1961).

79 STERLING, T. D., PERRY, H. and WEINKAM, J. J., "Automation of radiation treatment planning. III. A simplified system of digitising isodoses and direct print-out of dose distribution", *Brit. J. Radiol.*, **36,** p. 522 (1963).
80 LINDSAY, D. D. and STERN, B. E., "A new tissue-like material for use as bolus", *Radiol.*, **60,** p. 355 (1953).
81 INTERNATIONAL ATOMIC ENERGY AGENCY, *Applications of High Energy Radiation in Therapy*, Bibliographical Ser. No. 1, 1960 (Vienna, I.A.E.A.).
82 INTERNATIONAL COMMISSION ON RADIOLOGICAL UNITS AND MEASUREMENTS, *Clinical Dosimetry*, Recommendations of the ICRU, ICRU Report 10*d*, 1962, NBS Handbook 87 (Washington, D.C., Nat. Bur. Stand., 1963).
83 GREEN, A., JENNINGS, W. and BUSH, F., "Rotational Roentgen therapy in the horizontal plane", *Acta Radiol.*, **31,** p. 273 (1949).
84 SMITHERS, D. W., "Rotation therapy", *J. Fac. Radiol.*, **6,** p. 73 (1954).
85 MASSEY, J. B., "The role of moving field techniques in treatment with high-energy radiation. I. Some physical considerations", *Brit. J. Radiol.*, **35,** p. 90 (1962).
86 WACHSMANN, F. and BARTH, G., *Moving-field Radiation Therapy* (Univ. Chicago Press, 1962).
87 GOLDSCHEIDER, G. and STERN, B. E., "Rotational x-ray therapy planned to complement radium treatment of carcinoma of the cervix uteri", *Brit. J. Radiol.*, **31,** p. 88 (1958).
88 SNELLING, M. D. and STERN, B. E., "Arc therapy with 250 kV x-rays. IV. Treatment of carcinoma of the cervix by a combination of intra-cavitary radium and arc therapy", *Brit. J. Radiol.*, **31,** p. 298 (1958).
89 JENNINGS, W. A. and MCCREA, A. L., "Dose distribution in conical rotation therapy with a 2 MeV generator", *Radiol.*, **68,** p. 689 (1957).
90 KOLIBAY, W., "An x-ray tube mounting for convergent beam therapy giving high percentage depth dose", *Brit. J. Radiol.*, **23,** p. 653 (1950).
91 FLATMAN, G. E. and ELLIS, R. E., "A report on the use of convergent-beam x-ray therapy", *Brit. J. Radiol.*, **29,** p. 139 (1956).
92 GROOM, A. C., "The design of wedge filters for x-ray therapy", *Brit. J. Radiol.*, **24,** p. 676 (1951).
93 KEMP, L. A. W. and OLIVER, R., "The flattening of wedge isodose curves in the direction perpendicular to the wedge", *Brit. J. Radiol.*, **25,** p. 502 (1952).
94 COHEN, M., "Physical aspects of Roentgen therapy using wedge filters. I. Some new copper wedges and their applications", *Acta Radiol.*, **52,** p. 65 (1959).
95 COHEN, M., "Physical aspects of Roentgen therapy using wedge filters. II. Analysis of wedge field dosimetry". *Acta. Radiol.*, **52,** p. 158 (1959).
96 COHEN, M. and BURNS, J. E., "Physical aspects of Roentgen therapy using wedge filters. III. The effect of wedge fields on variations in kV and other factors", *Acta Radiol.*, **52,** p. 470 (1959).
97 COHEN, M., "Physical aspects of Roentgen therapy using wedge filters. IV. A new wedge principle using graphite and tin", *Acta Radiol.*, **53,** p. 153 (1960).
98 ROOSENBEEK, E. V. and GRIMM, J. H., "Wedge filters: their construction and use with the 22 MeVp betatron", *Amer. J. Roent.*, **85,** p. 926 (1961).
99 STEWART, J. G., "The clinical significance of dose distributing problems", *Brit. J. Radiol.*, **35,** p. 743 (1962).
100 TRANTER, F. W., "A wedge filter for use in treatment of carcinoma of the cervix uteri with 4 MV x-rays", *Brit. J. Radiol.*, **32,** p. 350 (1959).
101 PORTER, E. H., HALL, E. J. and ELLIS, F., "Point wedges: a development of wedge filter technique", *Brit. J. Radiol.*, **34,** p. 655 (1961).
102 JOLLES, B., *X-Ray Sieve Therapy in Cancer* (London, H. K. Lewis, 1953).

103 SEIDEL, K., "Depth doses in grid irradiation", *Strahlentherapie*, **99**, p. 549 (1956).
104 FURTH, J., "Conditioned and autonomous neoplasms: a review". *Cancer Res.*, **13**, p. 477 (1953).
105 MCWHIRTER, R., "The principles of treatment by radiotherapy of breast carcinoma", *Brit. J. Cancer*, **4**, p. 368 (1950).
106 ATKINS, H. L. and HORRIGAN, W. D., "Treatment of locally advanced carcinoma of the breast with Roentgen therapy and simple mastectomy", *Amer. J. Roent.*, **85**, p. 860 (1961).
107 BORGSTROM, S. and LINDFREEN, M., "Pre-operative Roentgen therapy of cancer of the breast", *Acta Radiol.*, **58**, p. 9 (1962).
108 BLOEDORN, F. G. *et al.* "Combined therapy: irradiation and surgery in the treatment of bronchogenic carcinoma", *Amer. J. Roent.*, **85**, p. 875 (1961).
109 SHERMAN, A. I., "The role of combined surgery and radiation in the treatment of cancer of the reproductive system", *Amer. Cancer Soc., Proc. Symp. on Radioth. Cancer*, 1960, p. 142 (1961).
110 RUFF, C. C. *et al.*, "Preoperative radiation therapy for adenocarcinoma of the rectum and rectosigmoid", *Surg. Gynec. Obstet.*, **112**, p. 715 (1961).
111 PATERSON, R. and RUSSELL, M. H., "Clinical trials in malignant disease. Part IV. Lung cancer. Value of post-operative radiotherapy", *Radiol.*, **13**, p. 141 (1962).
112 MITCHELL, J. S. and SIMON-REUSS, I., "Experiments on the mechanism of action of tetrasodium 2-methyl-1:4-napthohydroquinone diphosphate as a mitotic inhibitor and radiosensitiser, using the technique of tissue culture experimental methods and quantitative results", *Brit. J. Cancer*, **6**, p. 305 (1962).
113 MITCHELL, J. S., *Studies in Radiotherapeutics* (Oxford, Blackwell, 1960).
114 BINNIE, G. C., "Breast carcinoma treated with stilboestrol and deep X-ray", *Brit. J. Radiol.*, **22**, p. 714 (1949).
115 HOLME, P. B., "Treatment of the reticuloses by chemotherapy and radiotherapy", *Brit. J. Radiol.*, **34**, p. 569 (1961).
116 SCHWARTZ, E. E. *et al.*, "Combined irradiation and chemotherapy in ovarian carcinoma", *Radiol.*, **78**, p. 272 (1962).
117 ALEXANDER, L. L. *et al.*, "Bronchogenic carcinoma", *Amer. J. Roent.*, **87**, p. 375 (1962).
118 WALSTAM, R., "The dose distribution in the pelvis in radium treatment of carcinoma of the cervix", *Acta Radiol.*, **42**, p. 237 (1954).
119 GARCIA, M., "Further observations on tissue dosage in carcinoma of the cervix uteri", *Amer. J. Roent.*, **73**, p. 35 (1955).
120 WALTER, J. and MILLER, H., *A Short Textbook of Radiotherapy* (London, Churchill, 1959).
121 WINTERNITZ, J. G., "Carcinoma of the cervix: a discussion on the value and techniques of supplementary X-ray therapy", *Brit. J. Radiol.*, **21**, p. 27 (1949).
122 JOHNS, H. E., *The Physics of Radiology* (Springfield, Chas. C. Thomas, 1961).
123 COPCUTT, W. A., "The use of X-ray films in beam direction for X-ray therapy", *Brit. J. Radiol.*, **22**, p. 210 (1949).
124 MEAD, W. J. and COLLINS, V. P., "The principles of dilatancy applied to radiotherapy", *Amer. J. Roent.*, **71**, p. 864 (1954).
125 STRANDQUIST, M. and ROSENGREN, B., "Television-controlled pendulum therapy", *Brit. J. Radiol.*, **31**, p. 513 (1958).
126 JONES, D. E. A. and RAINE, H. C., (Correspondence), *Brit. J. Radiol.*, **22**, p. 549 (1949).

127 Stacey, A. J., Bevan, A. R. and Dickens, C. W., "A new phantom material employing depolymerized natural rubber", *Brit. J. Radiol.*, **34,** p. 510 (1961).
128 Ellis, F., Hall, E. J. and Oliver, R., "A compensator for variation in tissue thickness for high-energy beams", *Brit. J. Radiol.*, **32,** p. 421 (1959).
129 Watson, B. W., "A digital dosemeter for use in x-ray therapy", *Phys. Med. Biol.*, **8,** p. 161 (1963).

10 | *X-ray Crystallography*

Crystals have been objects of interest since antiquity. Indeed some writers have discussed crystals in almost mystical terms, having been impressed by their resemblance to living creatures in growth and reproduction. The term

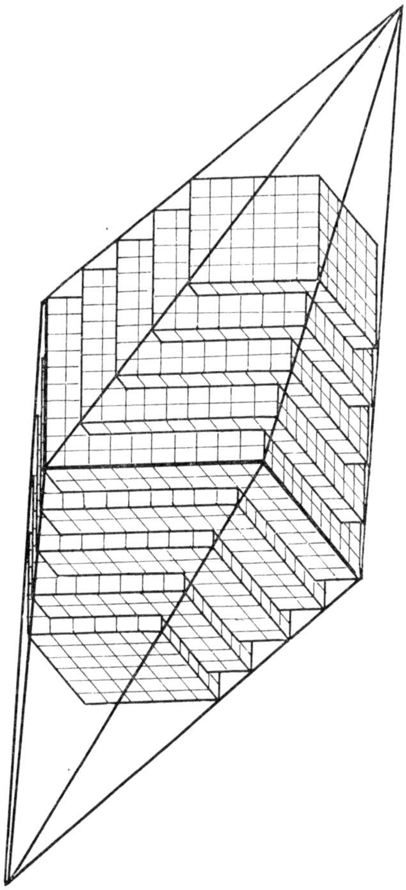

FIGURE 10.1 *Calcite Crystal on Haüy Model*

7—(T.989)

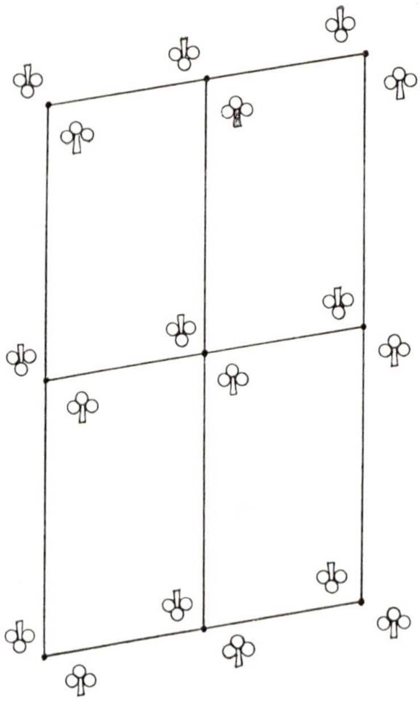

(a)

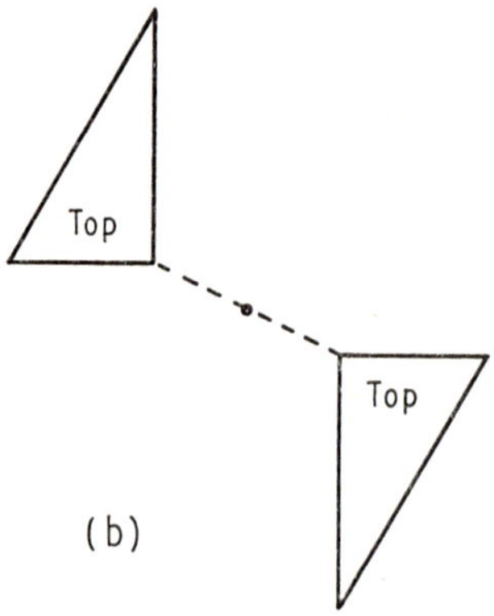

(b)

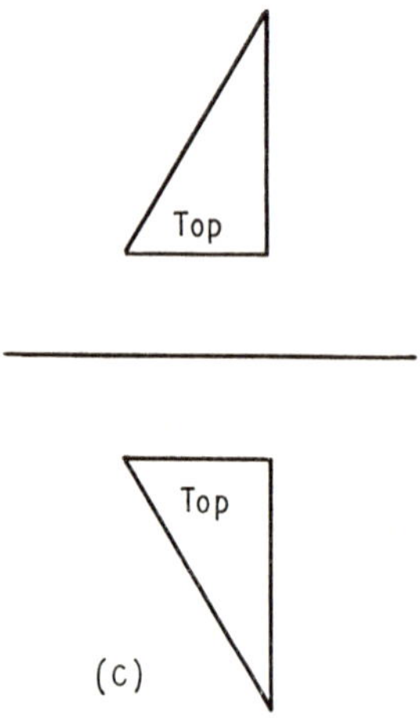

(c)

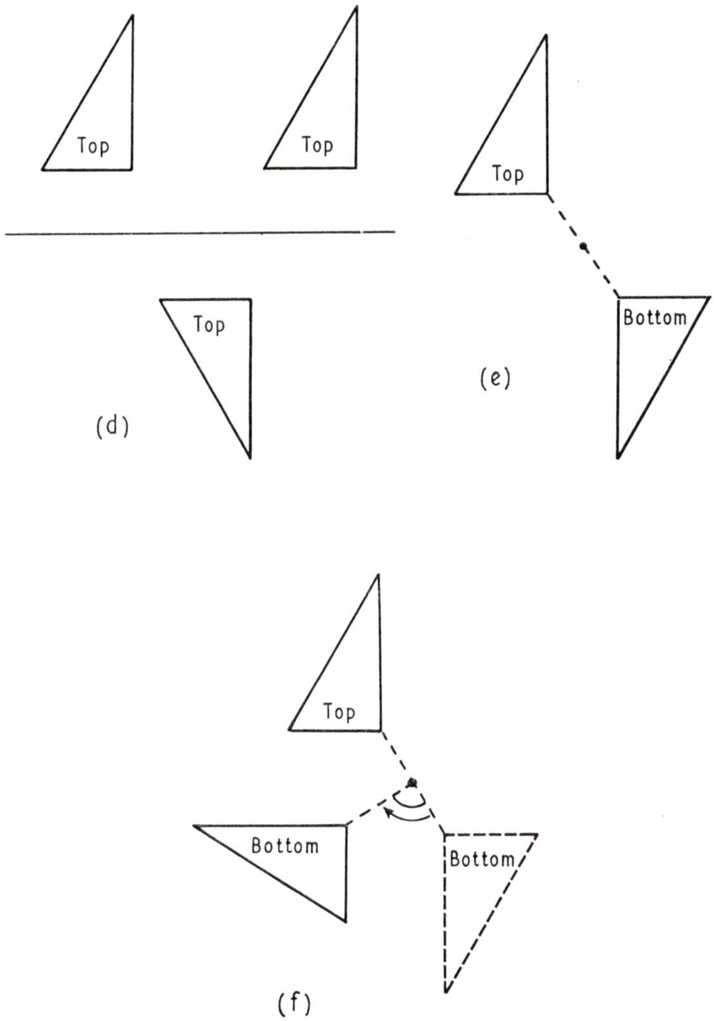

FIGURE 10.2 *Some Symmetry Elements*
(*a*) The pattern unit is rotated about an axis perpendicular to the plane of the pattern and then translated.
(*b*) Rotation about an axis. The pattern unit is a scalene triangle. The designated top and bottom surfaces of the triangle are to be regarded as different.
(*c*) Reflection across a plane.
(*d*) Reflection and translation.
(*e*) Inversion about a point.
(*f*) Inversion followed by rotation.

crystal itself is derived from the Greek word for ice, and was applied to the clear quartz found in the Alps, in the belief that it was ice transformed by intense cold.

The most striking aspect of crystals is their symmetrical shapes. However, one problem confronting early students of crystallography was that a single material could have more than one crystalline form. For example, calcite may occur in a long hexagonal prism forms or in a pointed form called dog-tooth spar. Both forms have the same angles between corresponding faces. To generalize, a particular material may crystallize in one of several alternative forms, but all the forms have the same angles between the corresponding faces. The different forms arise from the different relative areas of the faces. A particular form or "habit" will grow under particular conditions of temperature, concentration and purity. Even crystals of one habit, however, show small variations in relative face areas.

The shape of a house is not necessarily the same shape as a building brick. The same argument applied to crystal habits. In 1784 Haüy [1] demonstrated that all calcite crystal habits were constructed from a basic rhombic unit (Figure 10.1). Thus, the constancy of a crystal habit arises because identical basic units are assembled in a repetitive pattern.

Pattern Theory

The study of patterns is the study of regular arrangements. Patterns may be in one, two or three dimensions.

Patterns may be classified according to the operations performed on the pattern unit; the regular arrangement follows from ordered repetition of the operations.

For example, in a two-dimensional pattern (Figure 10.2*a*) the pattern unit is rotated about an axis perpendicular to the plane of the pattern, and translated. The angle of rotation and the distance translated are repeated. Wallpapers are familiar examples of two-dimensional patterns and an examination of such a paper will reveal the operations used to form the pattern.

Patterns may be classified according to the operators by which they are formed; the pattern unit may be translated, rotated, reflected or inverted in space. Most patterns contain combinations of these operators (Figures 10.2*b*, *c*, *d*, *e*, *f*). The two-dimensional pattern (Figure 10.2*a*) contains two such operators: rotation and translation.

The smallest complete pattern forms a parallelogram in two dimensions and a parallelepiped in three dimensions. It is known as a unit cell (Figure 10.3). The pattern operators are called the elements of symmetry of the pattern.

Patterns are grouped according to the symmetry of the pattern, and there is not an infinite number of possibilities. Consider a one-dimensional pattern. The pattern unit can only be translated, or reflected and translated. Thus there are two one-dimensional groups. Similarly, there are seventeen two-dimensional groups and 230 three-dimensional groups.

A continuous array of unit cells forms a space framework known as a space lattice or Bravais lattice, of which there are a possible fourteen. The corners of the framework are called Bravais points.

Space lattices may also be grouped according to crystal systems, of which there are seven (Table 10.1). Each system is characterized by one or two simple symmetry operations, such as rotation or reflection, and each system

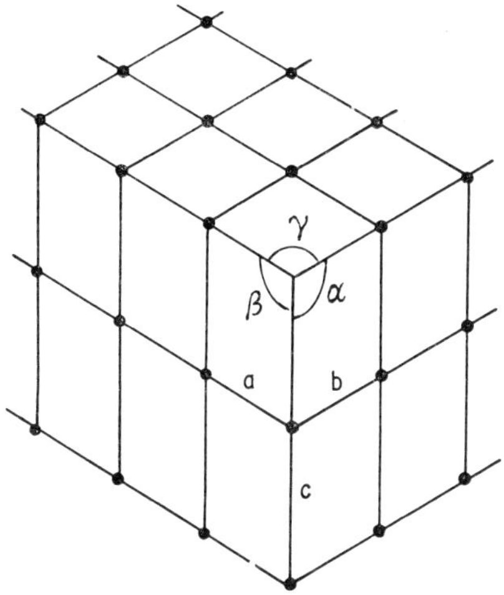

FIGURE 10.3 *Unit Cells in a Crystal forming a Bravais Lattice*

has a unit cell of a particular shape. For example the cubic system has a cube-shaped unit cell. If a cube is held with the cube corner pointing at the observer it is obvious that a rotation of 120° constitutes a symmetry operator, as such a rotation will restore the cube to an identical state.

There are four such axes of rotation in a cube and these four "threefold" symmetry axes characterize it. On looking at the face of a cube it may be

TABLE 10.1 *The Seven Crystal Systems*

System	*Symmetry*	*Unit Cell*
Triclinic	Nil	$\alpha \neq \beta \neq \gamma \neq 90°$ $\alpha \neq b \neq c$
Monoclinic	One twofold axis or mirror plane	$\alpha = \gamma = 90° \neq \beta$ $a \neq b \neq c$
Orthorhombic	Three mutually perpendicular twofold axes, or two mirror planes intersecting in a twofold axis	$\alpha = \beta = \gamma = 90°$ $a \neq b \neq c$
Trigonal	One threefold axis	$\alpha = \beta = \gamma \neq 90°$ $a = b = c$
Hexagonal	One sixfold axis	$\alpha = \beta = 90; \gamma = 120°$ $a = b \neq c$
Tetragonal	One fourfold axis or four fold inversion axis	$\alpha = \beta = \gamma = 90°$ $a = b \neq c$
Cubic	Four threefold axes	$\alpha = \beta = \gamma = 90°$ $a = b = c$

seen that a rotation of 90° is a symmetry operator, but this "fourfold" symmetry operation is not specific for cubes only. A rectangular block with only four sides equal out of the six, will have a fourfold axis of symmetry through the square face, but the unit cell is not cubic. Thus four threefold axes are required to specify cubic symmetry. Even a cubic-shaped unit cell need not imply cubic symmetry, as the crystal may have some property, such as piezo-electric effect, indicating a lower symmetry and not included in a threefold symmetry operator. Similarly one threefold axis only is not uniquely cubic. Group theory indicates that there must be four of them.

A crystal structure can be defined in terms of the primitive unit cell. But it may be more appropriate to define it in terms of a multiple unit cell

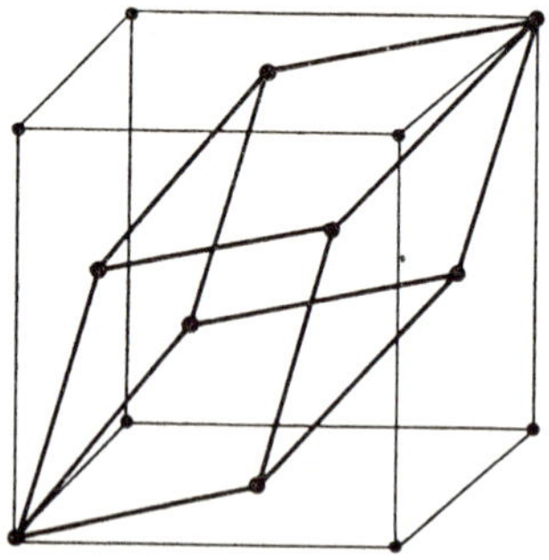

FIGURE 10.4 *A Face-centred Cubic Unit Cell showing the less Symmetrical Primitive Unit Cell*

of different shape which is more relevant to the crystal properties. This larger unit cell will contain more than one Bravais point (Figure 10.4). Each Bravais point is identical in atomic arrangement and surroundings with the others in the Bravais lattice.

Pattern theory is basic to the study of materials, because the internal arrangement of the atoms of a material is, in general, ordered. Even in liquids, the molecules exert sufficient influence on each other to prevent a completely random structure developing. Some solids, such as rubber, do not normally have a well-ordered structure, but may have this induced by, for instance, stretching. Others, such as metals, have a well-ordered structure which is not usually apparent to the naked eye, because growth conditions inhibit the formation of well-developed crystal faces. Thus, interest in crystals has moved from a study of their external appearance to the more fundamental study of atomic patterns.

Crystals are classified by their symmetry into seven crystal systems. Many common materials, such as the metals iron and copper, and also common salt, belong to the cubic system. The well-known crystals calcite and quartz belong to the trigonal system. The monoclinic system has most members.

X-ray Diffraction

During 1912, a group of scientists were working in Munich on apparently unrelated topics. Sommerfeld was working on X-rays, Laue was interested in interference and Groth was a leading authority on crystallography. Ewald, working under Sommerfeld, had prepared a doctor's dissertation on optical scattering phenomena in crystals, and this gave Laue the idea that if the wavelength radiation were short enough, the crystals would act as a diffraction grating in three dimensions, to give reinforcement maxima. Friedrich, who was also working in Sommerfeld's laboratory at that time, offered to verify this idea experimentally and after some failures, produced a diffraction pattern from copper sulphate [2]. This was an impressive verification of the suggestion that x-rays were a form of wave motion.

The mathematical treatment of the phenomena by Laue is difficult to apply in practical problems and a simpler treatment was developed by

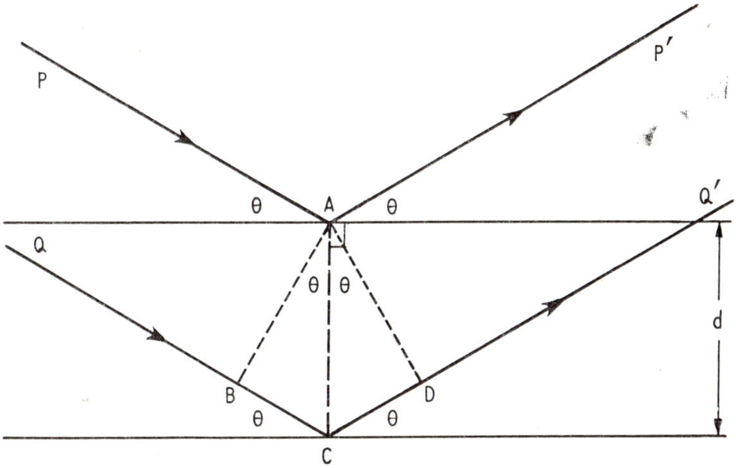

FIGURE 10.5 *Bragg's Law*
It is assumed that the diffracting material in the crystal lies on parallel planes, spaced a distance d apart. When reinforcement occurs, the planes "reflect" the x-rays, like mirrors. In the two adjacent planes shown, the ray QQ', reflected from the lower plane, travels a distance $2d \sin \theta$ i.e., $BC + CD$, further than the ray PP'.
For reinforcement: $2d \sin \theta = n\lambda$.

Bragg [3] at Cambridge. He proposed that the various atoms in a crystal could be thought of as lying on planes at different levels in the crystal. Each plane would act in a manner analogous to mirrors in optics and "reflect" the x-rays at all angles of incidence. However, the combined effect of all these reflecting planes regularly spaced would resemble an optical grating and give rise to intensity maxima. If PAP' and QCQ' (Figure 10.5) represent waves being reflected at a glancing angle θ from successive crystal planes of spacing d, the incident waves are in phase along the wave-front AB, and are also in phase along the reflected wave-front AD. If AC is perpendicular to the crystal planes then the extra path travelled by the lower portion of the wave front is

$$BC + CD = d \sin \theta + d \sin \theta = 2d \sin \theta \qquad (10.1)$$

For reinforcement this must equal an integral multiple of the wavelength λ. Hence

$$2d \sin \theta = n\lambda \qquad (10.2)$$

This is known as *Bragg's law*. Since λ is known and n follows from the order of diffraction, the interplanar spacing d can be determined from a measurement of the angular deviation, 2θ, of the beam.

This will give us the size and shape of the unit cell, and much information concerning the crystal or Bravais lattice, and its symmetry. It is not sufficient, however, to locate the position of the atomic arrangements. The crystal planes have their own atomic structure which affects the intensity of the

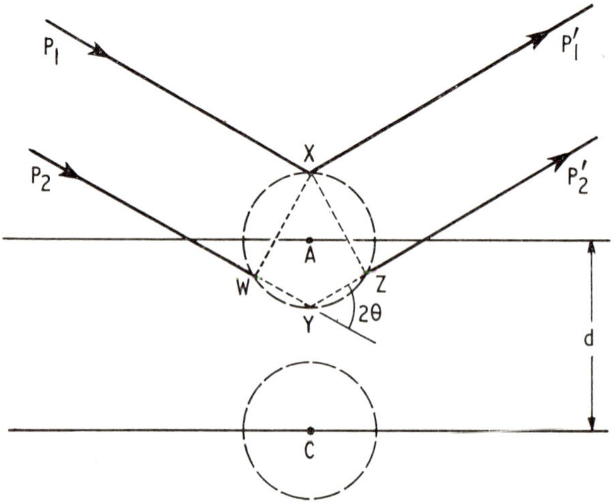

FIGURE 10.6 *Scattering of X-rays by an Atom*
The intensity of x-rays scattered by an atom is affected by the phase relationships between the beams scattered from the various parts of the atom, and the electron density in those parts.

diffracted beam. The intensity of the Bragg reflections thus depends on a large number of factors, including the number of atoms contributing to a reflection, their respective atomic numbers, and their arrangement in the diffracting unit.

Crystal diffraction is described in terms of Rayleigh scattering (Chapter 3), and it is the electrons within the atoms which are the effective agents.

If the x-rays scattered by all the electrons in the crystal unit cell were in phase, the intensity of the diffracted beams could readily be related to the crystal structure. The total intensity would be the intensity scattered per electron multiplied by the number of electrons in the unit cell. This is an oversimplification because of phase relationships. For example, consider the x-rays scattered by an atom on one reflecting Bragg plane (Figure 10.6). The x-rays scattered from point X (Figure 10.6) in the atom, when the beam has been deviated by the angle 2θ, are not necessarily in phase with those from point Y, as the lower part of the wave-front has travelled the extra distance WYZ; this effect will reduce the intensity of the beam. The phase relationship between the radiation scattered from different parts of the unit cell must also be considered. The conditions for the X-rays scattered from the Bragg plane at *C* to be in phase with those scattered from *A* will not be the same as for X-rays from Bragg planes between *A* and *C* to be in phase with *A*. Thus the fact that electrons are not all located on the Bragg planes *A* and *C* will reduce the intensity. It is usually not difficult to postulate a structure for an unknown crystal and then calculate the intensity of the expected reflections. The crystallographer is usually confronted with the converse problem of being presented with a list of intensities of radiation diffracted from many Bragg planes whose orientation with respect to the unit cell is known, and from thence he must find the structure. The difficulty lies in the fact that it is the intensity of the X-rays which he measures and not the amplitude of the vibrations. It is rather like measuring the force and angle of a football entering the goal and having to determine from this information about previous kicks. The problem is often indeterminate without additional information.

In crystal structure determination, the intensities of the diffracted beams are proportional to the square of vector sum of the radiation scattered from the atoms between the Bragg planes. The phase angle between the vectors is determined by the position of the atoms between the planes, and these angles cannot be directly determined because of the square relationship between amplitude and intensity. However, additional information about the crystal structure is generally available from its optical and other properties. Consequently the phase angles are sometimes known or otherwise inferred.

In this way many crystals have had their structures determined. Each structure presents its own problems. For structures of some complexity it is often possible to compute a Fourier series in two or three dimensions from the intensities, which will give a map of the electron density in the unit cell. When the unit cell is large and complex as with protein molecules, the number of Bragg reflections may be several tens of thousands and the labour of computation would be insuperable without electronic computers. Prolonged exposure of protein molecules to the X-radiation can affect their structure, making a further complication.

The structure of proteins contributes to the study of life itself. The importance of this work has been recognized by the award of Nobel prizes to several workers in this field.

Knowledge of the crystal structures with smaller molecules has greatly increased the understanding of phenomena such as the heat treatment of steel, the relationship between graphite and diamond, the elasticity of rubber, the flow of liquids and the strength of materials.

Experimental Methods

POWDER METHOD

Consider a collimated X-ray beam incident on a crystal. It will be deviated by an angle 2θ where θ is determined by the spacing of the atomic planes of the crystal and the X-ray wavelength. If there are many small crystals or *crystallites* of the same material with random orientations, the deviated beams will all be at an angle of 2θ with the incident beam, and consequently they will lie on the surface of a cone having a section of semi-angle 2θ. Thus, each different atomic plane in the unit cell will give rise to its own cone. This is the principle of the powder method. The crystals are crushed, filed or otherwise treated to reduce them to a uniform powder. The powder is usually

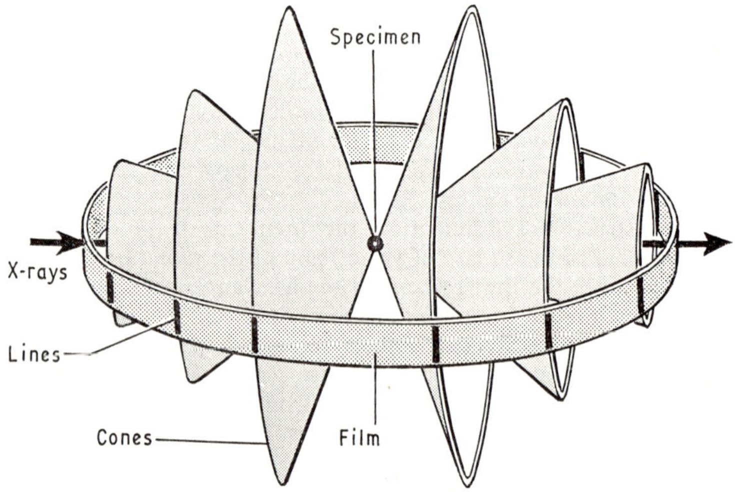

FIGURE 10.7 *Cones of Diffracted Rays formed in the Powder Method*
The Bragg planes in Figure 10.5, being oriented in all possible directions in the powdered crystals, give rise to cones of "reflected" rays.

made into a cylindrical shape by coating a glass fibre with adhesive and picking up some powder on the fibre, or by mixing the powder with adhesive and rolling it out into a cylinder having a diameter of approximately one third of a millimetre. The specimen is then mounted along the axis of a cylindrical powder camera, and a strip of film wrapped around the camera circumference (Figure 10.7). A motor may rotate the specimen to increase the randomness of the crystallite orientation.

The X-ray beam is filtered to give monochromatic radiation. Each of the various families of lattice planes will give rise to a "reflection" at its own angle θ. In a crystal belonging to the simple cubic Bravais Lattice in the cubic crystal system, the value of the interplanar spacing d of the h,k,l plane (*see* Fig. 10.10 for h, k, l plane 321) for a cube of side a, is given by

$$d = \frac{a}{(h^2 + k^2 + l^2)^{1/2}} \tag{10.3}$$

Using Bragg's Law,

$$\lambda = 2d \sin \theta \quad \text{(Consider } n = 1 \text{ throughout)} \tag{10.4}$$

From 10.3 and 10.4

$$\lambda = \frac{2a}{(h^2 + k^2 + l^2)^{1/2}} \sin \theta \tag{10.5}$$

$$(h^2 + k^2 + l^2)^{1/2} = \frac{2a}{\lambda} . \sin \theta = K . \sin \theta \tag{10.6}$$

In any particular experiment K is fixed by the crystal and the radiation, so that the values of $\sin \theta$ are determined by the sum of the squares of the three integers h,k,l. The face-centred and body-centred cubic Bravais lattices have a rather more complicated pattern of reflections owing to annihilation of some of the reflections appropriate to the simple cubic lattice by the centring lattice points. In less symmetrical crystal systems the reflections become more difficult to interpret, as allowance must be made for the non-integral angles between the unit cell vectors, and the fact that the vectors may not be equal in length.

When the diffracted beams are received on a flat plate normal to the x-ray beam, the intersections of the cones produce rings called Debye-Scherrer rings. In the powder camera the intersection of the cones and the cylindrical film form a sequence of slightly curved lines, which may be measured very accurately to give their spacing and hence the Bragg angles of the reflections.

The powder method is useful particularly for identifying materials, because every crystalline material will give a unique powder pattern, with reflections of constant angular position and intensity. A mixture of materials will give rise to patterns appropriate to the composition of the mixture, making quantitative analysis possible.

In the mechanical working of metals the crystallites often tend to assume special orientations; such a material is in a state of *preferred orientation.* This may have a profound effect on the mechanical properties of the material and it is important to be able to measure the extent to which it occurs. Preferred orientation is revealed by the fact that the Debye–Scherrer rings are reduced to arcs, or even spots.

When the crystallites are small or in a state of strain, the rings tend to become diffuse and this effect is used to investigate crystal size and strain phenomena [4].

Special powder cameras have been constructed for use at high and low temperatures [5].

THE LAUE METHOD

This method is named after the originator of the first experiments on crystal diffraction, in which a stationary single crystal of copper sulphate was irradiated with x-rays taken from a tube without any added filtration. This unfiltered white radiation contains a wide range of x-ray wavelengths and intensities, the minimum wavelength being determined by the voltage and the maximum by the filtration of the tube. For any particular Bragg plane the glancing angle with respect to the incident beam is fixed by the orientation selected for the crystal, so with θ and d fixed, the λ required to satisfy Bragg's equation is "selected" by the crystal from those present in the radiation. If the crystal orientation is altered the same Bragg planes may reflect, but for a different wavelength.

Consequently, the wavelength and the intensity of the reflected radiation are indeterminate, and because of this, the Laue method is not now used for structure determination. It does, however, reveal useful information on the symmetry of a crystal, eleven sets of Bravais point groups being distinguishable. It may also be used to study deformation of crystal structure, when bending of the crystal planes will result in elongation of the Laue spots. This is known as asterism. The widest use is probably to determine crystal orientations, found from the symmetry of the Laue picture. This is related to the symmetry of the crystal along the X-ray beam direction.

OSCILLATION AND ROTATION METHODS

When a small single crystal is placed in a monochromatic X-ray beam in a random orientation, the Bragg equation in general will not be satisfied for a

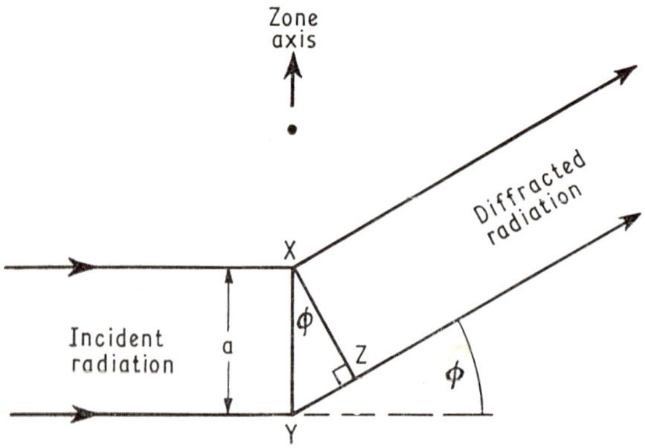

FIGURE 10.8 *Reinforcement of Diffracted Rays by a Zone (or Line) of Points* The condition for reinforcement is $a \sin \phi = n\lambda$. The diffracted radiation is not confined to the plane containing the zone axis and the incident radiation, but lies on a cone of semi-angle ϕ.

particular plane, as d and θ will have been fixed at arbitrary values. If, however, the orientation of the crystal is continuously varied, then the plane may momentarily pass through an orientation such that the value of θ required by Bragg's law is attained and a Bragg reflection will occur. In the usual experimental arrangement, the crystal is mounted on a fine glass fibre with a prominent family of planes (or zone axis) parallel to the axis of rotation or oscillation. The film is mounted on a cylinder with the same axis.

If the zone axis of the crystal has a repeat distance a, and Y and Z (Figure 10.8) represent two Bravais points on the zone, then the condition for reinforcement is that YZ shall be equal in length to one or more X-ray wavelengths, or

$$YZ = n\lambda$$

This condition does not apply only in the plane of the incident beam and the zone axis. The diffracted beams are constrained to lie on a cone of semi-angle $90 - \phi$ where ϕ is the angle of deviation. There is a family of such cones for various values of ϕ and the intersections of these cones with the cylindrical film will be lines. These are the *layer lines* (Figure 10.9) on which the Bragg reflections lie.

The value of the unit cell spacing a is easily calculated from the camera geometry:

$$\frac{n\lambda}{a} = \sin \phi \tag{10.7}$$

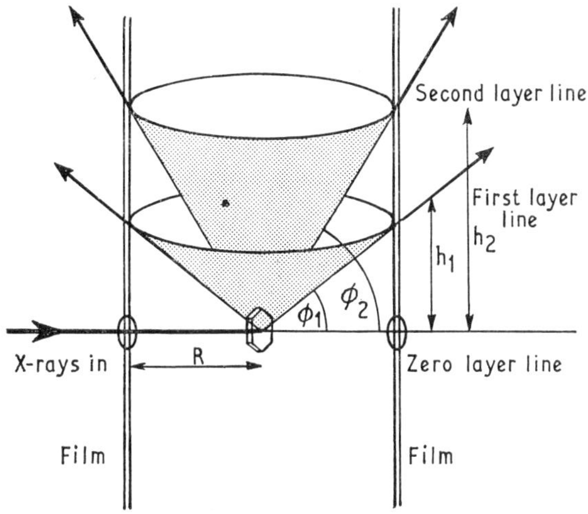

FIGURE 10.9 *Formation of "Layer Lines" by Single Crystal Diffraction*

In the second of the upper layer lines, the angle of deviation is ϕ_2, and the spacing of this layer line from the zero layer line is h_2. If the camera radius is R then

$$\frac{h_2}{R} = \tan \phi_2 \tag{10.8}$$

so that knowing λ and ϕ_2 and solving equations (10.7) and (10.8) a can be calculated. The other unit cell spacing can be found by rotating the crystal about the appropriate axis.

In order to determine structures it is necessary to know more than the geometry of the unit cell or the symmetry of the crystal. The Miller indices h,k,l, of the reflecting planes must also be found; this process is called indexing. It is performed by assuming a unit cell, and using it as a reference for the spacing of all the planes. The deviation of the Bragg reflections is known and each may be indexed from its position on the film. This can be a tedious process and it is easier to use the *reciprocal lattice* concept. The purpose of this concept is to describe the orientation, and the spacing of all the crystal planes. Consider a family of Bravais planes (Figure 10.10) mutually parallel. The origin O is an arbitrary point in the crystal. From it a line OX is constructed, perpendicular to the Bravais planes. If a point R on OX is defined such that the distance OR is proportional to the reciprocal of the spacing d of the Bravais planes, then R is the reciprocal point for those planes. Thus

$$OR = \frac{k}{d} \tag{10.9}$$

In geometrical problems, k is made numerically equal to one, whereas in diffraction problems, it is convenient to make k numerically equal to λ.

There is an unlimited number of Bravais planes, each with its own reciprocal point. The array of points in *reciprocal space* is called a *reciprocal lattice.* A change in the value of k merely alters the scale of the reciprocal lattice.

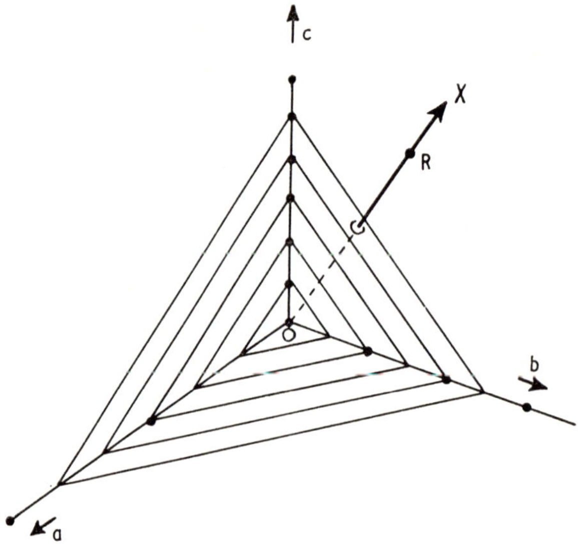

FIGURE 10.10 *Reciprocal Lattice Point at R for the 321 Planes*
The length OR is inversely proportional to the interplanar spacing, and in a direction parallel to the normal to the planes.

The reciprocal lattice is a convenient way of depicting all the Bragg planes of the Bravais lattice at once and assists in the indexing of rotation and oscillation photographs (Figure 10.10). A chart has been constructed [6] from which two of the three cylindrical co-ordinates of a reciprocal lattice point can be read from an oscillation photograph. A crystal with a complicated structure will give many overlapping Bragg reflections. The interpretation is facilitated by moving the film in conjunction with the crystal and photographing only one layer line at a time (Figure 10.11) as with the Weissenberg camera.

The reflections are spread over the whole film instead of being confined to a line. The photograph is a distorted picture of the reciprocal lattice, and makes possible indexing at sight.

LOW-ANGLE SCATTERING

The scattering process here is not Bragg reflection but a process resembling the scattering of light by dust particles, and the angle of deviation depends upon the size and shape of the particles. Diffuse intensity maxima are produced enabling estimates of particle size to be made. The main application is to study the structure of near-amorphous materials such as liquids and glasses.

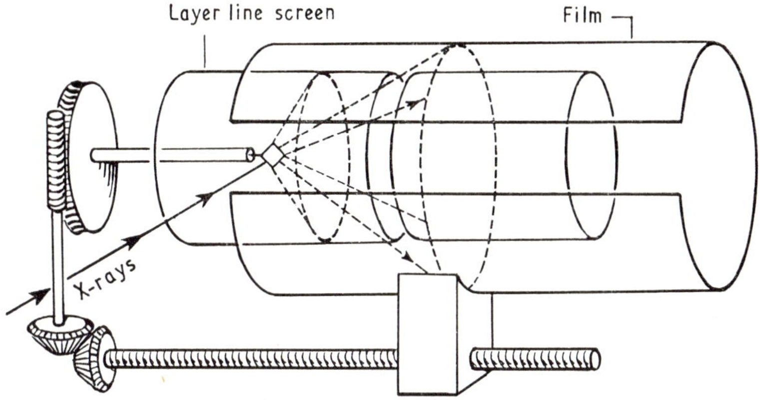

FIGURE 10.11 *The Weissenberg X-ray Goniometer*

Crystal Chemistry

The two main factors governing the crystal structure pattern are the shape and size of the atoms and molecules and the forces binding these units together. Much of the information on the binding forces in materials has been obtained from inter-atomic spacings obtained by x-ray diffraction.

CRYSTAL BINDING FORCES

There are five types of forces binding together the atoms in crystals, and more than one of these types of bonding may exist in one crystal. The division of solid-state science into organic chemistry, inorganic chemistry and metallurgy is largely a reflection of the various bond types.

(i) Van der Waals Forces

Van der Waals forces are weak universal forces. They are, for example, responsible for the changes of state of inert gas to liquid to solid, as the temperature falls. These gases are so stable that only the residual van der Waals forces are of importance. Other examples are found in organic chemistry, where the van der Waals forces weakly bind together the molecules of many organic compounds, the atoms of which may be tightly bound within the molecule by other forces.

(ii) Electrostatic Forces

Electrostatic forces are the binding forces in ionic crystals. There is no individual molecule in an ionic crystal, which is an assembly of ions, surrounded by neighbours of opposite sign, sufficient in number to neutralize the electrostatic forces. A crystal of sodium chloride consists of ions which are single ionized atoms. Each chlorine (or sodium) ion lies on a cubic face-centred lattice and has six neighbours of opposite sign. Ionic crystals also exist where radicals such as the SO_4 ion form groups.

(iii) Homopolar Forces

Homopolar forces are due to the sharing or linking of electronic or molecular orbitals. The structures of diamonds, silicon and germanium are formed in this way.

(iv) Metallic Forces

In metals, the atoms contribute electrons to a common electron "gas". The atoms resemble positively charged "bricks" surrounded by negatively charged electronic "mortar" and it is this mutual attraction which constitutes the metallic forces. The atoms can slip over one another without fracturing the metallic bond and this gives rise to ductility, a characteristic property of metals.

(v) Hydrogen Bond

Owing to the simple structure of hydrogen, which consists of a proton nucleus and one orbital electron, the hydrogen atom acts as a small force centre. It can bind together molecules with a force rather greater than a van der Waals force. It is important, for example, in forming inter- and intra-molecular bonds in proteins.

ATOMIC SIZES

In many crystals, the atomic size is an important factor in determining the atomic arrangement or "packing". However, the atomic size is not a constant for a given atom, but may depend, for example, on the nature of the chemical bond.

In ionic crystals, the ions pack so that the electrostatic charges are neutralized and so that the gaps between ions are as small as possible.

Application to Materials Science

Materials have such a large influence on the contemporary cultures that they have contributed names to certain epochs such as the stone age, the bronze age or the iron age.

Demands for materials of greater strength, higher melting point and particular electrical properties have stimulated much modern research into the development of new materials. Conventional materials may be combined in new ways, for example by dispersing fine alumina particles throughout a metal to harden it. Another procedure is to remove imperfections from small crystals to produce "*whiskers*" of very high strength.

Many of the well-known materials in common use, such as metals, ceramics and textile fibres, have a well-established basic crystal structure. Electrical, optical and mechanical properties are closely related to this structure. It is possible to devise materials with special properties after a detailed study of the effects of impurities and imperfections on the crystal structure, for example, in the development of semi-conductors, or carbon fibre crystals.

Kossel Lines

The atoms of a crystal can be excited to become independent sources of monochromatic radiation. This radiation will interact with other atoms in the crystal lattice, and reflection will occur at angles corresponding to Bragg's law. If a single crystal is used as the target in an x-ray tube, striking patterns are obtained from these interference effects. They were first observed by Kossel, Loeck and Voges in 1934 [7] and are called Kossel lines.

A rather similar effect is obtained if very divergent x-ray beams are emitted from a point x-ray source close to a stationary crystal. These have been used by Lonsdale [8] to investigate crystal texture. They may also be used to obtain accurate lattice parameters.

Kikuchi [9] line patterns in electron diffraction, obtained by diffracting electrons from single crystals, are analogous to these effects.

Literature References

1 Haüy, R. J., *L'Essai d'une Théorie sur la Structure des Crystaux appliquée à plusieurs genres de Substances Crystalisées* (Paris, Gogué et Neé de La Rochelle, 1784).

2 Friedrich, W., Knipping, P. and Laue, M., "Interferenz–Erscheinungen bei Röntgenstrahlen", *Sitzungsb. bayer. Akad. Wiss., math. phys. Kl.*, p. 303 (1912).

3 Bragg, W. L., "The diffraction of short electromagnetic waves by a crystal", *Proc. Cambridge Phil. Soc.*, **17**, p. 43 (1912).

4 Lewis, D. and Pearson, H., "X-ray line broadening in calcium fluoride", *J. Appl. Phys.*, **35**, p. 1939 (1964).

5 Peiser, H. S., Rooksby, H. P. and Wilson, A. J. C. (Eds.), *X-ray Diffraction by Polycrystalline Materials* (London, Chapman and Hall (for Institute of Physics), 1960).

6 Bernal, J. D., "On the interpretation of x-ray single crystal, rotation photographs", *Proc. Roy. Soc.*, A, **113**, p. 117 (1926).

7 Kossel, W., Loeck, V. and Voges, H., "Die Richtungsverteilung der in einem Kristall enstandenen charakteristischen Röntgenstrahlung", *Z. Physik*, **94**, p. 139 (1935).

8 Lonsdale, K., "Divergent-beam x-ray photography of crystals", *Phil. Trans. Roy. Soc.* A., **240**, p. 219 (1947).

9 Kikuchi, S., "Beugung der Materiestrahlen", *Physikal. Zeit.*, **31**, p. 777 (1930).

11 X-ray Chemical Analysis

X-ray chemical analysis is supplementary to, but not a replacement of, conventional methods.

X-ray spectroscopic analysis, X-ray fluorescence analysis and X-ray absorption analysis give information about the elements present in a sample, and depend upon electron transitions in the atoms examined. A fourth method, X-ray diffraction analysis is complementary to the other three, as it analyses chemical compounds and polymorphs. This is because the patterns obtained by X-ray diffraction methods are dependent on the arrangements of the atoms in the crystals of the materials examined, and these arrangements are unique to each compound or allotrope. The scattering of X-rays by the crystals to give the patterns relevant to X-ray diffraction analysis does not involve electron transitions, but depends upon coherent scattering.

X-ray chemical analysis has the merits of non-destructive testing. Preparation of samples is usually simple and only small quantities are required. All four methods are quick and labour saving, although careful standardization is required. A disadvantage is the high cost of the apparatus.

In all four methods the intensity of particular X-ray wavelengths must be measured. Bragg's law (Chapter 10) relates the wavelength λ of the diffracted radiation, the glancing angle θ and the spacing d, of the crystal planes:

$$\lambda = 2d \sin \theta \tag{11.1}$$

In the first three methods, a crystal with a known spacing d is used, and θ is measured, enabling λ to be calculated. In the fourth method a known value of λ and a measured value of θ are used to determine d.

The X-ray Spectrometer

The X-ray spectrometer resembles in principle an optical spectrometer [1]. The X-ray diffracting crystal is mounted at the centre of a graduated circle (Figure 11.1), so that the crystal can rotate about the axis of the circle. An X-ray beam, collimated by an arrangement of slits is incident on the crystal (Figure 11.2).

In some situations, the diffracting crystal is the specimen to be analysed. In others, the specimen produces radiations which are incident on, and analysed by, the crystal of the spectrometer. The angle of deviation between incident and diffracted beam is 2θ. In order to maintain alignment, the detector, which also rotates about the axis of the circle, moves at twice the angular velocity of the crystal.

The first X-ray spectrometer was constructed by the Braggs [2]. They discovered the X-ray line spectra, thus making possible modern crystal structure analysis and laying the foundation for X-ray spectroscopy.

FIGURE 11.1 *X-ray Spectrometer*
A, anode; *B*, divergence parallel slit assembly; *C*, divergence slit; *D*, centre of goniometer and specimen; *E*, focusing circle; *F*, receiving slit; *G*, receiving parallel slit assembly; *H*, scatter slit; *J*, counter tube.
(*Philips Service Manual for Wide-range Goniometer Type PW 1050*)

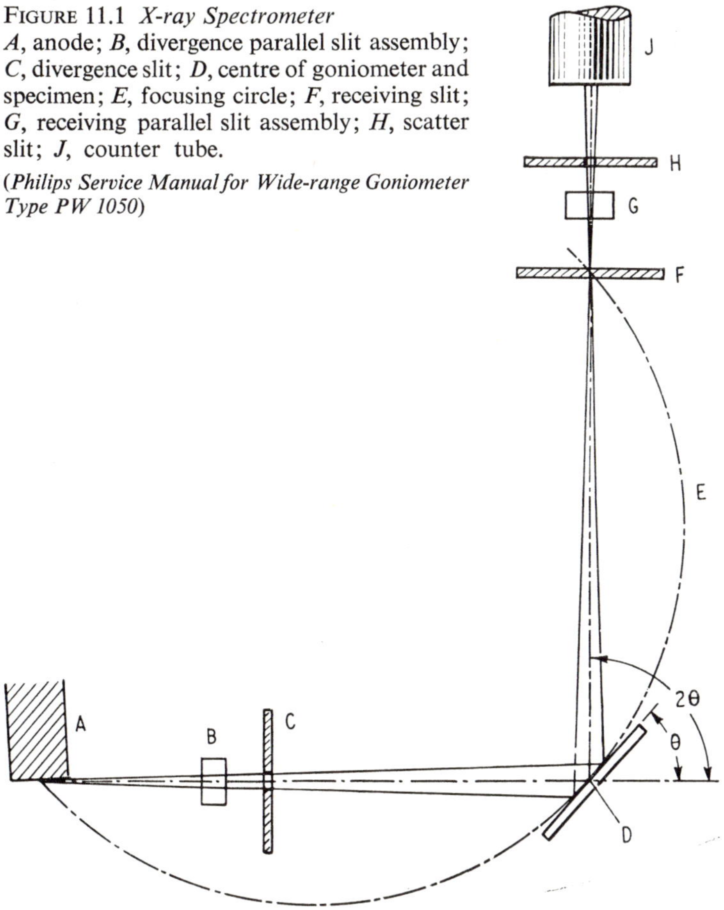

Some modern spectrometers have curved crystals to focus the diffracted beam. The geometry of the arrangement resembles that of the Rowland circle [1] in optics, so that the X-ray source, the crystal and the detector slit lie on a circle (Figure 11.3) called the focusing circle. To maintain alignment, the radius of curvature of the crystal planes is made equal to the diameter of the focusing circle.

Geiger, proportional or scintillation counters (Chapter 4) are used in modern spectrometers. A Geiger counter requires a simpler circuit and

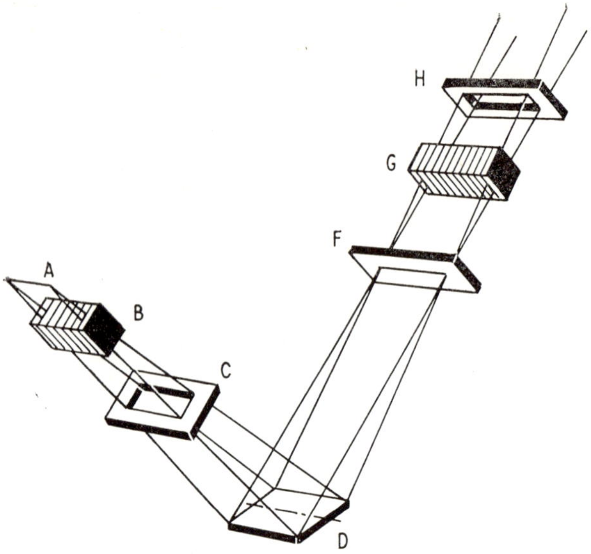

FIGURE 11.2 *X-ray Spectrometer Slit System*
A, line focus; *B*, divergence parallel slit assembly; *C*, divergence slit; *D*, specimen; *F*, receiving slit; *G*, receiving parallel slit assembly; *H*, scatter slit. (*Philips Service Manual for Wide-range Goniometer Type PW* 1050)

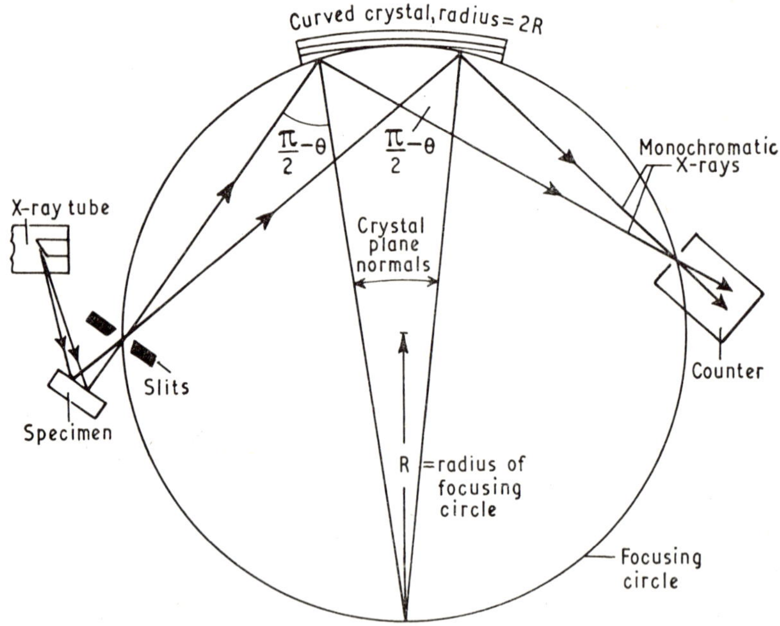

FIGURE 11.3 *Curved Crystal Focusing Spectrometer*

produces a larger pulse than the other two, but it can only be used for low count rates owing to its long paralysis time. The pulse height from the Geiger counter is independent of the energy absorbed in each triggering event and therefore also independent of X-ray photon energy. Pulse heights from proportional and scintillation counters are energy-dependent and so can be used for energy analysis. The proportional counter has a better energy resolution than the scintillation counter, but, being a gas counter, is not very sensitive at high X-ray energies. At low energies the sensitivity of a scintillation counter is limited by absorption in the protective coating on the crystal.

When the X-ray spectrometer is used to examine crystal structure, it is usually called a diffractometer [3].

Automatic spectrometers are often of the single channel type, in which the counter moves automatically from one angular position to another, the movement corresponding to the Bragg plane spacing of the X-ray wavelength to be examined. More rarely, they are of the multi-channel type with a separate spectrometer for each channel. This is the more sensitive type and is appropriate in X-ray spectral analysis when the spectrometer crystal is only an analyser and the specimen is separate and produces the incident beam.

For the rapid analysis of crystal structures, the output from an automatic spectrometer can be fed into a computer.

Non-dispersive Methods

It is possible to perform X-ray spectral analysis without an analysing crystal, or even without a spectrometer, if non-dispersive methods are used [11]. In these methods the radiation of differing wavelengths and intensities is sorted by absorption edge filters and pulse height analysers. The voltage of the X-ray tube may be varied to excite only the required characteristic wavelengths. This method is effective when the elements to be examined differ widely in atomic number.

X-ray Spectroscopic Analysis

If the sample is made the target in an X-ray tube and bombarded with electrons of suitable energy, it will emit characteristic X-rays. This is the basis of a method of chemical analysis. The emitted X-rays are analysed in an X-ray spectrometer, and the elements present in the sample qualitatively identified by their characteristic wavelengths. Quantitative estimation is also possible by measuring the relative intensities of the spectra. This method was first investigated by Castaign [4].

The specimen must be either electrically conducting, or made so by evaporating a metallic layer on the surface, otherwise surface stray electric fields may divert the incident electron beam. The surface should be smooth to prevent undue absorption effects from projections. For metallurgical specimens the ordinary polishing and etching procedure is usually adequate.

In the electron probe microanalyser the electron beam can be focused so that it irradiates an area of the order of a square micron. The beam may be made to scan a specimen to give an analysis averaged over a larger area.

In Duncumb's method [5], the sample is scanned and two cathode ray tubes are used. The first gives on its screen a recognizable visual image of the sample, achieved by making the electrons scattered from the sample produce, after amplification, an image on the screen. The second cathode ray tube shows the distribution of a particular element on its long-persistence screen, and the two screens are interpreted together. The distribution of the element is shown by the brightness of the screen at each point in question because the X-rays from the specimen, after analysis by a crystal spectrometer,

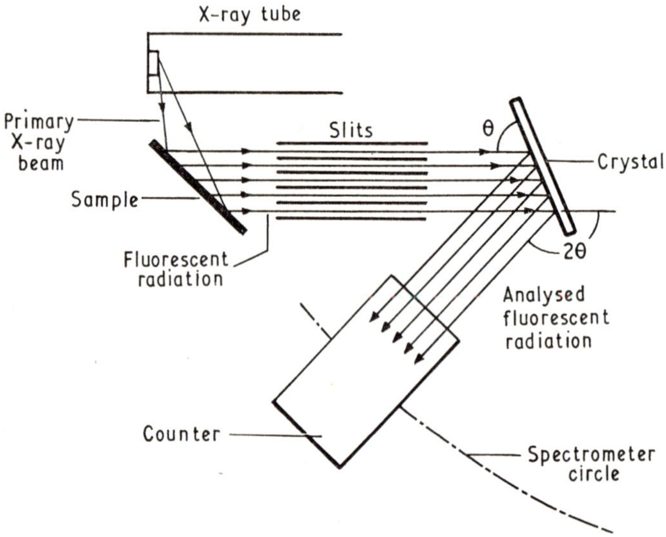

FIGURE 11.4 *X-ray Spectrometer for Fluorescence Analysis*

are incident on a proportional counter and the resulting signal from the counter is made to modulate the beam intensity of the cathode ray tube.

For quantitative analysis the proportional counter output is fed to a ratemeter. A standard sample is used for calibration.

The electron probe microanalyser has found considerable application in metallurgy for identification and analysis of precipitates, diffusion zones and inclusions. Interpretation can be complicated by enhancement due to fluorescence and by absorption effects. Carbon deposition from cracking of the vacuum oils and greases can create anomalies through absorption of the longer wavelength radiations.

X-ray Fluorescence Analysis

If a sample is irradiated with X-rays of sufficiently high energy it will emit fluorescent X-radiation. This radiation may be analysed in an X-ray spectrometer, and the elements present in the sample identified by their characteristic wavelengths (Figure 11.4). The relative proportions of the elements present are estimated by the respective intensities. When a sample contains at least one per cent of an element this method is more accurate than optical spectroscopy [6, 7, 8, 9, 10]. The X-ray fluorescent spectra are simple compared with the corresponding optical spectra.

The exciting radiation is taken from a tube with a tungsten target operating at about 50 kVp. The fluorescent radiation is analysed with a crystal, for example of pentaerythritol. Focused beams from curved crystals give comparatively high intensities.

In qualitative analysis, a spectrometer calibration of θ against λ is required. The radiation to be analysed includes the original exciting tungsten lines and the characteristic radiation from the specimen. Lines from sample impurities and target contaminating metals will also be present. Radiations of wavelengths from 0·05 to 0·25 nm may be easily analysed. The lower limit is fixed by the maximum tube voltage and the higher limit by absorption considerations. Elements of high atomic number may be analysed using their fluorescent L radiation. Absorption in air limits the atomic numbers to those above about 22 (titanium). However, boron ($Z = 5$) and carbon ($Z = 6$) are measurable using heavy metal stearate crystals *in vacuo* and counter windows made of Formvar a few tens of nanometres thick. Optical diffraction gratings with spacings coarser than conventional analysing crystals may extend the use of the X-ray analysis to still lower atomic numbers.

The measurement of light elements in a heavy element matrix is difficult because of absorption in the sample.

The intensity of the fluorescent radiation from an element in a mixture is usually not proportional to the concentration of the element. This is because of interaction with the radiation from the other elements and also because of differential absorption effects depending upon concentration. Enhancement is more pronounced with fluorescent analysis than with electron probe analysis because of the greater beam penetration.

X-ray Absorption Analysis

Materials may be analysed by their X-ray absorption properties. The method is usually based on an examination of absorption edge effects. Because of self-absorption, materials with a high absorption coefficient must be examined as foils or in dilute solution.

Elements present may be identified by their absorption edges. The sample is placed in an X-ray beam, the wavelength of which is varied continuously and the intensity of the transmitted beam is measured and plotted against corresponding wavelength. The resulting curve will have discontinuities at the absorption edges of the elements in the sample, showing an abrupt increase in transmitted intensity on the long-wavelength side of an edge. The X-ray beam of continuously varying wavelength may be obtained from white radiation, resolved into its components by a diffractometer; a disadvantage is the low intensity of the reflected X-radiation from the diffractometer.

There are a number of methods for quantitative absorption analysis.

In one of these the wavelength may be set to an absorption edge of a particular element, as for identification, and the change in mass absorption coefficient measured for wavelengths just longer and just shorter than the absorption edge wavelength. Another method is to irradiate the sample by a suitable intense characteristic line present in the incident irradiation and to measure the absorption; the analysis may be made directly, provided that the mass per unit area of the sample is known. A third method uses incident white radiation; more intense beams are thus available, but analysis must be made by calibration.

An important application of absorption methods of analysis is the determination of tetraethyl lead in petrol [2]. It has also been used for medical and biological material in thin sections.

When the spatial distribution of elements in a metallurgical specimen is required, but exact analysis is not necessary, informative radiographs using various characteristic radiations may be obtained from thin specimens. When the radiation is near the absorption edge of an element in the specimen, for example copper K_α radiation and a ferrous material, the distribution of the element in the mixture is indicated by its preferential absorption.

Chemical Analysis by X-ray Diffraction

Chemical analysis by x-ray diffraction is concerned with the arrangement of the atoms.

Consider a mixture of two substances. The relative intensities of the lines from one substance depend on the atoms in its diffracting planes, so that the diffraction pattern identifies the substance [13]. The intensity of one pattern relative to that of the other depends on the relative proportions of the two substances. Thus, a mixture may be analysed, qualitatively and quantitatively, by combining an examination of the diffraction patterns with intensity measurements of the diffraction rings.

For instance, when analysing a mixture of crystals such as those of iron and sulphur, each element will give its own diffraction pattern. In contrast, iron sulphide crystals will give a diffraction pattern different from that of the mixture. Similarly, a mixture of calcium carbonate and lead sulphate can be distinguished from a mixture of calcium sulphate and lead carbonate, because the atom arrangement is specific for each compound.

The method can be used to distinguish between allotropic forms of the same element, such as the diamond and graphite forms of carbon.

The diffraction pattern may be photographed in a powder camera. This is the method of choice for small samples such as fine wires and for the detection of weak lines. Alternatively, the diffraction pattern may be recorded using counting methods with a diffractometer. This method is suitable for larger samples in the form of sheet or discs and for poorly crystalline material giving diffuse rings.

Film densities are not proportional to exposures (Chapter 7) and in general, intensity measurements are more directly made on a diffractometer than with a powder camera. However, when the sample is a mixture of two substances, one coarsely and one finely crystalline, the former will give less continuous rings than the latter; this effect can be used to identify the rings originating from each compound and is more obvious using film than a diffractometer. When the crystals are pointed or flat, they will tend to lie in preferred orientations when packed in a diffractometer specimen holder, and this will modify the ring intensities. Packing anomalies are not important in the powder method, where the method of specimen preparation reduces this effect. In metallurgical analysis the exact size of the unit cell is used for quantitative analysis, requiring great accuracy in measurement on all the Bragg reflections. This is most conveniently carried out using a powder camera [13].

If the diffraction patterns of standard and sample prove to be identical, it is immediately evident that standard and sample are the same substance and it is unnecessary to index (Chapter 10) the lines unless the structure of the standard is unknown and is required. Indexing is easiest with single crystal spot patterns.

Qualitative Analysis by X-ray Diffraction

As every solid substance gives its own diffraction pattern characterized by the angular position of the diffracted lines and their intensity, these patterns are made the basis of systems of qualitative analysis.

Each laboratory should keep its own classified record of films and results. These records are most useful where the range of materials encountered is limited. Difficult analyses are also best carried out using the laboratory's own records. However, in time, the range of patterns to be analysed usually becomes too large to be handled by this direct method.

The basis of modern reference methods was devised by Hanawalt in 1936 [14]. His classification is intended to be independent of the wavelength of the analysing radiation, and is an advance over direct comparison of line positions, which depend on the radiation. Hanawalt's method compares the spacing, d, of Bragg planes and the intensities of their reflections.

Hanawalt, and later the American Society for Testing Materials [15] listed all known patterns on cards. Each card tabulates one pattern in order of decreasing values of d and records line intensities relative to the strongest (Figure 11.5).

To identify an unknown substance from its pattern, intensities and values of d are measured from the film. The strongest line is used for the first step in identification, and its value of d selected in a reference book. As the value of d is not usually known exactly owing to impurity and experimental error, the substance is not uniquely identified from the strongest line. Calculated values of d for the second and third strongest lines and their intensities relative to the strongest line, are used to help to complete the identification. The pattern can then be provisionally referred to one card, and all the lines are then used to complete the identification. There are more than ten thousand cards in the complete index; modern indexing systems [16, 17] are used in conjunction with it. This type of problem is suitable for solving by computer.

Practical difficulties arise in the identification of compounds by this method. The diffraction pattern may be unsatisfactory because, for example, poorly crystalline material may give diffuse reflections. Sometimes there is not enough material to give a good pattern. Often the sample is a complex mixture and its diffraction pattern is difficult to interpret. In such cases,

6-0392

d	2.98	2.79	2.04	3.88	$NaNO_2$
I/I_1	100	56	37	2	Sodium Nitrite

Rad. Cu λ 1.5405 Filter
Dia. Cut off Coll.
I/I_1 d corr. abs.?
Ref. Swanson et al., NBS Circular 539 Vol IV p 62-3 (1953)

Sys. Orthorhombic S.G. C_{2v}^{20} - Im2m
a_0 3.570 b_0 5.578 c_0 5.390 A C
α β γ Z 2
Ref. Ibid.

εα nωβ εγ Sign
2V Dx 2.135 mp Color
Ref.

Sample from the Fisher Scientific Co. Spect. anal.: <0.01 % Al, Ca; <0.0001 % Ag, Ba, Cu, Mg.
X-ray pattern at 26°C.

Replaces 1-0883, 3-0628

d Å	I/I_1	hkl	d Å	I/I_1	hkl
3.88	2	011	1.292	2	033
3.006	32	110	1.263	2	141
2.976	100	101	1.253	<1	231
2.789	56	020	1.238	5	042
2.695	5	002	1.235	4	213
2.035	37	121	1.213	1	024
2.007	29	112	1.1623	1	301
1.938	13	022	1.0728	2	321
1.784	14	200	1.0687	2	312
1.757	3	031	1.0525	3	143
1.710	8	013	1.0466	<1	233
1.650	4	130	1.0321	1	105
1.605	<1	103	1.0175	1	242
1.504	7	220	1.0027	1	224,330
1.488	<1	202	0.9689	2	044
1.406	5	132	.9679	1	125
1.391	6	123	.9392	<1	332
1.348	2	004	.9345	<1	323
1.313	5	222			

404

FIGURE 11.5 *Card from the A.S.T.M. Powder Diffraction File*
(Courtesy of The American Society for Testing and Materials)

before x-ray diffraction, it is helpful to remove any obviously identifiable components by hand-picking under a lower power microscope or by sieving, flotation or magnetic methods.

Quantitative Analysis by X-ray Diffraction

Quantitative analysis should be carried out by matching the pattern of the unknown mixture against that of a standard mixture. This is best performed on a diffractometer, but film methods using a microdensitometer can give nearly as accurate results. In favourable circumstances, matching by eye can give results to an accuracy of about 5 per cent. Sometimes the intensities from the mixture can be compared with that from an internal standard, being mixed with a standard substance in known proportions.

Literature References

1 Longshurst, R. S., *Geometrical and Physical Optics* (London, Longmans, 1960).
2 Bragg, W. H. and Bragg, W. L., "The reflection of x-rays by crystals", *Proc. Roy. Soc.* A, **88,** p. 428 (1913).
3 Klug, H. P. and Alexander, L. E., *X-ray Diffraction Procedures* (New York, Wiley, 1962).
4 Castaign, R. and Guinier, A., *Proc. 1949 Conference on Electron Microscopy, Delft* (Delft, 1950).
5 Cosslett, V. E. and Nixon, W. C., *X-ray Microscopy* (Cambridge University Press, 1960).
6 Gordon, F. W. and Davis, H. M., "Summarized proceedings of a colloquium on x-ray fluorescence, London, February and April, 1958", *Brit. J. Appl. Phys.*, **10,** p. 105 (1959).
7 Parrish, W., "X-ray spectrochemical analysis", *Philips Technical Review*, **17,** p. 269 (1956).
8 Jenkins, R., "Trace element analysis by x-ray fluorescent spectroscopy", *Conference on Limitations of Detection in Spectrochemical Analysis*, July, 1964 (Reprints supplied by Philips Scientific Equipment).
9 Booth, E., Wiltshire, N. and Braun, L., "The determination of the coating thickness of tinplate by x-ray fluorescence methods", *B.H.P. Tech. Bull.*, **23,** p. 15 (1965).
10 Jenkins, R. and de Vries, J. L., *Practical X-ray Spectrometry* (*Philips Technical Library*) (London, Cleaver-Hume, 1967).
11 Birks, L. S., *X-ray Spectrochemical Analysis* (New York, Interscience, 1959).
12 Clark, G. L., *Applied X-rays* (London, McGraw-Hill, 1955).
13 Peiser, H. S., Rooksby, H. P. and Wilson, A. J. C., *X-ray Diffraction by Polycrystalline Materials* (London, Institute of Physics, 1960).
14 Hanawalt, J. D. and Rinn, H. W., "Identification of Crystalline Materials", *Ind. Eng. Chem.*, **8,** p. 244 (1936).
15 American Society for Testing Materials, *Index to the X-ray Powder Data File*, A.S.T.M. Spec. Tech. Pub. 48-H. (Philadelphia, 1959).
16 American Society for Testing Materials, *Fink Index to the Powder Diffraction File*, A.S.T.M. Special Tech. Pub. 48 (Philadelphia, 1963).
17 American Society for Testing and Materials, *The Matthews Co-ordinate Index*. Foreword in Fink Index (*see* reference 16).

12 | *Uses of Large Doses*

The chemical yield of a radio-chemical reaction is described in terms of the *G*-value, defined as:

$$G = \text{number of molecules produced or converted per 100 eV absorbed} \tag{12.1}$$

Thus, a dose of 1 rad will produce or convert about $MG \times 10^{-12}$ g per gramme of material irradiated, where M is the molecular weight in grammes of the material. As *G*-values are numerically usually considerably less than a hundred unless there is a chain reaction, it follows that many megarads are required to transform a substantial fraction of a chemical.

Living organisms are damaged by smaller doses [1, 2]. 0·5 Mrad will kill most insects within twenty-four hours and 0·1 Mrad will kill them within a week. The latter dose is sufficient to produce a radiation lethargy which prevents the insects from doing much feeding damage. Lower doses of about 15 to 20 krads make the insects infertile and cause their death in a few weeks [3].

The heating effect of radiation is small. A dose of 1 Mrad (10^8 ergs/g) is equivalent to 2·4 cal/g so that the temperature rise in an aqueous medium is only 2·4°C from this dose. A few megarads is usually sufficient for radiation sterilization, which can thus be carried out without damage to heat-labile materials.

Suitable sources must deliver high dose rates. Van de Graaff generators or linear accelerators can provide high intensity beams of megavoltage X-rays, but kilocurie sources of radioactive nuclides, particularly cobalt-60, are also used.

Radiation as a Mutagenic Agent

A sudden change in hereditable characteristics of an individual may follow an alteration in the composition of a gene, known as a gene mutation. These mutations occur in the absence of any man-made source of interference, when they are said to be *spontaneous mutations*. The spontaneous mutation rate is low and has been estimated at 1 in 10^5 to 1 in 10^8 genes per generation. When a mutation confers some benefit on the individual, its incidence in the population is likely to increase by the process of natural selection. Therefore, when a stable environment has existed for many generations, the vast majority of genetic mutations are likely to be deleterious; it has been estimated [4] that only about 1 in 800 mutations is likely to be of value in plant breeding, and these may give properties which are useful to man rather than being of survival value to the plant. A minor change in environment can alter the balance for natural selection: in a smoky industrial district, survival is easier for moths with dark wings than with light, and a case is known where a dark-winged mutant form of previously light-winged moths has supplanted its forebears.

Radiation is a mutagenic agent. In microbiology and plant breeding, the incidence of mutations has been purposely increased by irradiation, the large number of deleterious ones being discarded and the very few that are beneficial retained for breeding [5]. The process is one of trial and error in the absence of complete knowledge of the genotype concerned [6]. Although sometimes useful, this type of plant breeding is unlikely to supplant conventional methods.

The LD_{50} or MLD (p. 289) varies widely in plants and seeds and is, for instance, about 1,000 rads for pine trees and about 100,000 rads for mustard seeds [7]. Obviously, doses well below the LD_{50} must be used in plant breeding. Mutation rates are probably of the order of 1 in 10^8 genes for each rad, for animals, lower organisms and plants. Many hundred chromosome loci may be involved in the determination of one attribute, such as resistance to a particular disease [8].

The radio-sensitivity of a particular plant material for a particular gene mutation depends on a number of factors: oxygen tension, temperature before, during and after irradiation, moisture content, chemical treatments, nutritional state, age and storage conditions [4].

Irradiation of barley has produced strains with stiff straw, others which mature early or which have a high protein content and yet others which thrive on poor soil [8].

When large-scale production of penicillin was introduced, the best available strain yielded 100 to 200 units of penicillin per ml of culture fluid. Conidia of this strain were irradiated with X-rays, and although most strains were less productive, one was found which produced 300 to 500 units per ml. Subsequently this strain was irradiated again, this time with ultra-violet light, and a second mutant selected which produced more than 1,000 units per ml. Thus, in all, nearly a ten-fold increase in productivity was obtained by the use of irradiation to induce mutations [9].

Entomological Control

The control of insect pests by insecticides may be technically difficult in some situations, such as grain and timber infestation. Radiation has been used in these cases as there is no chemical contamination and the insects do not have to be in contact with the destructive agent [10, 11].

The dose to kill 50 per cent of weevils in one day is 0·3 to 0·5 Mrads; to kill in two or three weeks, the dose is 50 krads, and this dose is lethal to most insects [12, 13, 14, 15]. The sterility dose for weevils is 12 to 15 krad, but there are technical problems in irradiating grain to even this lower dose, as the process is uneconomic unless very large tonnages can be irradiated to a near-uniform dose in a short time.

Insect pests can sometimes be controlled by the *sterile male technique*. Male insects or pupae are irradiated with a sterility dose and then released into the native population to mate with unirradiated females. The process has to be repeated a number of times. It is often economic to reduce the insect population with insecticides before applying the sterile male technique. The method has been used to control the white grub (*Melolontha vulgaris F*) [14], fruit flies, tsetse flies and screwworm flies [3, 15]. An exposure of 2·5 kR will sterilize male screwworm pupae. The sterile male technique was used to eradicate screwworm flies from an island and later from the whole of Florida and the south-eastern United States. This was a large operation in which 50 million flies were reared each week. Their weekly food requirements were 40 tons of meat, 35 gallons of honey and 4,500 gallons of beef blood. Eradication took eighteen months and cost about six million dollars, but screwworms were estimated to have cost the cattle industry over ten million dollars per year in Florida alone [4].

Preservation of Food

A major cause of foodstuff spoilage is the growth of bacteria, making the food unpalatable and sometimes poisonous, or with its nutritional value impaired. Other causes of spoilage are mould growth and the sprouting of vegetables. Irradiation can prevent all these [16, 17, 18] although it is not always practicable. Large doses cause the breakdown of organic substances in the food and the breakdown products often cause a change in taste.

There is no reason to suppose that irradiated foods are harmful, but as a precaution, long-term tests have been carried out on many generations of laboratory animals; human volunteers have eaten irradiated foods under close supervision. No ill effects have been detected in humans, but when animals have lived on diets with a large component of irradiated foods, they have sometimes developed symptoms thought to be due to vitamin deficiencies. Public health authorities in most countries consider that existing regulations prohibiting processes which render food injurious to health, are adequate for the protection of the public. But regulations in the Federal Republic of Germany prescribe the use of ionizing radiation for food treatment, while Canada and the U.S.S.R. specifically permit the irradiation of potatoes [4].

Cooked meats are liable to bacterial contamination. When exposed to 0·1 MR, the storage life of some cooked meats has been increased fourfold [13]. Foods of this kind can be pre-packed in plastic envelopes under clean but not aseptic conditions, leading to economies in transport and storage. Irradiation of imported meats has been advocated to reduce outbreaks of food poisoning.

An experimental dose of 1 Mrad of 3 MeV x-rays destroyed the orange-yellow colour of butter although there was only a slight increase in peroxide value [2]. This dose was probably unnecessarily high for preservation purposes.

A few thousand rads will delay or prevent the sprouting of certain vegetables, particularly potatoes, onions, carrots and mangolds [19, 20]. The optimum dose for potatoes depends on the variety and on storage temperature and is between 5,000 and 12,500 rads [4].

Most moulds are aerobic and therefore surface growths, so that they can be inhibited by non-penetrating radiation with no risk of altering the taste. However, it is often technically easier to reach all surfaces by using penetrating X- or γ-radiation. Bilberries have been treated in this way with a dose of 0·3 Mrads; there was no loss in flavour and only a small loss in vitamin C, but the storage life was increased three-fold.

Sterilization

Radiation sterilization has obvious advantages for materials such as rubber, plastics and some pharmaceuticals, which are damaged by heat. Another advantage is that this form of sterilization can be carried out after packing, making expensive sterile packing processes unnecessary.

A dose of 2 to 5 Mrads will reduce the probability of finding a living organism by a factor of at least 10^{10}, and is acceptable as giving complete sterility unless the initial contamination was excessive [12]. The radiosensitivity of bacteria is, however, markedly dependent on oxygen tension [21]. About 2 Mrads have been used to sterilize eye-drops, surgical sutures, dressings and disposable plastic catheters.

Partial sterilization at doses of about 0·1 to 0·2 Mrads is adequate for hospital blankets. Untreated blankets can be responsible for transmitting minor infections from patient to patient, even when care is taken to prevent serious diseases being spread in this way. Heat sterilization damages woollen blankets, and some hospitals have substituted cotton blankets which may be boiled. However, partial sterilization of woollen blankets by irradiation is now an economic proposition [13].

Initiation of Chemical Reactions

Energy absorbed from radiation may break a molecule into two or more fragments, some of which may be ionized. These fragments are likely to be highly reactive, and they may drift away to combine with other radicals or molecules to form new chemical material. When water is irradiated, the molecule is broken into charged ions, and these highly reactive and short-lived radicals may react with solute molecules, rather than recombining. Sometimes radiation initiates a chemical reaction which otherwise takes place only under special conditions, such as high temperature.

Even when chemical effects are useful, radiation as an agent is unlikely to compete economically with traditional methods in the chemical industry, unless there are special circumstances [22]. But there are exceptional cases where radiation has something to offer; it can promote the reaction between benzene and chlorine to form the insecticide benzene hexachloride, and can compete in this with the more traditional agent of ultra-violet light. Improved fuel oils have been obtained by radiation-induced reactions and cracking of hydrocarbons [12, 23].

Some long-chain molecules (polymers) consist of many sub-molecules (monomers) joined end to end. The process of forming the chains is known as *polymerization.* Irradiation sometimes promoted polymerization. Soft X-rays generated at 50 kVp from a beryllium-window tube have been used to polymerize acetylene and ethylene, both as pure gases and in mixtures of helium, argon and xenon. Noble gases accelerate the X-ray polymerization of acetylene. Other unsaturated hydrocarbons have been polymerized by this method [24].

Useful Effects on Materials

Degradation or polymerization may result in changes in the physical properties of a material rather than in chemical reactions. Thus, radiation doses of 5 to 100 Mrads may fracture the bonds of some polymers holding large molecules together, and the material may turn brittle and it may discolour;

TABLE 12.1 *Effect of Radiation on Polyethylene*
(From Carrol, J. G. and Bolt, R. O., *Radiation Effects on Organic Materials,* New York, Academic Press, 1963)

Dose (Mrads)	*Effect*
1 to 3	Grafting
5 to 7	Polymerization of monomer
5 to 25	Cross-linking
80 to 100	Degradation

it may undergo degradation. When the bonds of other polymers are broken in this way, they rejoin to form long chains or they may join to form cross-links between chains.

The processes of polymerization and cross-linking raise the distortion temperature and the melting-point and increase the strength of a material. Polymerization is the process underlying vulcanization and curing of rubber. Cross-linking can be achieved chemically as well as by irradiation, but in the former method the cross-links are formed by extraneous chemicals and form weak links. Radiation-induced cross-linking therefore results in a stronger material, although it may be more expensive [25, 26].

The effect of irradiation on plastics depends on temperature, but not, within wide limits, on dose rate [27, 28]. Polythene, cross-linked by irradiation, is useful as small inserts in valveholders, where they are able to withstand the high temperature associated with soldering [12].

Graft copolymerization can be achieved by irradiation [29, 30]. In this process, a monomer (one molecule of the chain forming a polymer) of one substance becomes incorporated in the polymer of another. The polymer is soaked in a solution of monomer and irradiated. Either a *block copolymer* or a *branch copolymer* may result. In the former, the polymer chain breaks and the monomer becomes inserted in it. In the latter, a bond is formed between a monomer molecule and one of the molecules of the polymer chain [25]. A few megarads is required for copolymerization (Table 12.1). The rate of grafting is positively correlated with dose rate, temperature and monomer concentration [31]. Radiation-induced graft copolymerization has been used to bond a layer of styrene to polytetrafluorethylene, to improve its adhesive properties [12].

Literature References

1 MOLE, R. H., "Physical parameters in radiobiology. Dose rate and time factor", in *Radiation Effects in Physics, Chemistry and Biology*, Ed. Ebert, M. and Howard, A. (Amsterdam, North-Holland Publ. Co., 1963).

2 DUNN, C. G. *et al.*, "Biological and photo-chemical effects of high energy, electrostatically produced Roentgen rays and cathode rays", *J. Appl. Phys.*, **19,** p. 605 (1948).

3 HORNE, T. and BROWNELL, L. E., "The use of radiation sources for insect control", in *Radioisotopes and Radiation in Entomology*, ST 1/PUB/21/9 (Vienna, International Atomic Energy Agency, 1963).

4 JEFFERSON, S. (Ed.), *Massive Irradiation Technique* (London, George Newnes, 1964).

5 FAVRET, E. A., "Contribution of radio-genetics to plant breeding", *Int. J. Appl. Rad. Isot.*, **13,** p. 445 (1962).

6 DAVIES, D. R. and WALL, E. T., "Artificial mutagenesis in plant breeding", *Nature*, **182,** p. 955 (1958).

7 EVANS, H. J., "Cellular and macroscopic damage in relation to cytological observations", in *Radiation Effects in Physics, Chemistry and Biology*, Ed. Ebert, M. and Howard, A. (Amsterdam, North-Holland Publ. Co., 1963).

8 CARTER, T. C., "Genetics—agricultural applications", in *Progress in Nuclear Energy* (Series VI, Biological Sciences, Vol. I), Ed. BUGHER, J. C., COURSAGER, J. and LOUTIT, J. F. (London, Pergamon, 1956).

9 STANIER, R. Y., DOUDOROFF, M. and ADELBERG, E. A., *General Microbiology* (London, Macmillan, 1962).

10 CORNWALL, P. B., "The disinfestation of foods, particularly grain", *Int. J. Rad. Isot.*, **6,** p. 188 (1959).

11 CORNWALL, P. B., CROOK, L. J. and BULL, J. O., "Lethal and sterilizing effects of gamma radiation on insects infesting cereal commodities", *Nature*, **179,** p. 670 (1957).

12 CROWLEY-MILLING, M. C., "The application of irradiation in industry" *J. Instn. Elect. Eng.*, **107,** Pt. A., p. 111 (1960).

13 PUTMAN, J. L., *Isotopes* (London, Penguin, 1960).

14 INTERNATIONAL ATOMIC ENERGY AGENCY, *Radiation and Radioisotopes Applied to Insects of Agricultural Importance*, ST 1/PUB/74 (Vienna, 1963).

15 KNIPLING, E. F., "The use and limitation of isotopes and radiation sterility in meeting insect problems", *Int. J. Appl. Rad. Isot.*, **13,** p. 417 (1962).

16 COLEBY, B. *et al.*, "Radiation processing of foods in Britain", *Nucleonics*, **16,** p. 188 (1958).

17 Silow, R. A., "The significance of nuclear science for food and agriculture", *Int. J. Appl. Rad. Isot.*, **13,** p. 313 (1962).
18 Food and Agriculture Organisation, *Report on the Meeting on the Wholesomeness of Irradiated Food* (Rome, 1962).
19 Burton, W., Horne, T. and Powell, D. B., "Effect of gamma radiation on stored potato tubers", *European Potato J.*, **2,** p. 105 (1959).
20 Cornford, C. E. and Powell, D. B., "Effect of gamma rays on stored mangolds", *Nature,* **180,** p. 996 (1957).
21 Hollaender, A., Stapleton, G. E. and Martin, F. L., "X-ray sensitivity of *E. coli* as modified by oxygen tension", *Nature*, **167,** p. 103 (1951).
22 International Atomic Energy Agency, *Industrial Uses of Large Radiation Sources,* ST 1/PUB/75. Proceedings of IAEA Conference 27th–31st May 1963 (Vienna, 1964).
23 Bolt, R. O. and Carroll, J. G., *Radiation Effects on Organic Materials* (New York, Academic Press, 1963).
24 Lind, S. C., *Radiation Chemistry of Gases* (New York, Reinhold).
25 Ellis, R. H., *Nuclear Technology for Engineers* (New York, McGraw-Hill, 1959).
26 Charlesby, A., *Atomic Radiation and Polymers* (Oxford, Pergamon, 1960).
27 Charlesby, A. and Davison, W. H. T., "Temperature effects in the irradiation of polymers", *Chem. and Ind.*, **8,** p. 232 (1957).
28 Todd, G. and Wild, G. A., "Temperature dependence of irradiation damage to polythene", *Nature*, **199,** p. 172 (1963).
29 Ballentine, D. G., "Role of radiation in graft and block copolymer synthesis", *Mod. Plastics*, **35,** p. 171 (1957).
30 Chandler, H. W., Henley, E. J. and Trachtenberg, E. N., "The mechanism of radiation-induced graft copolymerization", *Int. J. Appl. Rad. Isot.*, **13,** p. 239 (1962).
31 Restaino, A. J. in *Effects of Radiation on Materials*, Ed. Harwood, J. J. *et al.* (New York, Reinhold, 1958).

13 | *X-ray Equipment*

The basic operations in x-ray production are, firstly, the generation of free electrons, secondly their acceleration to a high velocity, and finally their stopping within a target. To accomplish this, the electrons should move in a vacuum, otherwise their velocity is diminished by collisions with the gas molecules. The evacuated vessel is known as the x-ray tube.

The usual method of electron production in modern x-ray tubes is by heating a tungsten filament to about 2,000°C in a high vacuum. Oxide-coated wires are not used in x-ray tubes because the oxide is easily stripped off by positive ion bombardment from residual gases. This is a consequence of the high tube voltage. However, an electron microscope[34] working at 50 kVp is available using either an oxide-coated platinum filament, or an oxide-filled platinum tube.

Early x-ray tubes were gas-filled. In a gas-filled x-ray tube the electrons are supplied by ionizing the residual gas in the tube and by bombardment of a cold cathode by the positive ions. *Cold emission* is used also in *flash radiography* [1]. This involves the emission of electrons from a cold metal surface under conditions of intense electric field strength, *in vacuo*.

In many tubes of a conventional type, the target consists of a tungsten button let into a copper block; copper is used because it is easy to fabricate and a good conductor of heat. Provision must be made for cooling the target as comparatively large amounts of heat must be dissipated, owing to the low efficiency of conversion of electrical energy to x-rays. The x-rays are emitted from the tube through a window which may be made of materials with low absorption coefficients, such as beryllium.

The electrons are accelerated by applying between filament and target a high voltage produced by a transformer or analogous device, e.g., a Van de Graaff generator, and the circuit may include other components, particularly rectifiers and capacitors to improve the wave-form of the applied voltage.

X-rays Generated Below 500 kVp

X-RAY TUBES

The early experiments of Langmuir on electron emission from hot tungsten cathodes led to the idea of a hot-cathode x-ray tube, although experience had suggested that the electron current would be erratic.

The first stable hot-cathode x-ray tube was described by Coolidge [2] in 1913. The tube was bulbous and made of glass, and the filament consisted of a flat spiral with a molybdenum focusing tube heated by an insulated storage battery. Great care was taken by Coolidge to evacuate the tube as perfectly as possible, and to degas the electrodes and tube by baking and by passing heavy currents; the result was an x-ray tube with novel advantages, such as a steady focal spot, sharp focus, stable voltage and high rating.

Pumping of X-ray Tubes and Getters

The manufacture of x-ray tubes is a remarkable process involving many elaborate techniques. It is not possible to mass-produce them and the labour is skilled and expensive.

The tube envelopes are made of a hard borosilicate glass, resistant to electrical and mechanical shock. The metal parts of the envelopes are made of special alloys, such as chrome–iron, which have the same coefficient of thermal expansion as the glass. After the tube is assembled, it is heated to remove absorbed gases in the envelope and in other components. After lengthy pumping, residual gases are removed by a getter, which is a heated reactive wire.

Any remaining gas, or traces of gas evolved while the tube is in use, may be reabsorbed by the metal components, thus preserving the vacuum. Once a tube has become appreciably "gassy", it is virtually impossible to restore the original vacuum. However, it is sometimes possible to effect a temporary improvement in a tube that has evolved a small amount of gas, by bringing the kilovoltage up very slowly with the filament switched off, so that the gas is ionized and the charged particles may then cling to the insulated parts of the tube.

Continuously Evacuated X-ray Tubes

Continuously pumped x-ray tubes are sometimes advantageous. For example, in x-ray diffraction work when various radiations may be required, it is not always economical to store a range of sealed-off tubes each with a different anode, particularly in view of the rather short shelf life of some tubes. Moreover, it is easy to repair a defect, such as a burnt-out filament, in a demountable tube, and this is not possible with a sealed-off tube. Contamination of tungsten on the target or window may be easily removed. Because they are easy to repair, the tubes may be run at higher ratings than comparable sealed-off tubes. On the other hand, continuously evacuated tubes have the disadvantage that they require more regular maintenance, especially with regard to the pumping gear. Also, they are rather less stable, as it is difficult to obtain a really good vacuum without prolonged pumping. The tube takes time to pump out, and when closing down, the system should be allowed to cool before the vacuum is reduced.

Continuously evacuated equipment has certain special applications, such as with fine-focus tubes and very high-voltage equipment. In general, however, it is not widely used owing to lack of mobility and the difficulty of obtaining a high enough vacuum without the use of a getter and other special degassing methods.

The Filament

The filaments of x-ray tubes are normally made of tungsten wire. The fabrication of tungsten into wires and ingots presents difficulties, and it was not until 1909 that Coolidge first made ductile tungsten, using very pure materials and a suitable ingot crystal structure. The metal is now made into bars by the process known as *sintering*: fine tungsten particles are intimately pressed together and then heated [3]. Modern tungsten wire is drawn through dies, has a fibrous structure and can be readily formed into helixes. On heating in the x-ray tube, however, the tungsten recrystallizes and becomes brittle owing to precipitation of impurities at the crystal boundaries. This makes it extremely difficult to adjust a used filament in a demountable tube.

Clean tungsten wire is not a very efficient source of electrons, but owing to the low vapour pressure of the metal it is possible to heat the filament to the required high temperature without excessive loss of metal. This high temperature also prevents gas absorption which might reduce electron emission. A constant temperature along the filament is essential for a uniform x-ray field. Filament failure is usually due to a "hot spot" developing, and this process is greatly accelerated if tungsten oxide has been allowed to form, as it has a higher vapour pressure than the metal.

In radiography the filaments are normally long thin coils, the tangential projection of which on the anode gives an effective small focal spot. Flat spiral filaments are sometimes used where a sharp focus is not as important as a high output, for example in radiotherapy. Hair-pin shaped filaments can produce an extremely sharp focus.

The filament is heated by a few amperes supplied from a step-down transformer. The anode (or tube) current is controlled by the filament temperature, which is sensitive to the supply voltage. The filament transformer is often at the full kilovoltage above or below earth potential.

Tube Foci

In industrial and medical radiography a sharp focus is essential and this is often obtained by using a cathode in the shape of a long wire spiral, from which the electrons are focused by a focusing hood to form a line source of x-rays on the anode.

The angular relationship between electron beam, anode and x-ray beam determine the apparent size of the source. By judicious choice of the variables, small, nearly square, sources can be obtained.

A compromise, however, may have to be made between size of source and other factors. One of these is uniformity across the beam, for which there is an optimum target angle (Chapter 5); another is high output. Some tubes have two foci, one fine and one broad. The former gives higher resolution and the latter higher exposure rates.

Another type of tube[35] with two foci has been designed for stereo-fluoroscopy. This is a rotating anode tube with the two foci suitably spaced for stereo projection.

A number of sealed fine-focus tubes have been produced with foci as small as 0·2 mm diameter; these tubes usually have magnetic or electrostatic focusing arrangements. A continuously evacuated tube[36] (Figure 13.1) designed by Ehrenburg and Spear [4] has an oil-cooled anode and a focusing hood with a variable biasing voltage. The focal spot is about 0·04 mm diameter. Another design used by Cosslett [5] consists essentially of an electron microscope assembly with a foil target, giving a spot with a diameter of 0·001 mm, the X-radiation emerging through the foil target.

The rod anode tube is an interesting modification designed for the radiography of confined spaces such as boilers. A coil focuses the electrons, which move along a narrow earthed metal tube on to a hooded anode (Figure 13.2), and the focal spot size will vary with coil current.

The focusing hood consists of a cup of thin metal, usually molybdenum, with an aperture corresponding to the shape of the filament. The filament is set just below the level of the flat surface so that the hood will repel the electrons, thus exerting a focusing effect upon them. The hood is usually at the

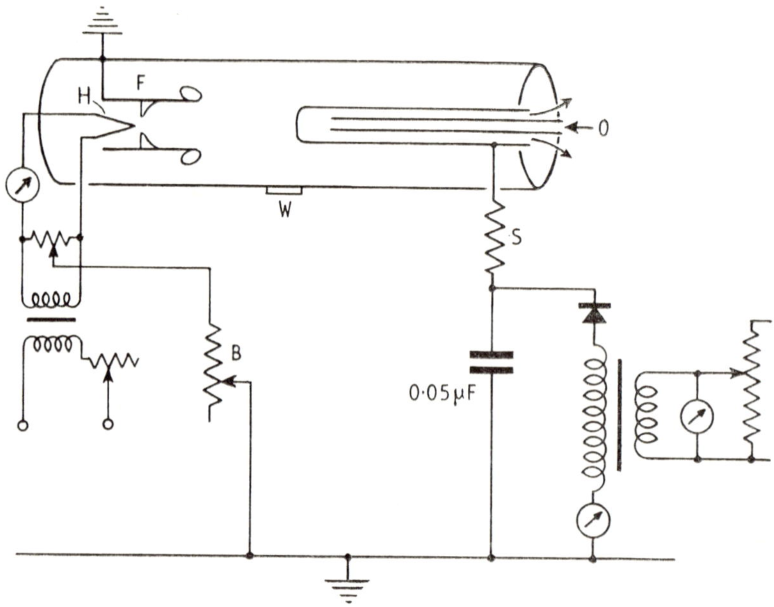

FIGURE 13.1 *Fine-focus Tube and Circuit*
B, biasing voltage; *F*, focusing hood; *H*, hairpin filament; *O*, oil cooling; *s*, surge-limiting resistance; *W*, window.
(From Ehrenberg, W. and Spear, W. E., *Proc. Phys. Soc. B*, **64**, p. 67, 1951. Courtesy of The Institute of Physics and The Physical Society)

same potential as the filament, although in some tubes the focusing effect is varied by applying a biasing voltage between the hood and the filament to vary the sharpness of the focus. The focusing effect depends on the tube design. For example, the tube may be made to focus more sharply by increasing the distance from the filament to the surface of the focusing hood, when a higher filament temperature will be required for a given anode current. Near the cathode the velocity of the electrons is relatively low and they move in accordance with the electric field direction, whereas near the anode the electrons have a higher velocity and consequently respond less to the same electric field acting along the same path distance.

As the focusing effect is a function of the velocity of the electrons, good focusing is easier to obtain when the electron velocity is constant, that is, with a constant kilovoltage supply to the tube.

The Anode and Hood

For general radiological use the characteristic components of the radiation are of little importance and the useful beam is composed of white radiation. The exposure rate is proportional to the atomic number of the target material and to the tube current (p. 221). Choice of target material is limited to metals of high atomic number and good thermal properties, such as platinum and tungsten. The material first in general use was platinum (atomic number 78) with a melting point of 1,173°C and thermal conductivity of 0·17 cal/cm sec °C at 18°C. Indeed, during the first world war, X-ray

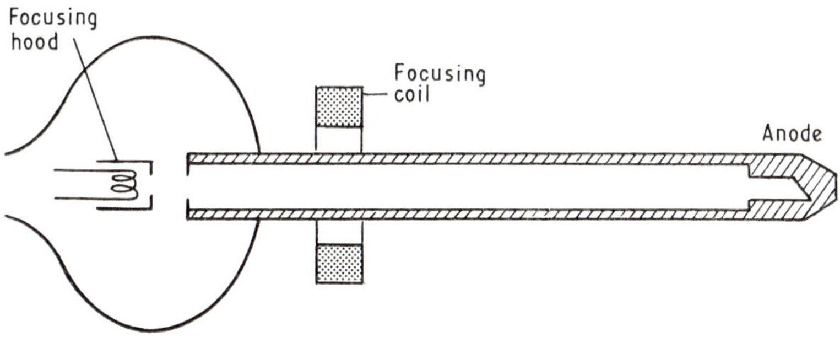

FIGURE 13.2 *Rod Anode Tube*

tubes were the largest users of platinum. Each anode then cost about twelve U.S. dollars.

Tungsten is now much cheaper than platinum; it has a much higher melting point (3,400°C) and an exceptionally low vapour pressure at high temperature. Its atomic number is 74 and its thermal conductivity is 0·48 cal/cm sec °C at 17°C.

In applications such as crystallography, it is important to be able to vary the quality of the radiation and in particular the characteristic components. Consequently, anodes are made of a variety of metals such as cobalt, chromium or iron. Some of these metals can be brazed directly on to the copper backing, but this is not always possible. Chromium, for instance, may be plated on to copper, using nickel as an intermediary keying metal. This can have the disadvantage that characteristic radiation of the underlying metal may appear as the chromium wears.

A metal hood may be fitted over the anode. The primary electron beam passes through a hole in the hood and the useful X-ray beam is emitted through a second hole at right angles to the first, and covered with a thin metal foil (Figure 13.3). The hood confines electrons reflected from the anode, which might otherwise strike the glass; it also acts as a primary X-ray beam limiter.

Another method of reducing scattered radiation is to place the cathode near the anode. However, this limits the voltage which may be applied to the tube and restricts the useful X-ray beam divergence. Furthermore, heat from the anode affects the filament temperature and therefore the anode current stability.

The energy of spurious radiation emitted from parts of the tube other than the anode is lower than that from the anode, because the scattered electrons causing this radiation have lower energies. These energies must be distinguished from those of X-rays arising from the anode and filtered by the casing, which are harder than the primary beam (Chapter 14).

Anode Temperature

The conversion of the electron energy to X-radiation has an efficiency of only about 1 per cent, and therefore high output has associated problems of heat dissipation.

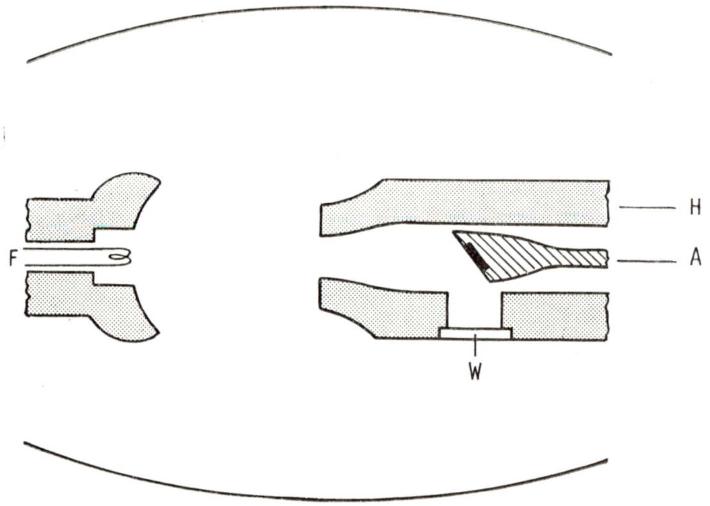

FIGURE 13.3 *X-ray Tube with Anode Hood*
A, anode; *F*, filament assembly; *H*, anode hood; *W*, window.

Consider a short time interval during an X-ray exposure. The heat produced at the anode is proportional to the product of the charge flowing to the anode and the tube voltage. Both charge and voltage may vary continuously and the total heat produced is proportional to these products summed over the whole exposure. The resultant anode temperature is a function of the heat produced, the thermal capacity of the anode and the heat loss.

The temperature of the anode is the main factor limiting X-ray output. Local overheating will cause target pitting, with loss in beam intensity and distortion of the focal spot. At very high temperatures, evaporation of the tungsten will cause evolution of gas and the deposit of tungsten on the window and other parts of the tube. Tungsten deposits on the glass cause stray electric fields which distort the focal spot. These mirror-like deposits should not be confused with the brown coloration due to effects of the radiation on the glass. The tungsten deposits may not be earthed and then associated electrostatic stresses can cause the tube to crack [6].

If the temperature of the anode becomes too high, electron emission will give rise to inverse current, if the tube is self-rectified. This current often converges on to the glass near the cathode, causing fluorescence and overheating. The glass may crack and the thin stream of air entering the tube will become ionized. This is visible as a spark, giving the false impression that the spark caused the puncture, instead of vice versa.

Tube Rating

Overheating is avoided by specifying the tube operating conditions: voltage, current and exposure duration. The values of these depend on the thermal capacity of the anode and the heat dissipation. When the tube is run continuously so that a steady state is reached, the limiting factor is the product of current and voltage. For instance, a tube may be run continuously at 5 mA and 200 kVp, or at 10 mA and 100 kVp with the same temperature effect.

In intermittent use, cooling between exposures allows higher instantaneous ratings. In the extreme case of flash radiography (p. 227), exposure times are so short that currents of thousands of amperes are safe.

The wave-form of the current affects both the tube rating and the useful X-ray output. At constant potential, the X-ray output for a given heat production is uniform. But for a fluctuating voltage, a proportion of the X-ray output is of unwanted lower energy, although it has associated heat production.

The filament of a tube must be run at a slightly higher temperature in a half-wave circuit than in a constant-voltage circuit to give the same mean current. This is because the same average current has to be produced in less than half the time, so that the peak current is more than twice as great.

Tube Cooling

Where the total heat production is small and exposure infrequent, the thermal capacity of the anode will preclude overheating. Usually, however, heat losses have to be enhanced by special design. This may be simply a finned radiator outside the tube and connected by a thick rod to the anode.

In modern shock-proof tubes, the space between the tube and the earth shield is completely filled with electrically insulating oil, and the heat is conducted into the oil using a copper ball as heat exchanger. An expansion chamber in the form of a bellows is provided for the hot oil. In some tubes, forced circulation of the oil is induced by pumping. The oil may be water-cooled.

Direct water-cooling may be used for tubes run at high continuous rating, although this is possible only where the anode is earthed. The water must not boil, as steam is inferior to water as a conductor of heat. Boiling will cause also scaling of the anode. In cold weather, it may be necessary to check that a frozen waste pipe does not stop the circulation of cooling water.

The Rotating-anode Tube

Another method of limiting anode temperature is to rotate the anode. This spreads the heat over an annulus while keeping the effective focal spot small.

The production of these tubes (Figure 13.4) is a remarkable industrial achievement. The anode is a dynamically balanced disc of tungsten rotating freely using a ball-bearing lubricated with a soft metal.

The disc is connected to a molybdenum shaft of low thermal conductivity, cast into a copper rotor activated by a coil outside the tube.

Rotation is at mains frequency, so that during an exposure of 1/50th second the anode completes one revolution. As the anode disc is made of tungsten, which has a high melting point, it can withstand higher temperatures than stationery anodes incorporating copper. In the former, anode cooling is primarily by radiation and in the latter by conduction.

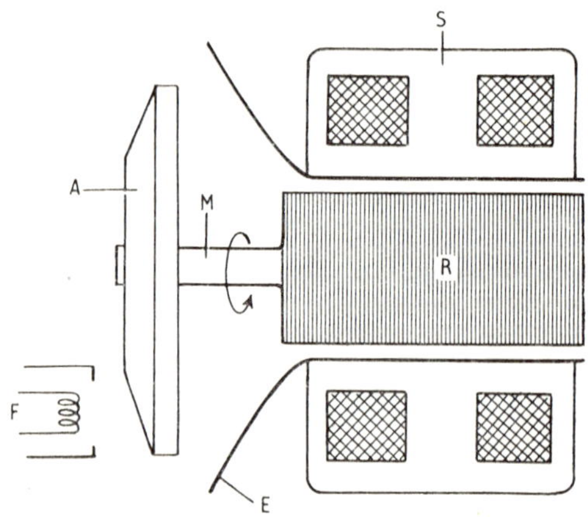

FIGURE 13.4 *Rotating-anode Tube*
A, rotating tungsten anode; *E*, glass envelope; *F*, filaments (fine and medium); *M*, molybdenum stem; *R*, copper rotor; *S*, three-phase stator.

If the anode fails to rotate at its correct speed during exposure, overheating will roughen the surface, or it may melt or crack the disc. Safety circuits should be included to prevent this.

Fine-focus Tubes

Small objects cool more rapidly than larger objects of the same shape. This is because the heat to be dissipated is proportional to the volume, but cooling is dependent on the surface area, and the ratio of surface area to volume decreases with increase in size.

Because of this, fine-focus tubes are possible with a small intense X-ray source, when high resolution and short exposure times are required. Compared with a conventional tube, a fine-focus tube operates at a lower anode current and therefore lower total X-ray output, but at a higher current per unit area of focal spot, and therefore higher X-ray output per unit area of focal spot. This allows for the more efficient heat dissipation from the small focal spot.

Operating conditions for fine-focus tubes are particularly critical as they are usually run at as high a rating as possible. Consequently, the anode may become locally overheated if the apparently low anode current is exceeded.

Field Emission Tubes

As the temperature of the cathode is lowered, electron emission requires a higher anode voltage. Intense emission can be obtained at room temperature if the electrostatic field is high enough. This is known as cold emission, and is the basis of the field emission tube. The field ion microscope operates on a similar principle. The acceleration of the emitted electrons depends on the potential gradient, which can be increased for a given potential difference by using highly curved surfaces.

The cathode may consist of a comb of needles surrounding a tapered anode (Figure 13.5). Fields of the order of 10^7 V/cm can produce continuous current densities of 10^7 A/cm^2, which is a million times higher than with conventional tubes. As the current response to the application of the voltage

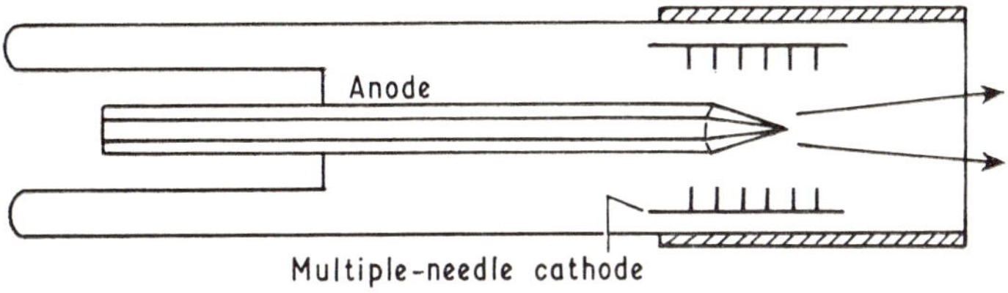

FIGURE 13.5 *Field Emission Cold Cathode Tube*

is nearly instantaneous, the chief use of field emission tubes is with pulsed circuits to give very short and intense exposures [1] in flash radiography.

Gas-filled X-ray Tubes

Besides the hot cathode and field emission methods, another way of producing an electron beam in an X-ray tube is gas ionization. This was the original method and was in common use before the first world war. The tube is evacuated to a pressure of approximately 10^{-4} mm of mercury. When a high voltage is applied between the cathode and the anode, the residual air becomes ionized and produces streams of electrons and positive ions. The positive ions produce further electrons on striking the cathode, and thus the ionization builds up, the final anode (electron) current being determined by the gas pressure, the applied voltage and the tube geometry. The concave curvature of the cathode produces an electrostatic focusing effect which may give a sharp focal spot. When run under correct conditions, a gas X-ray tube is self-rectifying, although a separate rectifier greatly improves stability.

The gas is slowly absorbed within the tube, and thus requires to be replaced by admitting more air through some leak arrangement.

The gas tube tends to be unstable and erratic in operation with a wandering focal spot. A further disadvantage is that the voltage and current are interdependent in a complex fashion. However, these tubes are sometimes used in X-ray diffraction work, as they give a beam free from tungsten characteristic radiation contamination, and can be made cheaply in a workshop with limited facilities.

The Finch [7] form of electron diffraction camera resembles the gas-filled X-ray tube in operation, in that the electron stream is produced by a gas discharge under similar conditions. In general, however, it would seem that the day of the gas-filled X-ray tube has passed, except for students as a demonstration of a wide range of principles and experimental methods.

TUBE CASING

The tube casing has a number of functions. It supports the X-ray tube, the cable terminations, applicators and collimators. It must be proof against leakage of insulating oil. It provides mechanical protection for the tube and an earthed shock-proof shield for the safety of the operator.

Radiation protection is ensured by a carefully designed lead lining, the housing itself being made of cast light alloy or welded sheet metal.

CABLES

Where some mobility of the tube is required, it is usual to lead the high voltage to it through a flexible cable (Figure 13.6). The two leads carrying the filament voltage are embedded in this cable and are insulated by rubber

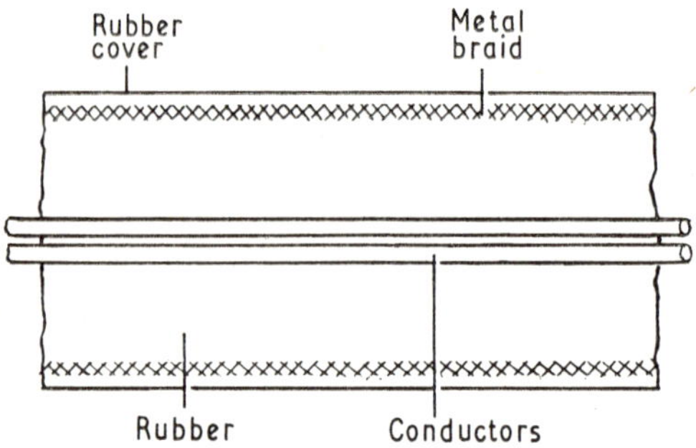

FIGURE 13.6 *H.T. Cable*

from the earthed braided sheath. If the cable is bent through an acute angle, the insulation may break down. Cables are frequently a cause of failure in X-ray apparatus.

A common site of cable faults is at the cable ends, or "terminations", which should be kept clean and lightly greased with a high-voltage grease to prevent corrosion or the ingress of moisture. The filament current is relatively heavy and the filament voltage low, so that contact resistances must be kept to a minimum to avoid sparking.

Cables give better performances with a constant voltage than with alternating voltages because, in the latter case, dielectric losses in the rubber cause heating and loss of insulation.

The capacitance of the cable depends on its length, the radius of its conductors and the dielectric constant of the rubber. At mains frequency, its contribution to the impedance of the high-voltage circuit is usually small compared with that from the inductance and resistance of the other components. However, an a.c. milliammeter in the anode circuit may demonstrate the presence of some current flowing through the cable insulation. When very long cables are used, the extra capacity introduced will alter the tube current wave-form and hence the exposure rate.

HIGH-VOLTAGE TRANSFORMERS

Below 500 kVp, the tube voltage is produced by means of a high-voltage transformer. This consists essentially of two coils of wire, the primary and the secondary. These are wound so that the magnetic field produced when a current passes through the primary winding also passes through the secondary. When the current in the primary alters, the primary magnetic field alters correspondingly, and this produces a change in the electromotive force in the secondary winding, which in turn produces a secondary current whose magnetic field opposes the change in the primary field. Thus, when an alternating voltage from the mains is applied to the primary winding so that the primary current is continuously altering, the secondary voltage will fluctuate in a corresponding manner. The greater the flux-linkage, the greater the induced electromatic force, and where the flux-linkage is complete the ratio of the maximum value of the E.M.Fs can be shown to be approximately

$$\frac{E_1}{E_2} = \frac{N_1}{N_2} \qquad (13.1)$$

where N_1 = number of primary turns
N_2 = number of secondary turns
E_1 = maximum primary EMF
E_2 = maximum secondary EMF

The efficiency of the linking process is greatly increased by the presence of a soft iron core which concentrates the flux lines. The flux linkage is greatest when the iron core is designed so that the flux lines in it form continuous loops, threading primary and secondary. The core is made of thin insulated laminations of special iron alloy, arranged to reduce eddy currents to a minimum.

The windings are of insulated wire. The primary is wound on a limb of the core and the secondary is arranged to fit over it, and is separated from the primary by a tube of insulating material. The secondary is wound in layers, the wire being varnished to retain it in position and improve insulation. The sections are separated by sheets of insulating paper. The ends of the secondary are led out through stand-off insulators. The windings are usually completely immersed in transformer oil to prevent *silent discharge*, and also to dissipate heat. Where weight is important, a high-pressure insulating gas may be used instead of oil. It is important to preserve the insulation of the separators by keeping them cool.

Breakdown in insulation is seldom sudden, provided that the transformer is run within the correct rating; it usually occurs following gradual deterioration due to the entrance of moisture or dust, or by corona discharge. Corona discharge will produce ozone and nitrogen oxides which attack many insulating materials, in particular rubber, which therefore is not used for the insulation of leads in x-ray transformers. Oil and gas have an advantage over solid substances for insulation, in that they are self-healing if momentarily punctured by a voltage surge. The insulation performances of an x-ray transformer may be checked by measuring the primary current under

On the other hand, very small doses have sometimes been reported to have stimulating effects [15, 16]. For example, when hens' eggs were exposed to 2·9 R, given over the whole incubation period, the hatching of chicks was increased by 2·7 per cent and the egg-laying capacity of the young hens rose by 7 to 10 per cent compared with the non-irradiated controls [17]. There is much discussion about the interpretation of this kind of experiment, and many workers hold the view that the stimulating effect of small doses has not been proved. Control animals may die from diseases different from those

TABLE 14.2 *Annual Maximum Permissible Doses for Adults Exposed in the Course of their Work*
(From ICRP Publn. 9, Oxford, Pergamon, 1966)

Organ or tissue	*Class of person*	*Rems (annual)*
Gonads, red bone-marrow (and whole body)	All except pregnant women*	5
Skin, bone, thyroid	All	30
Hands and forearms, feet and ankles	All	75
Other single organs	All	15

* After a pregnancy has been diagnosed, the exposure of the woman should be such that the dose to her foetus, accumulated during the remaining period of pregnancy, does not exceed 1 rem.

TABLE 14.3 *Annual Dose Limits for Members of the Public*
(From ICRP Publn. 9, Oxford, Pergamon, 1966)

Organ or tissue	*Class of person*	*Rems (annual)*
Gonads, red bone-marrow (and whole body)	All	0·5
Skin, bone	All	3
Thyroid	Over 16 years of age	3
Thyroid	Under 16 years of age	1·5
Hands and forearms, feet and ankles	All	7·5
Other single organs	All	1·5

killing the irradiated group, or the irradiation may partially suppress some infection.

It is therefore accepted that all casual radiation exposure is undesirable, and that vigorous steps shall be taken to reduce such exposures to a minimum. The International Commission on Radiological Protection (ICRP) recommends upper limits of dose [18] which should not be exceeded, and keeps the situation continuously under review. Their recommendations are amended when necessary in the light of new knowledge. Upper limits of dose must be acceptable to the individual and to society, in the light of the benefits derived

[18]. In this sense, the dose is "acceptable" or "permissible". The ICRP [18] recommends that the term "maximum permissible dose" (MPD) should be used for exposure of radiation workers to controllable sources. For planned exposures of the general public, the term should be "dose limit" and for unplanned exposures from uncontrolled sources the term "action level" is recommended. The magnitude of the dose varies with circumstances and is discussed in detail in the *Recommendations* [8, 18, 19] (Tables 14.2, 14.3). The MPDs have tended to decrease throughout the years, but further decreases will be limited by the approach to background dose.

Recommendations of the ICRP are incorporated into national legislation or into recommendations and codes of practice (p. 255 *et seq.*).

Biological Effects of Radiation

The biological effects of radiation at dosages within maximum permissible levels are, by intent, marginal. They may be so in the sense that the chance of showing some effect is small, when this effect is of an all-or-none nature as in the induction of leukaemia; or they may be so in the sense that the effect itself is small, as in a reduction of the life-span.

Biological effects of radiation are often delayed, and it may be years between exposure and evident damage. This is one factor complicating the problems of radiation protection, as it may make cause and effect difficult to establish.

Radiation effects may be either somatic, when they affect the irradiated individual, or genetic, when they affect his descendants.

SOMATIC EFFECTS OF RADIATION

1. Cataract

Cataract may occur after a dose which, rarely, can be as low as 200 rads received during a few weeks [3]. Its induction is not therefore a risk for radiation workers, except after an accident. Such an accident is a possibility in X-ray diffraction work, where narrow beams of high intensity have to be accurately aligned in the apparatus.

Radiation-induced cataracts have followed exposure to both neutrons and X-rays. Although neutron irradiation is the greater hazard, some two hundred cases of cataract induced by X- or gamma-radiation have been reported in the literature [20].

2. Fertility

Loss in fertility is a sensitive indicator of damage by irradiation [3]. The effect in mice is markedly dependent on the age at irradiation, and at the most sensitive age of two or three weeks, the total reproductive capacity of female mice may be reduced to zero, after an exposure of only 20 R [21]. Exposure rate, as well as total exposure may determine the degree of infertility when mice are irradiated with gamma-rays [22].

Nevertheless, neither loss in fertility nor radiation-induced menstrual disturbances are thought to occur in women after doses within maximum permissible levels [3]. Temporary loss in fertility occurs after a gonad exposure of about 170 R in the female and about 300 R in the male; castration exposures are approximately 300 R and 700 to 1,000 R respectively, and depend on the age of the individual.

3. Effect on the Foetus

Low doses have led to development abnormalities in mice. Although these have been found in man only after large doses, it is prudent to assume that the human foetus may be readily damaged by irradiation, particularly at some stages of development. Certain tissues, such as the retina and the brain, are probably particularly radiosensitive [3]. Irradiation of the foetus may be followed by the development of childhood malignancies (p. 253).

4. Longevity

Exposure to ionizing radiation has reduced the life span in animals [23, 24, 25]. Lindop and Rotblat [24], working with mice irradiated with megavoltage x-rays, found that the shortening of life was due to the earlier incidence of all causes of death in irradiated animals. This earlier incidence was not of the same magnitude for different diseases, nor was it the same for mice of different ages [26]. Under controlled conditions, the amount of life-shortening was proportional to the dose. Other workers have found that the degree of fractionation has some effect on the shortening of the life span [27].

Less is known about life-shortening in man. It is, however, considered likely that any life-shortening from doses within maximum permissible dose levels will be small. Neary [23] has calculated that a radiation worker who accumulated a MPD of 200 rems in a working life, might thereby have his life shortened by 16 days or about one year, according to the assumptions on which the calculations were based. Failla and McClement [28] have suggested a period of eight months. No shortening of life is apparent amongst British radiologists during the years 1897 to 1956 [29], although United States radiologists were thought to die about five years earlier than their medical colleagues who were not exposed to radiation [30]; some workers [31] do not accept this conclusion, and others consider the question still to be open [20].

5. Leukaemia

Leukaemia may be radiation-induced, and the probability of developing the disease is measurably increased after a dose of about 100 rads [18, 32]. If there is no threshold dose, an increase in leukaemia incidence will be demonstrable at all dose levels, provided that the numbers under investigation are sufficiently large. It has been estimated that diagnostic x-ray examinations of the trunk are associated with a significant excess of the myeloid-monocytic group of leukaemias, and that 8 per cent of all leukaemias, other than lymphatic, are caused by diagnostic x-radiation, and a further 3·6 per cent by radiotherapy [33].

Cases of leukaemia occurred in Hiroshima and Nagasaki which could be attributed to the radiation from the atomic bombs: there were higher incidences at distances nearer to the explosions [3]. Radiotherapy patients given a particular treatment to the spine have a higher incidence than expected, and the incidence increases with increasing dose [34].

After much argument as to whether or not irradiation of the foetus during diagnostic radiology of the mother, can be responsible for the later development of leukaemia and other childhood malignancies [3, 35, 36, 37], the best evidence now supports such a cause and effect [38]. An exposure of about 5 R *in utero* probably increases by 50 per cent, the chance of the child subsequently developing leukaemia or another malignant disease [11]. However, to put the risk in perspective, it should be emphasized that the number of irradiated children developing one of these diseases is less than five in 10,000 live births. Put in another way, in a population of children equal to that of England and Wales (but actually from MacMahon's survey in the United States), five or six hundred cases of malignant disease occur "spontaneously" by the age of ten [11]. Antenatal radiography to ten or eleven per cent of the mothers adds perhaps twenty-six extra malignancies.

There is obviously no case for excluding all such radiography, but only for restricting it to cases where it is likely to forestall other dangers. When radiography is necessary, every effort should be made to reduce radiation exposure of the foetus to a minimum.

The ovaries should not be irradiated for treatment of infertility, nor to induce a temporary artificial menopause [5].

In the general population, the mortality from acute and chronic myeloid leukaemia is increasing, and these are forms which can be radiation-induced [3]. But leukaemia remains a rare disease, and it has not been proved that the increase is due to radiation. On the contrary, Court Brown and his co-workers [39] have concluded that exposure to ionizing radiation at the environmental level is unlikely to account for more than one per cent of the excess in mortality due to leukaemia, if this dose is capable of being leukaegenic at all.

It is uncertain whether radiation exposure at MPD levels are capable of being leukaegenic, or whether there is a threshold effect [40]. In making its recommendations, the ICRP has taken a conservative view and has fixed the dose to the blood-forming organs at a level low enough to ensure that any possible increase in leukaemia incidence is likely to be very small indeed.

6. Cancer

Ionizing radiations, if the dose is high enough, are known to be carcinogenic. In a review of radiation-induced cancer of the skin, Peterson [41] concluded that cancer developed in areas which showed clinical signs of radiation damage, but that there was no demonstrable relationship between severity of the reaction and latent period. Nineteen of his cases had undergone radiotherapy and two had accidental radiation injuries. All were irradiated during the early history of x-rays between 1902 and 1923; several had been given radiotherapy as children. Cade [42] reviewed 34 cases of irradiated patients who subsequently developed cancer; he concluded that (1) cancer develops exactly at the site severely damaged by radiation, (2) latent periods ranged from 8 to 56 years, and (3) all varieties of carcinoma and sarcoma may be radiation-induced. Mitchell and Haybittle [43] reported a case of carcinoma of the skin appearing 49 years after a single exposure of about 1,500 R.

There is no conclusive evidence that radiation at current occupational levels are carcinogenic, although this was not the case in the early days of x-rays [3] when doses were higher.

GENETIC EFFECTS OF RADIATION

It is well established that ionizing radiations can cause genetic effects [44, 45]. For this to occur, the radiation must have affected the germ cells or the cells ancestral to them in the reproductive organs. Radiation can cause chromosome "stickiness", breaks or aberrations or gene mutations. These often lead to a loss in reproductive capacity, so that damaged cells may die and prevent the development of affected individuals. However, gene mutations in particular may be non-lethal, resulting merely in changes in the hereditary material. If consequent effects appear in the animal or plant in which the mutation occurs, they are said to be dominant. If the particular mutation has to be carried by both parents before its result can be manifest, the mutation is said to be recessive; consequently, many generations may pass before the result of a recessive mutation appears in a descendant. The absence of abnormal individuals in his immediate descendants is no indication that the parent does not carry an increased load of recessive genes. Recessive mutations are the more common type.

Agents other than radiation may also act as mutagens. Different gene mutations are not equally sensitive to a particular mutagen, whether this is a chemical substance or radiation [46, 47]

As man has been exposed from time immemorial to ionizing radiations and other mutagens, it is not surprising that novel types of mutation have not been produced by man-made radiations. Nor is it surprising that most mutations are harmful, since those having survival value have been selected over previous ages. It is conceivable that a few detrimental effects could become beneficial, if living conditions were to change radically. In general, the more rapidly times are changing, the higher the optimum mutation rate, but enough is not known to make this generalization quantitative.

The genetic effects of radiation have been summarized by Penrose and Stevenson [3]. There is no known threshold dose below which gene mutations do not take place, nor is there thought to be a recovery process. Hence, to a first approximation, the number of mutations caused is proportional to the total dose. This has been demonstrated down to an exposure of 25 R for the fruit fly [31], and down to higher exposures for mammals. However, there is evidence [44, 48, 49, 50] that the genetic effect depends on dose rate as well as on total dose, a low dose rate being less effective than a high one.

In spite of its importance in fixing MPD levels, the absence of a threshold dose has not been proved. When working with experimental animals, statistical problems are formidable in assessing the genetic effects of small doses. For man, there are even greater difficulties, as his life span is long and exposure conditions comparatively uncontrolled. As findings are consistent with the absence of a threshold dose, this absence has been assumed in determining radiation protection recommendations. The margin for error is thus on the safe side, as in considering radiation-induced leukaemia.

In view of this assumption that even very low doses can occasionally produce genetic effects, and that such changes are irreversible and usually detrimental and may not become apparent for many generations, great efforts must be made now to minimize the radiation exposure of the population. There must be constant effort to reduce the dose from man-made sources, not merely to within MPD levels, but as far below as possible.

Laws and Recommendations for Radiation Protection

The ICRP *Recommendations* [8, 18] are acknowledged as embodying the considered opinions of the world's experts. These recommendations form the basis of numerous international (Table 14.4) and national publications.

TABLE 14.4 *Short List of International Publications Concerning Radiation Protection*

Manual of Industrial Radiation Protection. Part 2. *Model Code of Safety Regulations (Ionizing Radiations)*
(Geneva, International Labour Office.1959. London Office: 38 Parliament Street, S.W.1.).

International Atomic Energy Agency, Safety Series
No. 8, The Use of Film Badges for Personnel Monitoring (Vienna, I.A.E.A., 1962).
No. 9 Basic Safety Standards for Radiation Protection (Vienna, I.A.E.A., 1962).

World Health Organization Technical Report Series
No. 196, Medical Supervision of Radiation Work. Second Report of the Expert Committee on Radiation (Geneva, W.H.O., 1960).

International Atomic Energy Agency
Medical Supervision of Radiation Workers (Vienna, I.A.E.A., 1968).

The Evaluation of Risks from Radiation. A Report by a Task Group of Committee
I.C.R.P. Publication 8 (Oxford, Pergamon Press, 1966)

International Atomic Energy Agency
Basic Safety Standards for Radiation Protection (Vienna, I.A.E.A., 1967).

It is the responsibility of each country [18, 51] to introduce or encourage legislation and codes of practice, as it thinks fit.

The precise level at which MPD levels are fixed is a matter of judgment based on incomplete scientific evidence. It can be argued that it is unwise to accept the same levels in all circumstances, and that the economic benefits of nuclear power and the industrial use of ionizing radiations can raise the expectation of life to such an extent where poverty is widespread, that higher MPD levels might be acceptable than is the case for industrialized nations, where benefits are smaller. Even in the latter, some argue that too much time and money is devoted to radiation protection; others hold that this is not so, because radiation hazards are different in kind and more uncertain in degree than more conventional hazards.

LEGAL ASPECTS

The legal position in the United Kingdom, particularly with respect to the United Kingdom Atomic Energy Authority, has been summarized by Sim [52].

Apart from specific legislation, in this country common law gives some legal protection against injury by dangerous materials. Common law has been established by custom and usage and rests on precedent. The doctrine of strict liability implies that, before goods are transferred, they must be made as safe as possible, and attention must be drawn to any remaining danger. Both employer and employee are legally responsible for damage caused by the latter in the course of his employment. It is, however, usual for the master rather than his servant to be sued by an injured third party, as it is more likely that the master will be in a position to pay damages or compensation, although the employee may, in turn, sue his employer.

The master is obliged to provide a safe system of working and proper equipment as well as to supervise and control his undertaking properly. If he fails to do these things, and the servant is injured as a result, the master will be held to be negligent. But if the servant is injured because he has failed to use safety equipment provided, judgement will depend on circumstances.

Since 1946, a number of laws concerning radiation hazards have been passed in this country. These take the form of Acts of Parliament and legislation made under powers given by Act of Parliament. Copies are available from Her Majesty's Stationary Office. Some are concerned with x-radiation, and these are summarized here.

1. Radioactive Substances Act, 1948

This Act covers certain apparatus producing radiations as well as radioactive materials. The "appropriate minister" is empowered to control the use of irradiating apparatus and to make safety regulations for their use.

2. Atomic Energy Authority Act, 1954

The A.E.A. is put under an absolute duty to ensure that no ionizing radiations from any of their premises cause any injury to any person, or any damage to any property, whether or not he or it is on such premises or elsewhere.

3. Ionizing Radiations (Sealed Sources) Regulations and Orders

1961, No. 1470. Factories. The Ionizing Radiations (Sealed Sources) Regulations, 1961.

1961, No. 1710. Factories. The Ionizing Radiations (Sealed Sources) (Radiation Dosemeter and Dose Rate Meter) Order, 1961.

1961, No. 1711. Factories. The Ionizing Radiations (Sealed Sources) (Leakage Test) Order, 1961.

1961, No. 1712. Factories. The Ionizing Radiations (Sealed Sources) (Laboratory Certificate) Order, 1961.

1961, No. 1713. Factories. The Ionizing Radiations (Sealed Sources) (Radiation Dose Record) Order, 1961.

1961, No. 1714. Factories. The Ionizing Radiations (Sealed Sources) (Health Register) Order, 1961.

1961, No. 1715. Factories. The Ionizing Radiations (Sealed Sources) (Transfer Record) Order, 1961.

This is a series of statutory instruments designed to minimize radiation hazards for those within the jurisdiction of the Factories Acts. The term "sealed sources" in this context includes x-ray machines.

RECOMMENDATIONS AND CODES OF PRACTICE

In fields not covered by legislation, organizations have issued codes of practice or similar documents designed to control radiation hazards. Although not legally binding, the standards laid down in them and the way in which their recommendations were carried out might well influence a Court of Law in deciding whether or not a person or organization had been negligent. They are often comprehensive publications and may well include matter irrelevant in a particular situation. Consequently, it is advisable to issue a group of workers with a simplified and brief list of safety requirements, based on the Code in point and bearing on the particular hazards to which the group is subjected. Some Codes explicitly require local rules to be issued.

1. *Code of Practice for the Protection of Persons Against Ionizing Radiations Arising from Medical and Dental Use* (London, H.M.S.O., 1964).
   This Code deals with hazards to both staff and patients.
2. *Code of Practice for the Protection of Persons Exposed to Ionizing Radiations in Research and Teaching* (London, H.M.S.O., 1968).
   This Code is published by the Department of Employment and Productivity, who will advise on its provisions and visit establishments in which the Code applies.
3. *The Use of Ionizing Radiations in Schools, Establishments of Further Education and Teacher Training Colleges*, Department of Education and Science, Administrative Memorandum 1/65.
   This Memorandum applies to educational establishments having equipment which might generate x-rays and with an accelerating voltage greater than 5 kVp. It controls also the use of radioactive substances.
4. *Code of Practice for the Display of Ionizing Radiations at Exhibitions* (London, H.M.S.O., 1961).
5. *Home Office Requirements for the Control of Shoe-fitting Fluoroscopes* (London, H.M.S.O.)
6. *Ionizing Radiations: Precautions for Industrial Workers* (London, H.M.S.O., 1961).
   This is based on the *Ionizing Radiations (Sealed Sources) Regulations and Orders.*
7. *Model Code of Practice for Ports and Harbours*, Ministry of Transport (London, H.M.S.O.).
8. *X-Ray and Gamma Ray Safety Precautions* (British Welding Research Association, 29, Park Crescent, London W.1., 1957).

RESPONSIBILITY

The ICRP [18] has laid down that the authority in charge is "responsible for the protection of persons working in the establishment or on an operation and for limiting the exposure of members of the public so as to comply with the relevant national or local requirements".

In this context, it is not always obvious who is the authority in charge, nor which person is responsible for radiation protection. Thus, the Department of Education and Science has placed responsibility for safe working in maintained and assisted establishments with local education authorities, and it has made governing bodies responsible for other establishments. The United Kingdom Atomic Energy Authority requires that heads of Establishments ensure that all members of their staff comply with the provisions of their Code. In places of employment under the jurisdiction of the Factories Acts, the employer of the classified worker is the person responsible for his safe working conditions, whether or not he is the owner or occupier of the factory.

The Controlling Authority is ultimately responsible for the implementation of the two Codes of Practice for medical and dental work and for research and teaching (p. 257). In the case of hospitals, this Authority is the Board of Governors or the Hospital Management Committee (Board of Management in Scotland). For research and teaching, the Authority is "the body, person or persons ultimately responsible for the control of the establishment".

SAFETY OFFICERS

The person or persons responsible for radiation safety should delegate radiation protection duties to experts. Excluding medical work, two levels of competence are required: a "qualified expert" or health physicist who can advise on radiation protection, and a radiation protection officer who can apply radiation protection regulations. It is desirable for each department to have its own radiation protection officer, who can supervise closely day-to-day working conditions, and who works under the general direction of a health physicist or other safety officer. In a small establishment, however, the duties of the responsible person, the qualified expert and the health physicist may all be vested in one person.

Success of protection measures depends to a large extent both on the personalities of those responsible and on details of organization. Safety regulations may easily become identified as mere inconveniences imposed from above, but this can be avoided if they are seen to be accepted by members of the safety organization. This in turn implies that safety officers should work alongside the rank-and-file, and that all should be exposed to similar hazards. Furthermore, it is always difficult for a radiation worker to keep a balance by respecting the dangers without becoming unduly nervous of them. He can best acquire this outlook if he can discuss his doubts and difficulties as they arise, with an expert whom he feels to be not too remote. In his turn, a safety officer will gain much from attempting to carry out his own regulations, and any weakness in them should speedily become apparent. There is therefore much to be said against the appointment of a safety officer whose duties confine him largely to an office.

MEDICAL OFFICERS

Medical officers play a highly specialized role in radiation protection work; they should be familiar with the biological effects of radiation. All medical questions arising in protection work are referred to a medical officer. These questions include medical examinations with particular reference to haematology, before and during occupational exposure, as well as advice on the action to be taken in the event of overexposure. Many organizations set up radiation protection committees which co-ordinate radiation protection work, review cases of high exposure and give general support and advice to safety officers; a medical officer specializing in the subject plays an essential part on such a committee. Controlling authorities are obliged, in the Codes of Practice, to appoint a supervisory medical officer. Under the Factories Acts, an "appointed doctor" has to be approved by certificate of the Chief Inspector of Factories, or where there is no such medical practitioner, the appointed factory doctor for the district should act in radiation safety.

TRAINING

The ICRP [8] has observed that "it is the duty of the workers, upon the instruction of a qualified expert or responsible supervisor, to follow the recommendations and instructions which have been drawn up in the interests of radiation protection . . .".

A worker is able to carry out this duty in so far as he is trained to do so; he will be willing to do so if he recognizes its value. Training should therefore include both the physical principles and an indication of the social and biological implications of radiation hazards. In the words of the Code of Practice in Research and Teaching "it must be impressed on every individual working with ionizing radiations . . . that he has a duty to protect both himself and others from any hazard arising from his work and that he must not expose himself or others to ionizing radiations to a greater extent than is reasonably necessary for the purpose of his work". If a worker is unable to accept this duty, he is unsuitable for radiation work.

The Veale Committee [53] has reported on the need for training in radiological health and safety in this country. Courses in radiation protection are available (Table 14.5), although some large establishments train their own staff.

TABLE 14.5 *Publications listing Courses in Radiation Protection*

Title of publication	*Publisher*
Nuclear Studies	Science Department, The British Council, Albion House, 59, New Oxford Street, London, W.1.
Catalogue of Courses on Nuclear Science and Technology	European Nuclear Energy Agency, Organization for Economic Co-operation and Development, 38, Boulevard Suchet, Paris, XVIe, France
Directory of Courses in Occupational Safety and Health	International Labour Office, CH 1121, Geneva 22, Geneva, Switzerland. (London address: 38–39, Parliament St., S.W.1.)
List of Conferences, Meetings and Training Courses in Atomic Energy	International Atomic Energy Agency, Vienna, 1, Kaerntnerring, Austria.

CATEGORIES FOR RADIATION EXPOSURE

In fixing MPD levels, the ICRP [18] has considered both somatic and genetic effects of radiation.

Genetic effects are considered to provide the more stringent limitations; in this respect, the important parameter to consider is the total gonad dose received by that section of the population below the mean child-bearing age, 30 years. It follows that some groups in this section may receive higher gonad doses than others, provided that both the size of the group and its MPD are limited with the overall condition in mind. The MPD to other parts of the body can also be higher in some groups, for instance where the workers are under medical surveillance, than in others, provided that the risk of a serious radiation effect is always acceptably small for an individual. Accordingly, the ICRP has divided the population into two categories: those exposed in the course of their work, and members of the public.

Occupational Exposure

Adults exposed to radiation in the course of their work should receive less than certain stipulated annual maximum permissible doses (Table 14.2), and not more than half (rounded upward to the next whole number), the annual MPD in any one quarter. Thus, the MPD in one quarter is 3 rems for the gonads and the red bone-marrow. It may be justifiable [18] to allow a dose of 3 rems in consecutive quarters, provided that the accumulated dose does not exceed

$$5(N - 18) \text{ rems} \tag{14.1}$$

where N is the age in years.

Women of reproductive capacity [18] should accumulate dose at levels within those for annual, not quarterly, limits, so that they should receive less than 1·3 rems in each quarter. After a pregnancy has been diagnosed, the dose to the foetus accumulated during the remaining period of pregnancy should be kept below 1 rem.

Adults exposed in the course of their work who are most unlikely to receive as much as 3/10 of the MPD need not usually be monitored nor under special medical supervision, although environmental monitoring is desirable.

When it is considered that workers might receive doses in excess of 3/10 of the annual MPD, they should be monitored continuously and be under medical supervision. Such workers are known as *occupational*, *designated* or *classified* workers.

Health Surveillance

Health surveillance is not essentially different from that in general industrial medical practice [18], but it may in addition provide information on, for instance, an individual's "normal" blood count, which could be useful in the event of subsequent overexposure. Some recommendations require the result of a medical examination to determine the acceptance or rejection of a potential employee.

Medical examinations should be repeated at a frequency depending on the circumstances. The Codes of Practice for both Research and Teaching and that for Medical and Dental Work require designated workers to have a medical examination within four months immediately prior to designated employment and subsequently annually. Classified workers subject to the Factories Acts are required to undergo a medical examination by a specially appointed doctor, at least every fourteen months. Regulations are laid down regarding those competent to carry out blood tests and regarding the nature of the test. Comprehensive records must be kept in a health register (1961, No. 1470. The Ionizing Radiations (Sealed Sources) Regulations).

Age

The ICRP formula (equation 14.1, p. 260), relating dose and age, implies that occupational workers should be over 18 years of age. But exposure is explicitly permitted below this age [18], provided that the annual dose is less than 5 rems, and that the dose accumulated by the age of 30 years is not greater than that (60 rems), from the formula.

Some countries do not permit occupational exposure below the age of 18 years. In the United Kingdom, the position is somewhat anomalous: no person under 18 coming within the jurisdiction of the Factories Acts, may be a classified worker. Other bodies allow occupational exposure between the ages of 16 and 18 years, in some cases with limits lower than 5 rems/yr.

Although the ICRP [18] does not mention a lower age limit, it has been generally accepted that young persons under 16 should receive no significant radiation exposure, other than that due to medical procedures and natural background. Thus, in United Kingdom schools no pupil below that age is permitted to work with x-rays or materials which are significantly radioactive. Over the age of 16, a pupil should receive less than 50 rems/yr as a result of experiments in schools.

Controlled Areas

A controlled area [18] is "any area to which access is controlled for the purpose of protecting persons from exposure to radiation or radioactive materials. Access can be controlled in a variety of ways, the minimum being by the use of warning signs. The extent of a controlled area is a matter for professional judgement, but in all cases the extent should be such that it is extremely unlikely that workers outside the controlled area will receive doses in excess of 3/10 of the appropriate maximum permissible doses. Other considerations may require an enlargement of the controlled area".

This definition of a controlled area differs in emphasis from earlier ones. Like other terms in radiation protection, definitions vary from one publication to another as well as in different editions published by the same bodies.

Abnormal Exposures of Radiation Workers

Abnormal exposures include voluntary ones, undertaken in special circumstances, for example, to rescue somebody, and involuntary ones due to accident.

All abnormal exposures must be recorded. If the dose exceeds twice the annual limit, the competent medical authority must be consulted.

Personal Monitoring

The ideal personal monitor would give a warning indication before any serious damage had been done to the individual.

For instance, in considering the risk of leukaemia, an ideal monitor would measure some change in the blood as a precursor to leukaemia, while the change is reversible. Such a biological effect would be influenced by a number of factors such as the amount of bone marrow irradiated and the dose rate, as well as by the magnitude of total dose.

Other effects would demand different indicators.

Merely to state the problem is to demonstrate how far it is from solution; such biological indicators are not available.

Physical methods are accepted as the most practicable for personnel monitoring. They are sensitive and, in general, specific, but most measure only the integrated dose over a small area. Some types can be used to estimate the nature and quality of radiation; some are fairly insensitive to the direction of the incident radiation. All physical dose indicators can give misleading results, because what is measured is not necessarily the important parameter for a particular biological effect.

Occupational workers receiving significant doses (*vide ultra*) should be continuously measured so that MPDs are not exceeded and so that individual records may be kept. When, in the past, this has not been done, it must be assumed that the worker received the MPD during the unmonitored period, and this should be entered on his cumulative record.

Monitoring, to be effective, must be accompanied by some mechanism whereby the record of each worker is transferred to his new place of employment. Thus, for factory workers, the Ionizing Radiations (Sealed Sources) Regulations, 1961, provide for detailed records to be kept of all personal monitoring, and for a summary of these records to be sent on a "transfer record" to an employee's new place of work. Designated persons coming within the terms of reference on the Codes of Practice must be similarly documented.

A worker is entitled to inspect his dose record. Much can be said for interpreting this proviso in a positive way, and, as a matter of routine, for sending monitoring records to the workers concerned. The records can be of educational value to the individual, and should help him to recognize and deal with hazardous procedures when they occur in the future. They also give encouragement to the careful worker. Furthermore, it is bound to be a tedious matter for an individual to carry out protective measures, and these duties become less of a labour when it is evident that dose records are not confined to an office pigeonhole.

Personal monitors should be worn continuously during working hours. No single test should extend beyond four weeks and it is usually desirable to change film badges weekly or fortnightly. Long periods of use increase the possibility of erroneous results due to the fogging of films and other technical faults; perhaps of more importance, events resulting in high doses become increasingly difficult to identify.

Film badges [54] are recommended for continuous monitoring. The amount of blackening on the photographic film depends on the energy and type of radiation as well as on the dose. Therefore, in order to measure the latter, it is necessary to estimate the former (Chapter 7). It is also useful to know the former when assessing the significance of the dose. Film badges should be supplied by a recognized laboratory[38] whose reports are likely to be accepted in a Court of Law. Centralization of monitoring establishments has the additional benefit that the work load is large enough for automation to be appropriate.

A number of types of personal monitor are available which give an immediate measurement of the dose. Some of these comprise a form of ionization chamber and are designed to be energy-independent over a limited range, for which they may be calibrated. Quartz fibre dosemeters[39] [55] incorporate an optical system for reading the position of the fibre, and the exposure can be read by holding the dosemeter up to the light. Small capacitor ionization chambers[2] made of Bakelite-graphite are X-ray energy-independent above about 5 mm Al HVL. They are read in conjunction with an electrometer. Other types[40,41] [56], usually based on a Geiger counter, give visible or audible warnings when the dose rate is high.

More recently, solid-state devices have been developed for low dose measurements (p. 71). Some of these [57] are suitable for long-term monitoring to a degree unattainable with older methods. Thermoluminescent dosemeters (Chapter 4) are often used for personal monitoring, and are available from the Radiological Protection Service (*see* p. 123) for finger-tip and other monitoring.

The International Organization for Standardization has co-ordinated specifications for film badges and pocket electroscopes (quartz fibre dosemeters) for personal monitoring.

Exposure of Individual Members of the General Public

People in the population at large will receive some radiation exposure in addition to that from the natural background and medical procedures. This excess may come from luminous watches, television sets and environmental contamination. Some individuals who cannot be included as adults exposed in the course of their work may receive exposure from certain other man-made sources; examples are children performing school experiments with ionizing radiations.

Individual members of the public should receive, in general, less than one-tenth of the MPD levels for radiation workers (Table 14.3). For school children, the Department of Education and Science has set lower dose limits (p. 261). Experiments in schools are permitted only if they are carefully controlled. The exposure rate 1 metre from an X-ray target must be less than 0·5 mR/h, and the X-ray set must be totally enclosed in a lead box. As any one child will spend only a short time in the neighbourhood, his exposure from this source should be negligible.

Exposure of Population

Medical procedures are the main source of population exposure. Hazards from these procedures are, in general, small and can be kept within acceptable limits by careful attention to techniques. Radiological examinations of the lower abdomen and pelvis of women of reproductive capacity call for special care, because of the radiosensitivity of the embryo and the foetus. Except when there are urgent reasons, these examinations should be limited to the ten days following the onset on menstruation, or, during pregnancy, until after its conclusion or at least until its latter half [18].

The ICRP [18] recommends that the genetic dose to the population should be kept to a minimum. It should not exceed 5 rems from all sources additional to the dose from natural background and from medical procedures.

This implies that the contribution from the general public must be less than 5 rems, to allow for the higher contribution from radiation workers. As the mean child-bearing age is taken to be 30 years, a genetic dose of 5 rems corresponds to an average of 0·17 rem/yr, neglecting any weighting for the expected number of children conceived subsequent to the exposure. Somatic effects are expected to be very small. For instance, 0·17 rem/yr to the red marrow might give rise to about 3 cases of leukaemia per year per million persons exposed.

It is not the intention, neither is it practicable, to monitor the general public, although Henson [58] has developed a dosemeter capable of measuring the dose accumulated in two weeks, at dose rates of the same order of magnitude as background, with a standard error of ±10 per cent. However, he considered this dosemeter to be unsuitable for measuring mean population dose rates.

Installations and Shielding

Radiation can be minimized when adequate shielding is provided on installation of the X-ray equipment. To this end, the ICRP [8] has laid down general

TABLE 14.6 *Occupancy Factors*
(From ICRP Pubn. 3, *Report of Committee III on Protection against X-rays*, p. 21. Oxford, Pergamon, 1960)

Full Occupancy ($T = 1$)	Control space, offices, corridors and waiting space large enough to hold desks, darkrooms, workrooms and shops, nurse stations, rest and lounge rooms routinely used by occupationally exposed personnel, living quarters, children's play areas, occupied space in adjoining buildings.
Partial Occupancy ($T = \frac{1}{4}$)	Corridors too narrow for desks, utility rooms, rest and lounge rooms not used routinely by occupationally exposed personnel, wards and patients' rooms, elevators using operators, unattended parking lots.
Occasional Occupancy ($T = \frac{1}{16}$)	Closets too small for future occupancy, toilets not used routinely by occupationally exposed personnel, stairways, automatic elevators, sidewalks, streets.

principles for planning new or modified installations. Protection may be achieved by distance and by protective barriers; careful design can lead to economical installations with minimal barriers.

Shielding must be designed so that maximum permissible exposure levels are not exceeded (Tables 14.2, 14.3). The thickness necessary for a particular barrier depends primarily on the incident radiation intensity and the transmission factors for the radiation and barrier material. These determine the exposure rate on the "safe" side of the barrier. An acceptable rate is governed by the distance to persons at risk, whether or not they are only occasionally present, how much the set is used and the category from which the individuals come. "Occupancy" and "use" factors are laid down by the ICRP (Tables 14.6 and 14.7).

Choice of material for a barrier depends on convenience and cost. For X-ray energies where Compton attenuation processes are predominant, the mass of shielding required is nearly independent of the atomic number of the material, so that thickness varies inversely with density. However, for low- or high-energy radiation where photoelectric absorption or pair production respectively is important, absorption increases with increase in atomic number. This is why lead is often the material of choice: it is a cheap material for one with a high atomic number.

Broad beam attenuation is "obtained when the maximum amount of scattered radiation is included in the transmitted beam" [8]. Broad beam

TABLE 14.7 *Use Factors*
(From ICRP Pubn. 3, *Report of Committee III on Protection against X-rays*, p. 21. Oxford, Pergamon, 1960)

Full Use ($U = 1$)	Floors of radiation rooms except dental installations, doors, wall and ceiling areas of radiation rooms routinely exposed to useful beam.
Partial Use ($U = \frac{1}{4}$)	Doors and wall areas of radiation rooms not routinely exposed to the useful beam, floors of dental installations.
Occasional Use ($U = \frac{1}{16}$)	Ceiling areas of radiation rooms not routinely exposed to the useful beam.

data [8], and not attenuation coefficients are usually relevant for radiation protection calculations. The difference in calculated barrier thickness may be large. For instance, when 100 mm steel is used to attenuate photons of about 1·2 MeV, about 1·4 per cent is transmitted under narrow beam conditions, and about 7 per cent for a broad beam [59].

PRIMARY PROTECTIVE BARRIERS

Primary protective barriers are those "sufficient to attenuate the useful beam to the required degree" [8]. Thickness required may be obtained after calculating B, the allowable transmission (in R/mA-min at 1 metre) given by [8]

$$B = \frac{Pd^2}{WUT} \tag{14.2}$$

where P = maximum permissible exposure for design purposes
d = distance in metres from source to position occupied
W = weekly work load in mA-min/week
T = occupancy factor (Table 14.6)
U = use factor (Table 14.7)

The output of the X-ray set is required and, if this is unknown, approximate values of R/mA-min for various conditions can be obtained for published data (Figures 14.1, 14.2 and 14.3).

Values of B for a range of X-ray energies are available with lead or concrete barriers (Figures 14.4 to 14.9).

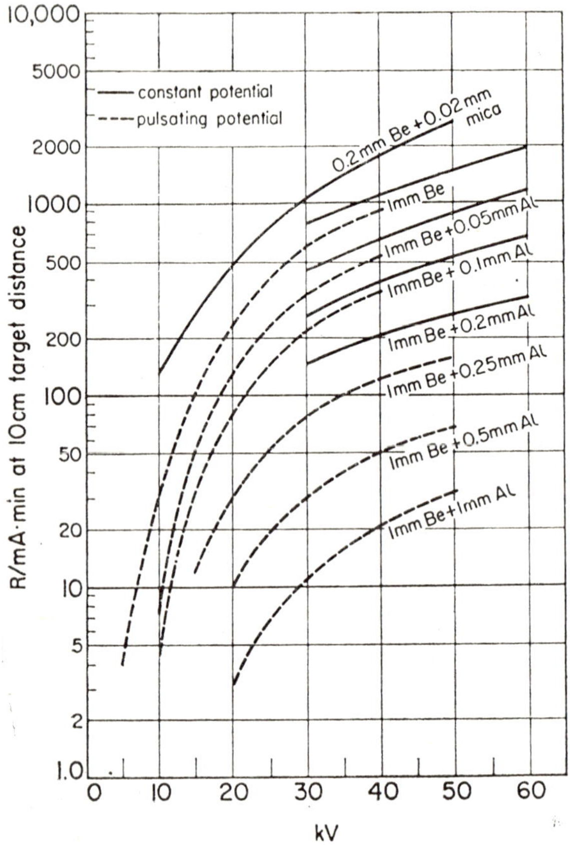

FIGURE 14.1 *Approximate radiation output in R/mA-min at a distance of 100 mm, measured in air, of the primary radiation from a tube with tungsten target and total filtrations from 0·2 mm Be to 1 mm Al.*
To get the dose rate for target materials other than tungsten multiply the values for the output by the factor $Z/74$, where Z is the atomic number of the target material in question and 74 the atomic number of tungsten.

(Recommendations of the International Commission on Radiological Protection. ICRP Publication 3, Report of Committee III. Oxford, Pergamon Press, 1960)

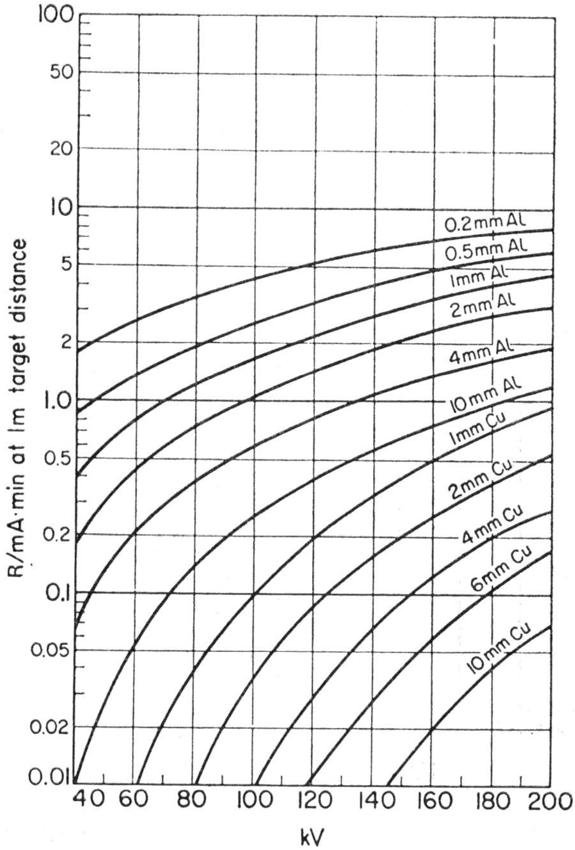

FIGURE 14.2 *Approximate radiation output in R/mA-min at a distance of 1 m, measured in air, of the primary radiation from a tube with tungsten target and total filtrations from 0·2 mm Al to 10 mm Cu at 40 to 200 kV constant potential.*
(ICRP Publication 3, Report of Committee III)

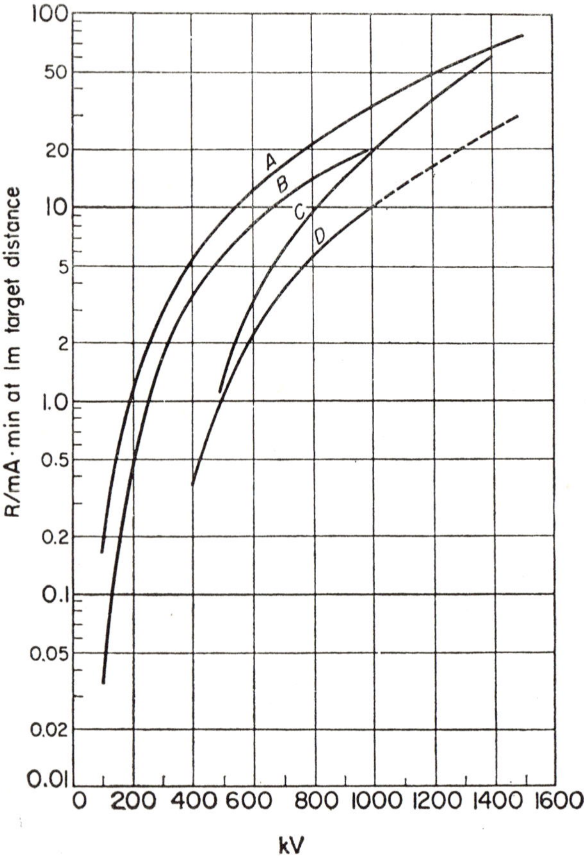

FIGURE 14.3 *Approximate radiation output in R/mA-min at a distance of 1 m, measured in air, of the primary radiation at 100 to 1,500 kV constant potential.*

A. Total filtration 0·5 mm Cu, tungsten-reflection target.
B. Total filtration 2·5 mm Cu, tungsten-reflection target.
C. Total filtration 2·8 mm W + 2·8 mm Cu + 2·1 mm brass + 18·7 mm water, tungsten-transmission target.
D. Total filtration 2·0 mm Pb + 4·1 mm Fe, gold-reflection target.

(ICRP Publication 3, Report of Committee III)

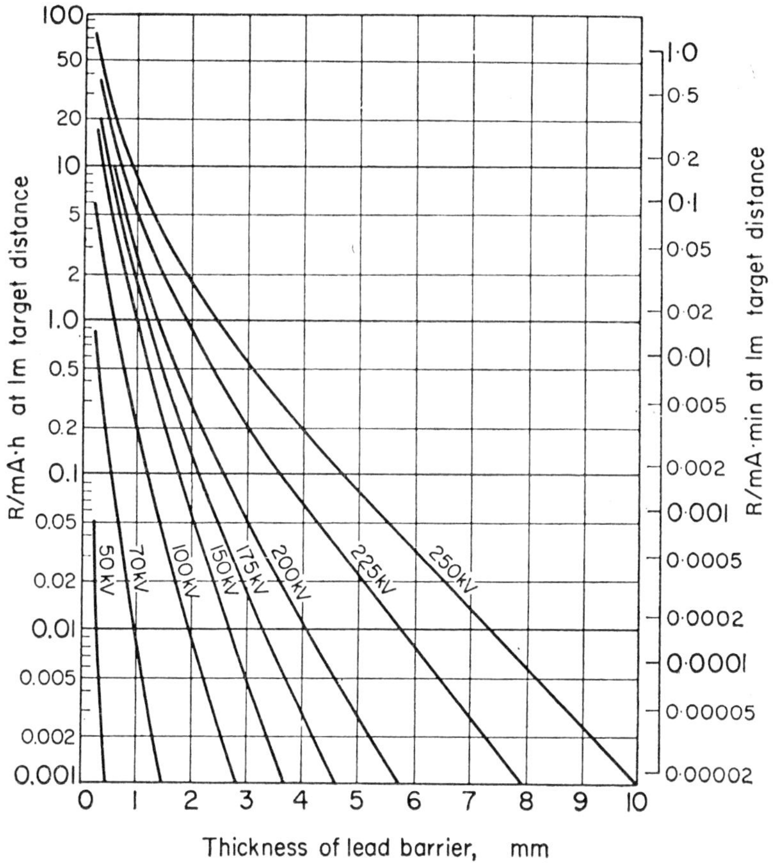

FIGURE 14.4 *Attenuation in Lead X-rays Generated at 50 to 250 kV*
The curves were obtained with a half-wave generator and a 90° angle between the electron beam and the axis of the x-ray beam. The inherent filtration was 3 mm of aluminium for the 150 to 250 kV curves and 0·5 mm of aluminium for the other curves.
Direct-current potentials require 10 per cent thicker barriers than for the pulsating potentials given above.
(ICRP Publication 3, Report of Committee III)

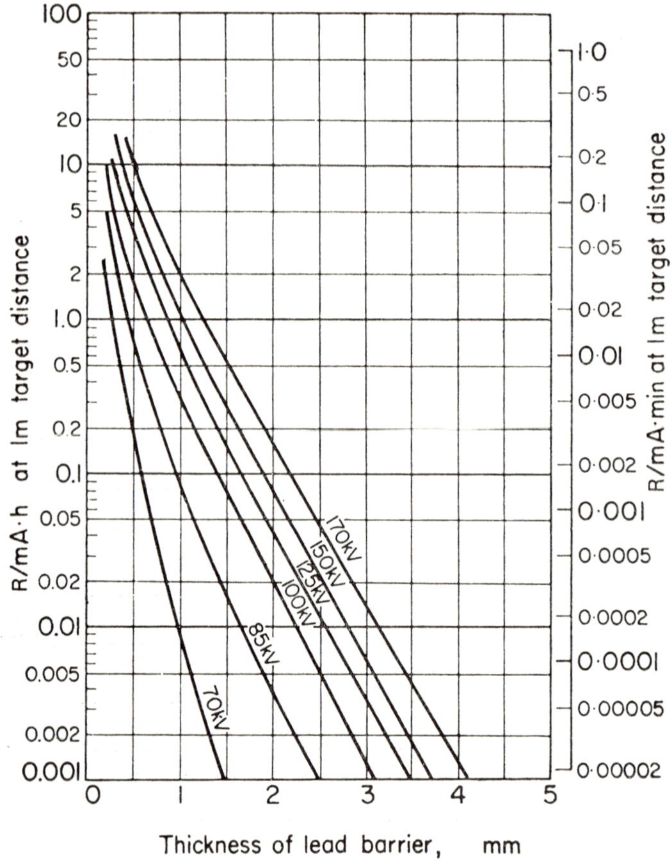

FIGURE 14.5 *Attenuation in lead of X-rays Generated at 70 to 170 kV*
The curves were obtained with a constant potential generator and a 90° angle between the electron beam and the axis of the x-ray beam. The inherent filtration was approximately 1 mm of aluminium.

(ICRP Publication 3, Report of Committee III)

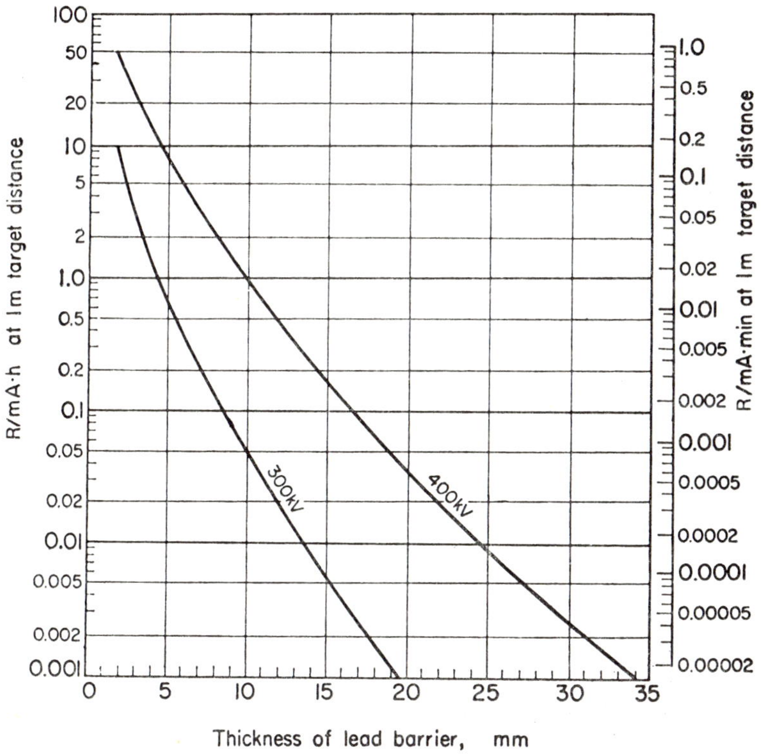

FIGURE 14.6 *Attenuation in Lead of X-rays Generated at 300 and 400 kV*
The curves were obtained with a constant potential generator and a 90° angle between the electron beam and the axis of the X-ray beam. The inherent filtration was approximately 3 mm of copper.

(ICRP Publication 3, Report of Committee III)

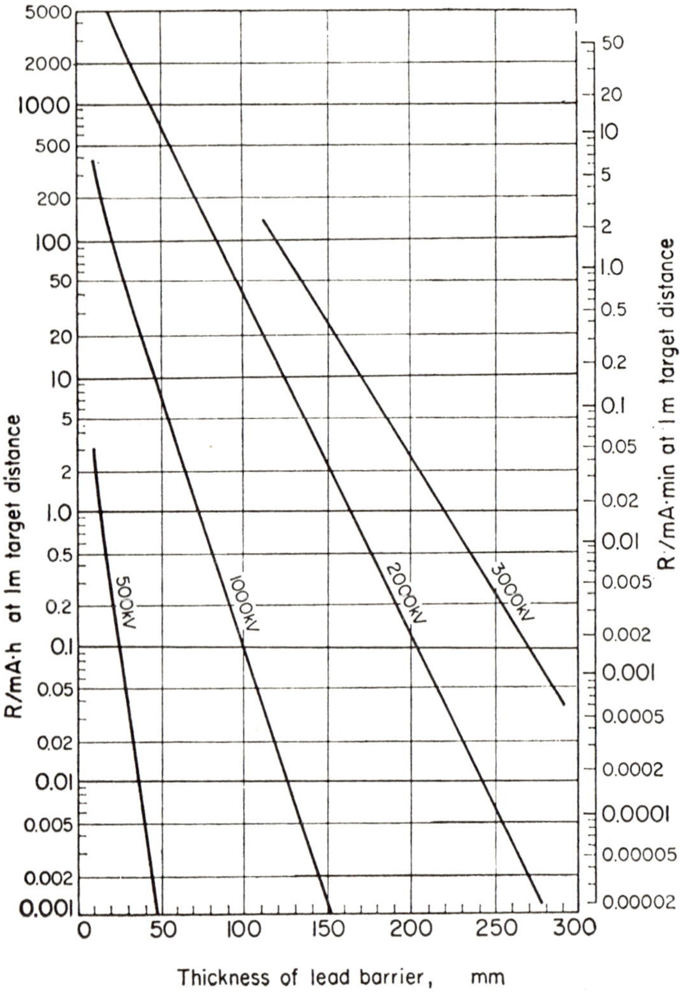

FIGURE 14.7 *Attenuation in Lead of X-rays Generated at 500 to 3,000 kV* The curves for 500 and 1,000 kV were obtained with a constant potential generator and a 0° angle between the electron beam and the axis of the X-ray beam. The inherent filtration was 2·8 mm of tungsten, 2·8 mm of copper, 2·1 mm of brass and 18·7 mm of water. The curve for 2,000 kV was obtained by extrapolation to broad-beam conditions from measurements made with a constant-potential generator and a 0° angle between the electron beam and the axis of the X-ray beam. The inherent filtration was equivalent to 6·8 mm lead. The 3,000 kV curve is based on values for large fields. The mean value of the many half-value layers, which must be considered when calculating the protection, is 15 mm lead.

(ICRP Publication 3, Report of Committee III)

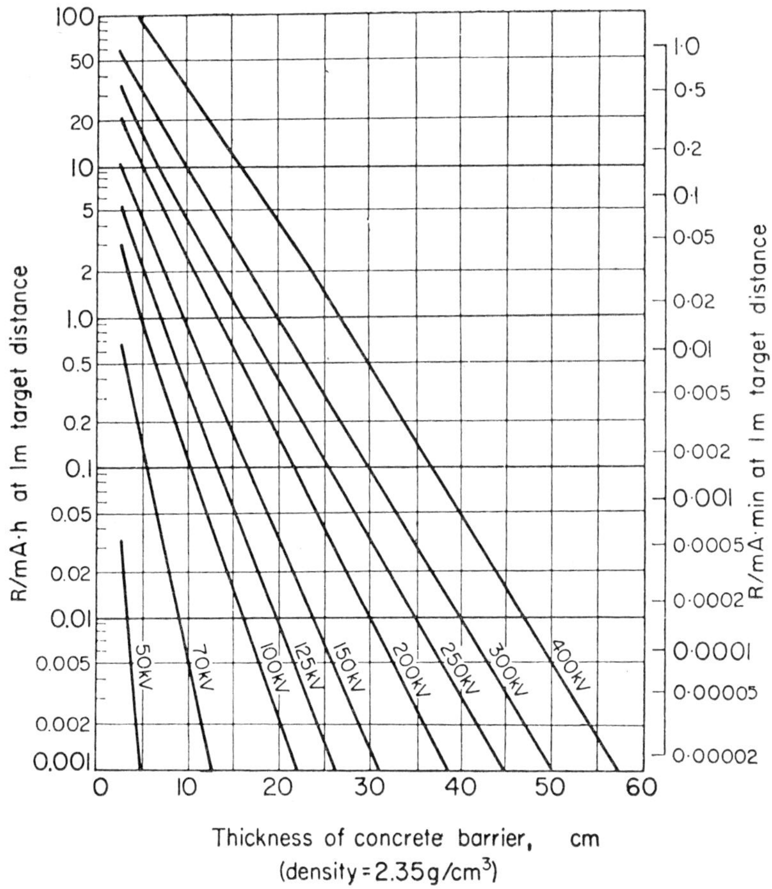

FIGURE 14.8 *Attenuation in Concrete of X-rays Generated at 50 to 400 kV*
The measurements were made with a 90° angle between the electron beam and the axis of the x-ray beam. The curves for 50 to 300 kV were obtained with a half-wave generator. The total filtration was between 1 and 3 mm of aluminium. The curve for 400 kV was interpolated from data obtained with a constant-potential generator. The inherent filtration was approximately 3 mm of copper.

(ICRP Publication 3, Report of Committee III)

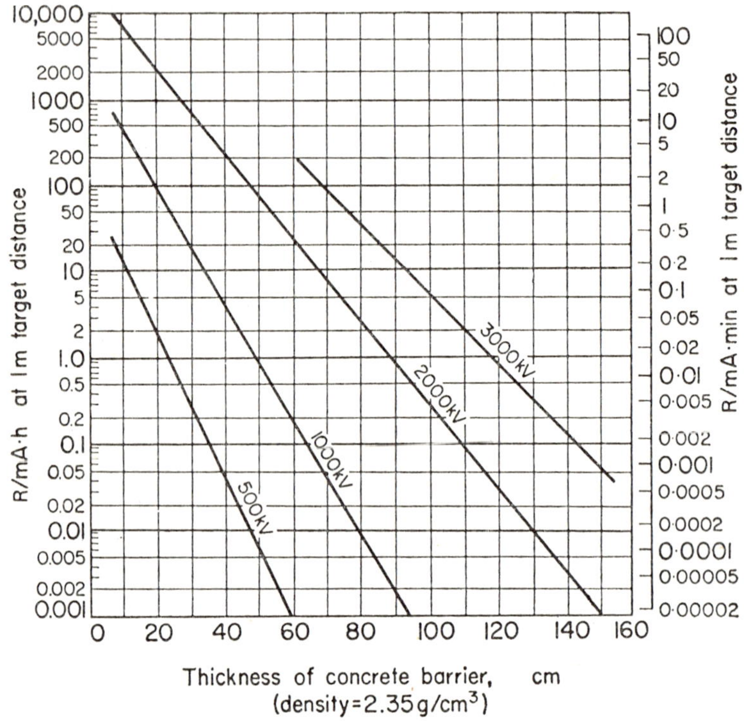

FIGURE 14.9 *Attenuation in Concrete of X-rays Generated at 500 to 3,000 kV* The curves for 500 and 1,000 kV were obtained with a constant-potential generator and a 0° angle between the electron beam and the axis of the x-ray beam. The inherent filtration was 2·8 of tungsten, 2·8 mm of copper, 2·1 mm of brass and 18·7 mm of water. The curve for 2,000 kV was obtained by extrapolation to broad-beam conditions from measurements made with a constant-potential generator and a 0° angle between the electron beam and the axis of the x-ray beam. The inherent filtration was equivalent to 6·8 mm lead. The 3,000 kV curve is based on values for large fields. The mean value of the many half-value layers, which must be considered when calculating the protection is 75 mm concrete.

(ICRP Publication 3, Report of Committee III)

Greene and Massey [60] have published data for the attenuation of 4 MV x-rays in concrete; broad-beam conditions were reached for an area$^{1/2}$ of about 700 mm.

Plasters and concretes incorporating barium salts are sometimes used for protective barriers because barium has a high atomic number; these materials may have twice the density of brick. Iron may be introduced into concrete with similar results: Bashandy [61] has measured the mass absorption coefficients as well as the energy distribution of gamma-rays in three different types of concrete, the densest containing 43 per cent magnetite and 43·6 per cent scrap iron. When Compton absorption is important, compressed concrete of normal constituents and high density may be the material of choice. Special materials[42,43] for radiation protective barriers are obtainable commercially.

Lead linings for walls and floors are useful particularly when converting existing buildings or providing enclosures for x-ray work. Lead easily loses its shape and is difficult to work. It is obtainable,[9,44,45,46] however, in a variety of thicknesses, sandwiched between plywood.

Sheet iron is suitable for screening low-voltage x-rays.

In building protective barriers, radiation-transparent cracks must be avoided. This may be achieved in a solid structure by overbonding between units or by using chevron-shaped bricks. The last have the additional advantage that the bricks interlock to form a rigid structure; both lead and barium concrete bricks are available in this shape.

The high-intensity, low-energy x-ray beams of x-ray diffraction apparatus must be provided with beam traps which are adequate to absorb the primary beam [62].

BARRIERS FOR LEAKAGE RADIATION

When x-rays are used in open installations, it is essential that the emission shall be limited to the useful beam. Leakage radiation, defined as "all

TABLE 14.8 *Permissible Leakage Radiation*
(From information in ICRP *Report of Committee III.* Oxford, Pergamon, 1960)

Type of equipment	*Maximum permissible leakage radiation at specified distances from the equipment*
Television set	0·5 mR/h at 50 mm, where accessible
X-ray analysis	25 mR/h at 50 mm, where accessible
Diagnostic X-ray set	100 mR/h at 1 metre
X-ray medical diagnostic high-tension generator containing rectifying valves	20 mR/h at 50 mm 2 mR/h at any readily accessible place in occupied space within a controlled area
X-ray therapy set	1 mR/h at 1 metre 30 mR/h at 50 mm, where this point is accessible to the patient

radiation, except the useful beam, coming from the tube or source housing" [8], must be below certain limits of exposure rate (Table 14.8). Some manufacturers aim to make any leakage radiation less than one-tenth of the maximum permissible rate. It is, however, a duty of the radiation protection officer to arrange for a survey of every new or modified installation, and for its periodical monitoring.

Shielding required for leakage radiation may be calculated [8] by

$$B_L = \frac{Pd^2 60I}{LTW} \tag{14.3}$$

where B_L = allowable barrier transmission, expressed as a fraction of unity

I = maximum rated continuous current in mA under conditions of maximum leakage

L = permissible leakage radiation in R/h at 1 metre.

Other symbols are as for equation (14.2).

Leakage radiation is highly filtered and consequently often harder than the useful beam; appropriate data must be used in calculating the thickness of shielding required.

As it is usually convenient to use thin shielding for leakage radiation, lead, tungsten, uranium or certain heavy alloys may be used. The last are obtainable[47] in densities up to 18·0 g/cm³.

BARRIERS FOR SCATTERED RADIATION

The primary beam, in order to be useful, must be incident on some object, and consequently to some extent, scattered by it. This results in a diffuse radiation likely to be further scattered by other objects and by walls, floors and ceilings of the building. The scattered radiation is less intense and, in general, softer than the useful beam, but protection from it often has to be provided in all directions.

If the scattering surface is in the useful beam [8] and 1 metre from the source, then

$$\frac{B_s D_s}{D_u} = \frac{Pd_s^2}{WT} \tag{14.4}$$

where B_s = allowable barrier transmission of the scattered radiation, expressed as a fraction of unity

D_s = attenuated dose in the scattered beam 1 m from the scatterer

D_u = incident dose on the scatterer

d_s = distance in metres from the scatterer to the position in occupied space

If the scatterer is located f metres from the source, the equation becomes

$$\frac{B_s D_s}{D_u} = \frac{Pd_s^2 f^2}{WT} \tag{14.5}$$

Values of D_u/D_s for a range of X-ray energies are available (Table 14.9). For all X-ray energies, the useful beam attenuation curve may be used.

In the range 50 to 250 kVp, the energy of scattered radiation is at a maximum for material of medium atomic number. For 4 MV X-rays, the energy increases with atomic number of the scatterer. The amount of scattered radiation also depends on the material, and is higher for lower atomic number scatterers [63].

When a secondary protective barrier is to shield against both leakage and scattered radiation, and the thickness required for either is about the same, the total transmitted radiation will be appreciably greater than *P*. Therefore, one-third of a "tenth-value layer" (TVL), should be added to the larger thickness of barrier. If the calculated thicknesses differ by more than one TVL, then the thicker barrier is adequate [8].

TABLE 14.9 *Ratio of the Dose Incident on the Scatterer* (D_u) *to the Unattenuated Dose in the Scattered Beam at 1 metre from the Scatterer* (D_s)
(From ICRP, *Report of Committee III.* Oxford, Pergamon, 1960)

X-rays generated at	D_u/D_s
500 kV or less	1,000
1,000 kV	48
2,000 kV	4
3,000 kV	0·8

DESIGN OF INSTALLATION

Whenever possible, the X-ray beam should be totally enclosed and the enclosure inaccessible during the irradiation [8]. Then it is only necessary to provide suitable interlocks and other warning devices to ensure that there is no accidental entry, provided also that the level of radiation outside the enclosure is below maximum permissible levels.

In many cases, however, open installations are unavoidable. These are designated radiation areas [8]. Authorized persons have access to these areas, but usually may not remain there during irradiation. Protection must be adequate so that MPD levels are not exceeded. For radiotherapy at over 100 kVp, only the patient may be in the treatment room during exposure [8], and interlocks must be provided so that the set is switched off before another person enters, unless the radiation level in the room is reduced below 10 mR/h at 1 metre from the tube housing, or below 2 mR/h on average. For radiotherapy at under 100 kVp, audible or visible signals must indicate when the tube is in operation.

Entry to open installations may be by suitably protected doors, or by a maze made from protective barriers and including a safety alarm to prevent unauthorized entry. The latter arrangement is usually chosen for very high-energy X-rays, when suitable doors would be very heavy. Greene and Stephenson have published details of the design of a treatment room using a 4 MV linear accelerator [64], and Jones [65] has studied general protection problems in the planning of new radiotherapy departments.

Guidance in planning a diagnostic X-ray department is provided by the Ministry of Health [66]. Their report advocates a properly protected fixed control enclosure provided with a fixed control unit and a lead glass observation window. General design requirements have also been considered by Westman [67], who recommends limiting the size of a department, and if necessary building a second department elsewhere in the hospital.

Dealler [68] has studied the design of new departments with protection aspects particularly in mind; he has suggested building baffle walls in front of doorways, so that the door itself has to be only a scatter barrier and consequently does not need heavy shielding. He has pointed out the advantage in siting x-ray departments on upper floors, as this reduces the likelihood of radiation reaching occupied areas outside. However, in directing useful beams through windows or ordinary walls of high and isolated buildings, there is a risk of irradiating workers occasionally outside, such as window cleaners and painters, who could receive very high doses.

Sometimes, as in radiotherapy, it is essential to be able to see into an x-ray room (Chapter 9). A window may be built into a wall provided that it does not transmit radiation above the permitted level, and this usually implies that it must provide the same amount of radiation protection as its surrounding wall. For x-rays generated at up to a few hundred kVp, lead glass is suitable. For higher energies, the window may consist of transparent walled tanks filled with a salt solution of high density: a 42-inch thickness of zinc bromide solution with a density of at least 2·35 g/cm^3 has been used in a tank to form a window for viewing patients undergoing treatment with a 4 MV linear accelerator [64]. In this energy range where the atomic number of the screening material is unimportant, thick sheets[48] of good quality glass with oil films of the same refractive index as the glass between the sheets, may make an economical window. An alternative is to use closed circuit television [69].

The Useful Beam

Control is necessary to ensure the correct exposure. The size and energy of the beam and its direction and the focal distance must also be carefully adjusted.

TIMERS AND DOSEMETERS

By themselves, timers do not ensure that the correct exposure is made, because fluctuations in electrical supply may lead to output variations. However, their use prevents gross accidental over- or under-exposure, and is required by the ICRP [8] for medical work. When dose is controlled by an automatic timer, the exposure rate may be kept constant by close attention to the meters on the control panel, or by a stabilizing circuit [70]. Alternatively, an integrating dosemeter[49] may terminate the exposure after a predetermined dose has been delivered [71] (Chapter 13).

During ciné radiography, very short exposure times may have to be accurately controlled, and special apparatus has been designed for this [72], as well as for general diagnostic work [73].

FIELD-SIZE LIMITERS AND SHUTTERS

X-ray beams should be of the smallest practicable area in order to reduce scatter, and in medical and veterinary work to avoid unnecessary irradiation of the patient [5]; in diagnostic work, limitation of beam size is the most productive way in which to reduce the radiation dose to the population at large [74].

When the high voltage across an x-ray tube may be raised only slowly, a shutter operated from the control desk can be opened after the running-up of the voltage.

In x-ray diffraction work, even short accidental exposures may lead to disastrous consequences, because the exposure rate may be as high as 10^5 R/min. Physical constraints are required to make accidental exposure impossible. Many ingenious and sometimes complicated devices have been installed in attempts to provide full safety without unduly frustrating the user. Protection from the primary beam may be by a flap which covers the tube housing as soon as the camera is withdrawn [75, 76]. A warning light should indicate that the x-rays are on. Warning of stray radiation is an additional safeguard, but is not easy to arrange because the primary beam is very narrow, and therefore any stray radiation tends to vary in intensity from point to point. End-window Geiger tubes are not entirely satisfactory in a radiation warning system, because of their confined geometry, and thin-walled glass Geiger tubes may be preferable, particularly if the x-ray set is housed in a cupboard, whose walls scatter back any stray radiation. A high signal rate may be made to break the high-voltage circuit. In addition, a loudspeaker[50] indicating the count rate gives a warning just before the x-rays are switched off.

BEAM DIRECTION

The direction of the useful beam is important from two points of view regarding protection.

First, with regard to radiation levels outside the x-ray enclosure, protection is greatly simplified if the beam can be fixed in direction, and to a lesser extent if its movement is limited, so that some protective barriers have to shield only stray radiation. When there is limited movement, this must be by mechanical restraints.

Secondly, minimum-sized fields can be used only when beam direction is accurate. Beam direction entails the relative positioning of (1) x-ray tube, (2) specimen or patient and, when relevant, (3) fluorescent screen or film.

The specimen or patient must not move. In medical work with adults or older children, it is usually necessary only to make them comfortable and ask them to remain still; they should be under observation. The period of immobilization must include setting-up as well as treatment. Babies and animals, unless they are anaesthetized, usually require some form of restraint such as a harness [77]. When they have to be held, this should be done by someone not occupationally exposed; he should be outside the useful beam and should wear protective clothing. In dental radiography, the film may be held by the patient if it cannot be fixed in position. Exceptionally, it may be held by another person provided that he is not a radiation worker, but never by the dentist or any of his staff.

In x-ray diffraction work, special care must be taken to avoid exposure while lining-up. For the alignment of the specimen, the fluorescent screen should be mounted on a rod and the screen viewed indirectly using inclined mirrors or a television screen. If this is impossible, the screen may be viewed directly, but from the side. Special safe mounting systems have been devised [78].

Medical fluoroscopy should be carried out only when other methods will not suffice. The screen must be protected by a lead glass sheet having a lead equivalent of at least 1·5 mm for 75 kVp x-rays, 2·0 mm for 100 kVp x-rays and an additional 0·01 mm per kVp for higher voltages. Tube, aperture and fluorescent screen should be aligned together so that the useful beam cannot fall outside the screen [8].

FOCAL DISTANCE

With protection aspects in mind, the ICRP [8] has recommended certain focal distances: in medical fluoroscopy, the focal–skin distance should be between 300 mm and 450 mm; in medical radiography with mobile equipment, the focal distance should be at least 300 mm. Shorter focal–skin distances would tend to deliver an unduly high skin dose on the proximal side of the patient. The output of an x-ray set is unlikely to be high enough to give a satisfactory fluorescent image at a focal–skin distance of greater than 450 mm.

FILTERS AND MONITORS

X-ray sets emit, in general, photons of a wide range of energies. The softer parts of the beam may well contribute only to scatter, and in medical use, to irradiation of the patient's skin, as they are insufficiently penetrating to contribute to the film blackening. Low-energy photons will be preferentially absorbed by a filter, and for this reason medical diagnostic equipment should have permanent aluminium filters fitted under the tube [8]; they should be at least 2 mm thick for operation above 70 kVp and at least 1·5 mm thick for lower operating voltages.

In orthovoltage radiotherapy it is common practice to obtain a range of beam qualities by inserting different filters, while maintaining the same operating tube voltage. Such filters affect the exposure rate, and any confusion could lead to grave errors in dosage. Therefore, each accessible filter must be clearly marked with its thickness and material [8], and an interlock system is desirable whereby the high voltage can be switched on only when the selected filter is in position [79].

A transmission monitoring chamber, fitted below the filters, will indicate that the exposure rate is as expected [8, 80], and provides an additional safeguard that the correct filter is in position. It does not indicate the skin dose, as this depends also on the field size and focal–skin distance.

Ionization chambers have been designed especially for monitoring diagnostic sets [73, 81].

THE RADIOGRAPHIC IMAGE

An obvious way to give only a low dose during radiography is to use sensitive photographic emulsions. Fast films[10,11] are available with which, for example, a chest x-ray may be taken for a subject of average build, with an exposure of only 6 mA-s at 65 kVp.

Intensifying screens [82, 83] or image intensifiers[4,15,20,22] [84, 85, 86] also serve to reduce the dose (Chapter 8).

In all systems, a point will be reached when further reduction of dose to the patient can be achieved only at the expense of a decline in image quality [87], and a compromise must be made.

In assessing patient dose, something allied to the total energy absorbed is the parameter of interest. It is possible to measure approximations to this: one instrument can measure R × cm^2, another can measure mW-sec. Both instruments have their merits [88], but require calibration and to be used with judgement.

Protective Clothing

Protective clothing can reduce the dose to both worker and patient. It must be carefully maintained and frequently examined to ensure that it has not deteriorated. Gloves should have no covering over the lead, so that any cracks may be revealed. Cracks frequently develop in sheet lead subject to bending, but materials[9] are available where this is not a risk, made for instance, of overlapping lead scales.

The equivalent lead thickness in clothing must be selected with the X-ray hazard in mind. For instance, for fluoroscopic examinations with X-rays up to 100 kVp, gloves [8] should have a lead equivalent of at least 0·25 mm. Body aprons must have a lead equivalent of at least 0·25 mm for X-rays of up to 150 kVp.

Spectacles of lead glass and other protective clothing are available[9,51] commercially.

Protective clothing for patients is primarily for local shielding of radio-sensitive organs such as the gonads and the eyes. The nation-wide survey carried out under the auspices of the Adrian Committee [4, 5] investigated dose to the skin, the gonads and the blood-forming organs from every type of diagnostic radiology, by a comprehensive system of measurements. A special ionization chamber[1] was developed for this purpose [89] which is substantially energy-independent for X-rays of HVL between 1 mm and 4 mm Al. This chamber is used with a portable electrometer [90] which can measure exposure rate or exposure over a wide range, so that measurements can be made either inside or outside the useful beam. A number of other special investigations have been published, particularly with reference to the gonad dose [91, 92, 93]; this can often be reduced by lead body shields[9,51] [94], and should be so reduced when practicable [8].

Warning Signs and Safety Devices

In addition to safety switches, appropriate warning of a radiation hazard must be provided. A red light, sometimes accompanied by an audible warning, may be used; they should operate only when the X-rays are switched on.

There is a British Standard for a sign of radiation hazard; replicas are available[52,53] commercially.

It should be possible to ensure that no one is advertently present in an enclosed X-ray installation when the set is in operation. A switch in the enclosure should be provided to break the high-tension circuit, and this should be used by anyone in the enclosure. As he will close the switch before he comes out, there should be an associated time delay so that x-rays cannot be generated until he is outside. Door locks should be provided with handles on the inside, so that the lock may be opened from within the enclosure.

Safety devices are particularly important in x-ray diffraction work, where dose rates are so high that exposure for even a second may be catastrophic.

Radiation Surveys

Radiation surveys are the final test of the adequacy of protective measures. A survey must be carried out before an x-ray set is put into use, after a modification likely to increase the hazard, and periodically to ensure that, both inside and outside controlled areas, radiation is below maximum permissible levels. Because the beam is narrow in x-ray diffraction work, a general low level of exposure rate cannot be taken as evidence that there is no dangerous leakage. Careful monitoring with suitable instruments or large pieces of film must be carried out on installation and at frequent intervals. Any leakage must be promptly eliminated [95].

It is often useful to find areas of leakage radiation with a sensitive monitor such as a Geiger or scintillation counter and then to measure the exposure rate using an ionization chamber which is insensitive to the energy and direction of the radiation.

Surveys must be undertaken by a qualified expert; the Radiological Protection Service (*see* p. 123) will survey any installation on request.

A survey will include examination of the protection afforded by building structures and other protective barriers, and of warning signs and signals, interlocks and similar devices and protective clothing, as well as a study of personnel monitoring techniques and of work-load [8].

The results of a survey should be submitted in a formal written report which should include any recommendations [8].

Literature References

1 Kaye, G. W. C., "The story of protection", *Radiography*, **6**, p. 59 (1940).
2 Kerley, P. "The x-ray martyrs", *Brit. Med. J.*, **ii**, p. 368 (1961).
3 Medical Research Council, *The Hazards to Man of Nuclear and Allied Radiation*, Second Report (London, H.M.S.O., 1960).
4 "*Radiological Hazards to Patients*". Interim Report of the (Adrian) Committee, 1959, London, H.M.S.O.
5 "*Radiological Hazards to Patients*". Second Report of the (Adrian) Committee, 1960, London, H.M.S.O.
6 Spiers, F. W., "Interim report on the determination of dose to bone marrow for radiological purposes", *Brit. J. Radiol.*, **36**, p. 238 (1963).
7 Osborn, S. B., "Variations in the radiation dose received by the patient in diagnostic radiology", *Brit. J. Radiol.*, **36**, p. 230 (1963).
8 International Commission on Radiological Protection, *Report of Committee III on Protection against X-Rays up to Energies of 3 MeV and Beta- and Gamma-rays from Sealed Sources*, Recommendations of the ICRP, ICRP Pubn. 3 (London, Pergamon, 1960).

9 McCarthy, R. and Mejdahl, V., "The genetically significant dose from luminous wrist watches", *Phys. Med. Biol.*, **8,** p. 279 (1963).
10 Mole, R. H., "Possible consequences of radiographic tissue dose", *Brit. J. Radiol.*, **36,** p. 241 (1963).
11 United Nations, *Report of the U.N. Scientific Committee on the Effects of Atomic Radiation,* Suppl. No. 16 (A/5216) (New York, United Nations, 1962).
12 Mayneord, W. V., "Some problems of radiation protection", *Brit. J. Radiol.*, **24,** p. 525 (1951).
13 Shiels, D. O., "Some effects on lymphoid cells of occupational and accidental exposure and ionizing radiations", *Second Australasian Conference on Radiation Biology* (London, Butterworth, 1959).
14 Evans, H. J., Court Brown, W. M. and McLean, A. S. (Eds.), *Human Radiation Cytogenetics* (Amsterdam, North Holland Publ. Co., 1967).
15 Carlson, L. D., Scheyer, W. J. and Jackson, B. H., "The combined effects of ionizing radiation and low temperature on the metabolism, longevity and soft tissues of the white rat", *Radiat. Res.*, **7,** p. 190 (1957).
16 Cork, J. M., "Gamma-radiation and longevity of the flour beetle", *Radiat. Res.*, **7,** p. 551 (1957).
17 Kostin, I. G., "Experience with the use of microdoses of gamma-irradiation in poultry breeding", *Biophys.*, **5,** p. 572 (1960).
18 International Commission on Radiological Protection, *Recommendations of the ICRP,* ICRP Publ. 9 (London, Pergamon, 1966).
19 International Commission on Radiological Protection. *Protection against Electromagnetic Radiation above 3 MeV and Electrons, Neutrons and Protons,* Recommendations of the ICRP, ICRP Pubn. IV (1953–1959) (London, Pergamon, 1964).
20 National Academy of Sciences, *Long-term Effects of Ionizing Radiations from External Sources,* Committee on Pathological Effects of Atomic Radiation. Publ. 849 (Washington, D.C., Nat. Acad. Sci. Nat. Res. Council, 1961).
21 Ebert, M. and Howard, A. (Eds.), *Radiation Effects in Physics, Chemistry and Biology* (Amsterdam, North Holland Publ. Co., 1963).
22 Carter, T. C., Lyon, H. F. and Phillips, R. J., "Induction of sterility in male mice by chronic gamma irradiation", *Brit. J. Radiol.*, **27,** p. 418 (1954).
23 Neary, G. J., "Ageing and radiation", *Nature*, **187,** p. 10 (1960).
24 Lindop, P. J. and Rotblat, J., "Shortening of life and causes of death in mice exposed to a single whole-body dose of radiation", *Nature,* **189,** p. 645 (1961).
25 Loutit, J. F., *Irradiation of Mice and Men* (Chicago, Univ. Press, 1962).
26 Crosfill, M. L., Lindop, P. J. and Rotblat, J., "Variation of sensitivity to ionizing radiation with age", *Nature*, **183,** p. 1729 (1959).
27 Lawson, B. G. *et al.*, "Effect of single and divided doses of x-irradiation on longevity of rats", *Radiat. Res.*, **18,** p. 255 (1963).
28 Failla, G. and McClement, P., "The shortening of life by chronic whole-body radiation", *Amer. J. Roent.*, **78,** p. 946 (1957).
29 Court Brown, W. M. and Doll, R., *Proc. Second U.N. Internat. Conf. on Peaceful Uses of Atomic Energy, Geneva,* **23,** p. 179 (1958) (New York, U.N., 1958).
30 Warren, S., "Longevity and causes of death from irradiation in physicians", *J. Amer. Med. Ass.*, **162,** p. 464 (1956).
31 Bacq, Z. M. and Alexander, P., *Fundamentals of Radiobiology* (London, Pergamon, 1961).
32 National Academy of Sciences, *Effects of Ionizing Radiation on the Human*

Hemapoietic System, Committee on Pathologic Effects of Atomic Radiation, Publ. 875 (Washington, D.C., Nat. Acad. Sci., Nat. Res. Council, 1961).
33 STEWART, A., PENNYBACKER, W. and BARKER, R., "Adult leukaemias and diagnostic x-rays", *Brit. Med. J.*, **ii**, p. 882 (1962).
34 COURT BROWN, W. M. and DOLL, R. *Leukaemia and Aplastic Anaemia in Patients Irradiated for Ankylosing Spondylitis*, Medical Research Council Special Report 295, (London, H.M.S.O., 1957).
35 STEWART, A., WEBB, J. and HEWITT, M. A., "A survey of childhood malignancies", *Brit. Med. J.*, **i**, p. 1495 (1958).
36 COOPER, A. G. S. and STEINBECK, A. W., "Leukaemia following irradiation *in utero*", *Brit. J. Radiol.*, **32**, p. 265 (1959).
37 LANCET, "Harmful effects of diagnostic irradiation", *Lancet*, 1963, **i**, p. 254.
38 MACMAHON, B., "Prenatal x-ray exposure and childhood malignancies", *J. Nat. Cancer Inst.*, **28**, p. 1173 (1962).
39 COURT BROWN, W. M. *et al.*, "Geographical variations in leukaemia mortality in relation to background radiation and other factors", *Brit. Med. J.*, **i**, p. 1753 (1960).
40 LAMERTON, L. F., "An examination of the clinical and experimental data relating to the possible hazard to the individual of small doses of radiation", *Brit. J. Radiol.*, **31**, p. 229 (1958).
41 PETERSON, O., "Radiation cancer. Report of 21 cases", *Acta. Radiol.*, **20**, p. 221 (1941).
42 CADE, S., "Radiation-induced cancer in man", *Brit. J. Radiol.*, **30**, p. 393 (1957).
43 MITCHELL, J. S. and HAYBITTLE, J. L., "Cancer of the skin appearing 49 years after a single diagnostic Roentgen exposure", *Acta. Radiol.*, **44**, p. 345 (1955).
44 CARTER, T. C., LYON, M. F. and PHILLIPS, R. J. S., "Genetic hazard of ionizing radiations", *Nature*, **182**, p. 409 (1958).
45 PURDOM, C. E., *Genetic Effects of Radiation* (London, Newnes, 1962).
46 STEVENSON, A. C., "Comparison of mutation rates at single loci in man", *Effect of Radiation on Human Hereditary*, p. 125. (Geneva, W.H.O., 1957).
47 FAHMY, O. G. and FAHMY, M. J., "Cytogenic analysis of the action of carcinogens and tumour inhibitors in *Drosophila melanogaster*. V.", *J. Genet.*, **54**, p. 146 (1956).
48 CARTER, T. C., "Radiation-induced gene mutation in adult female and foetal male mice", *Brit. J. Radiol.*, **31**, p. 407 (1958).
49 RUSSELL, W. L. and RUSSELL, L. B., *Proc. Second U.N. Internat. Conf. on Peaceful Uses of Atomic Energy*, **22**, p. 360 (New York, U.N., 1958).
50 RUSSELL, W. L. and KELLY, E. M., "Comparison between mutation rate induced by chronic gamma- and acute x-irradiation in mice", *Science*, **127**, p. 1062 (1958).
51 WORLD HEALTH ORGANIZATION, *Protection Against Ionizing Radiation. A Survey of Existing Legislation*, (Geneva, W.H.O., 1964).
52 SIM, D. F., "Atomic energy legislation and codes of practice", *Atom*, **59**, p. 7 (1961).
53 UNITED KINGDOM ATOMIC ENERGY AUTHORITY, *Training in Radiological Health and Safety*, Report of a Committee appointed by the U.K.A.E.A. (London, H.M.S.O., 1960).
54 LANGMEAD, W. A. and ADAMS, N. (Eds.), *Proc. Symposium on Film Badge Dosimetry, Harwell*, 1961, A HSB(RP) R 19 (Harwell, U.K.A.E.A., 1962).
55 BRITISH STANDARDS INSTITUTION, *Direct-reading Personal Dosemeter for x- and γ-Radiation*, British Standard 3385: 1961 (London, H.M.S.O.).
56 TAYLOR, D., "Radiac instrumentation", *J. Sci. Inst.*, **29**, p. 315 (1952).

57 Miyanagi, I. and Yamamoto, H., "Studies in silver-activated metaphosphate glass as a personal dosemeter", *Health Phys.*, **9,** p. 965 (1963).
58 Henson, P. W., "A photographic dosemeter for the measurement of personal doses of environmental radiation", *Phys. Med. Biol.*, **8,** p. 423 (1963).
59 Price, B. T., Horton, C. C. and Spinney, K. T., *Radiation Shielding* (London, Pergamon, 1957).
60 Greene, D. and Massey, J. B., "Some measurements on the absorption of 4 MV x-rays in concrete", *Brit. J. Radiol.*, **34,** p. 389 (1961).
61 Bashandy, E., "Shielding properties of different types of concrete", *Int. J. App. Rad. Isot.*, **13,** p. 87 (1962).
62 Kennard, O., Martin, A. J. P. and Woodger, L., "Labyrinth radiation trap for use with standard x-ray diffraction equipment", *J. Sci. Inst.*, **36,** p. 48 (1959).
63 Martin, J. H. and Muller, G. M., "Quantity and quality of scattered x-rays", *Brit. J. Radiol.*, **34,** p. 227 (1961).
64 Greene, D. and Stephenson, S. K., "The design of a treatment room to house a gantry-mounted 4 MeV linear accelerator", *Brit. J. Radiol.*, **34,** p. 640 (1961).
65 Jones, B. E., "Some protection aspects in the planning of new radiotherapy departments", *Brit. J. Radiol.*, **35,** p. 517 (1962).
66 Ministry of Health, *Hospital Building Note No.* 6 (London, H.M.S.O., 1961).
67 Westman, C. E., "Planning of new diagnostic departments", *Brit. J. Radiol.*, **35,** p. 265 (1962).
68 Dealler, J. F. B., "Design of new diagnostic x-ray departments—protection aspects", *Brit. J. Radiol.*, **35,** p. 269 (1962).
69 Van der Plaats, C. J., *Medical X-Ray Technique* (Eindhoven, Philips Technical Library, 1962).
70 Clayton, C. G. and Farmer, F. T., "Stabilization of x-ray therapy apparatus by ionization current", *Brit. J. Radiol.*, **22,** p. 669 (1949).
71 Watson, B. W., "A digital dosemeter for use in x-ray therapy", *Phys. Med. Biol.*, **8,** p. 161 (1963).
72 Connelly, F. C., "An apparatus for the accurate control of x-ray exposure times", *Brit. J. Radiol.*, **31,** p. 510 (1958).
73 Reinsma, K., *Dosemeters for X-Ray Diagnosis* (Eindhoven, Philips Technical Library, 1962).
74 World Health Organization, *Public Health and the Medical Use of Ionizing Radiation*, Fifth Report WHO Expert Committee on Radiation (Geneva, W.H.O., 1965).
75 Hughes, W. A., "A simple shutter-closing device for x-ray tubes", *J. Sci. Inst.*, **39,** p. 93 (1962).
76 Barnes, D. C. and Franks, A., "A safety shutter system for crystallographic x-ray tubes", *J. Sci. Inst.*, **39,** p. 648 (1962).
77 McFarlane, R. K., "A baby-holder for erect oesophageal chest and barium meal examination", *Brit. J. Radiol.*, **28,** p. 54 (1955).
78 Hughes, W. A. and Taylor, C. A., "A safe universal mounting and alignment system for x-ray goniometers", *J. Sci. Inst.*, **38,** p. 493 (1961).
79 Bower, J., Jones, J. C. and Orton, K. F., "An interlock system for a superficial x-ray set", *Brit. J. Radiol.*, **29,** p. 407 (1956).
80 Henry, W. H. and Aitken, J. H., "Monitoring ionization chamber for an x-ray generator", *Brit. J. Radiol.*, **34,** p. 516 (1961).
81 Borthwick, R. A., "An ionization chamber with short collecting time for routine testing of diagnostic x-ray equipment", *Brit. J. Radiol.*, **28,** p. 386 (1955).

82 JOHNS, H. E., *The Physics of Radiology* (Springfield, Chas. C. Thomas, 1961).
83 HANSELL, P. and OLLERENSHAW, R. (Eds.), *Longmore's Medical Photography* (London, Focal Press, 1962).
84 ROYLANCE, J., "The value of image intensification in opaque meals examination", *Brit. J. Radiol.*, **35**, p. 409 (1962).
85 NEMET, A. and HEWETT-EMETT, A., "Applications of x-ray image intensification to industrial problems", *J. Brit. Inst. Radio Engineers*, **17**, p. 523 (1957).
86 SCHWEIR, V., "One year's experience with the BV 20 combination of image intensifier and TV in casualty surgery", *Medicamundi*, **7**, p. 133 (1961).
87 SPIEGLER, G. and KEANE, B. E., "Image contrast and radiation protection: a figure of merit", *Brit. J. Radiol.*, **38**, p. 771 (1965).
88 ADRAN, G. M. and CROOKS, H. E., "The measurement of patient dose", *Brit. J. Radiol.*, **38**, p. 766 (1965).
89 OSBORN, S. B. and BURROWS, R. G., "An ionization chamber for diagnostic x-radiation", *Phys. Med. Biol.*, **3**, p. 37 (1958).
90 STUART, S. and OSBORN, S. B., "A wide-range direct-reading x-ray dosemeter", *Phys. Med. Biol.*, **3**, p. 255 (1959).
91 KEANE, B. E. and SPIEGLER, G., "Assessment of dose to the gonads outside diagnostic x-ray beams", *Brit. J. Radiol.*, **34**, p. 362 (1961).
92 EPP, E. R., WEISS, H. and LAUGHLIN, J. S., "Measurement of bone marrow and gonadal dose from chest examinations as a function of field size, field alignment, tube kilovoltage and added filtration", *Brit. J. Radiol.*, **34**, p. 85 (1961).
93 TWINN, K. and ALLSOPP, C. B., "Some problems encountered in the investigation of gonad doses received by patients undergoing dental x-ray examinations", *Brit. J. Radiol.*, **35**, p. 366 (1962).
94 BENEŠ, V., KOŠEK, P. and MICHAL, V., "Reduction of gonad dose in the radiography of the hips in infancy", *Brit. J. Radiol.*, **34**, p. 818 (1961).
94 ABRAHAMS, S. C. and BLACKMORE, W. R., "A shielding device for x-ray diffraction cameras", *Rev. Sci. Inst.*, **24**, p. 885 (1953).

Glossary

SOURCES OF DEFINITIONS

1 British Standards Institution, Glossary of terms used in radiology. B.S. 2597: 1955 (London, B.S.I., 1955).

2 International Commission on Radiation Protection, *Recommendations: Report of Committee III on Protection Against X-rays up to Energies of* 3 MeV *and Beta- and Gamma-rays from Sealed Sources* (London, Pergamon, 1960).

3 MacNalty, A. S. (Ed.), *The British Medical Dictionary* (London, Caxton, 1961).

4 International Commission on Radiological Units and Measurements (ICRU), Report 10*a*, 1962. NBS Handbook 84 (Washington, D.C., Nat. Bur. Stand., 1962).

5 Abercrombie, M., Hickman, C. J. and Johnson, M. L., *A Dictionary of Biology* (London, Penguin Books, 1951).

6 Ditchburn, R. W., *Light* (London, Blackie, 1963).

7 Merriman, A. D., *Concise Encyclopaedia of Metallurgy* (London, Macdonald and Evans, 1966).

Word	*Source o definition*	*Definition*
Annual genetically significant dose to a population	2	The average of the individual gonad doses each weighted for the expected number of children conceived subsequent to exposure.
Ankylosing spondylitis	3	Poker spine: a condition in which the ligaments of the spine ossify and the lateral joints become ankylosed. The sacro-iliac joints are also affected.
Chromosome	1	A rod-shaped or thread-like body of chromatin, in the nuclear structure of a cell, which splits longitudinally as the cell divides, one half going to each daughter cell.
Controlled area	2	Where persons occupationally exposed could receive doses in excess of 1·5 rems/yr.
Cancer	3	1. A malignant tumour arising from epithelial cells. 2. More loosely, any malignant growth.
Carcinoma	3	A malignant epithelial tumour which tends to spread locally and generally throughout the body, destroying normal tissues as it progresses and eventually proving fatal. It is most common after middle age.
Coherence	6	Two beams of light are said to be mutually coherent when the phase difference between the waves by which they are represented is constant during the period normally covered by observations.
Cold shut	7	A casting defect that arises from the freezing-over of the surface of the ingot or casting before the mould has been filled.
Fettle	7	"To put right," applied to cleaning and dressing operations carried out on castings and forgings.
Gene	1	The biological unit of heredity. Genes are generally considered as occupying specific sites on the chromosomes.
Genetic dose to a population	2	The dose which, if it were received by each person from conception to the mean age of child-bearing, would result in the same genetic burden to the whole population as do the actual doses received by the individuals.
Half-value layer (HVL) or Half-value thickness (HVT)	2	The thickness of a specified absorbing material which, when introduced into the path of a beam of radiation, reduces the dose-rate to one-half of its original value.
Haversian canals / Havers' canals	3	The fine canals in bone which contain blood vessels and are surrounded by concentric lamellae of bone.
Leukaemia	3	A fatal disease of unknown aetiology, showing gross disturbance of the leucocytopoietic tissues, first in the bone marrow and then affecting the blood and other organs. Any or several of the types of white blood cells may be involved, thus characterizing the particular leukaemia . . .
Leukaemia, acute myeloid	3	Acute myelosis, an acute leukaemia in which the main cellular disturbance concerns the myeloblasts and myelocytes.
Leukaemia, myeloid	3	Chronic myelosis, a chronic form of leukaemia with mild to very great splenomegaly and hepatomegally, and usually of relatively slow progress (from three to sixteen years); it usually shows a

Word	*Source of definition*	*Definition*
		very high myeloid leucocytosis in blood and bone marrow when first diagnosed. It may follow exposure to toxic and radioactive agents, and certain other diseases of the blood, but is usually of unknown aetiology.
Linear energy transfer	4	The linear energy transfer (L) of charged particles in a medium is the quotient of dE_L by dl where dE_e is the average energy locally imparted to the medium by a charged particle of specified energy in traversing the distance dl. It may conveniently be expressed in keV/mμ.
Locus	5	A particular position in a particular chromosome which always contains one kind of gene, or one of a particular set of allelomorphic genes. Homologous chromosomes contain identical sets of loci.
Mutation	1	A change in the characteristics of an organism produced by an alteration in the hereditary material.
Mean lethal dose	1	That single dose of whole body irradiation which will cause the death within a specified period of 50 per cent of those receiving it. Abbreviation: MLD
Occupational exposure	2	Exposure of an individual who normally works in a controlled area constitutes occupational exposure
Occupancy factor	2	The factor by which the work-load should be multiplied to correct for the degree or type of occupancy of the area in question. This factor may be based in general on the degree of occupancy averaged over a year.
Orthovoltage radiation	2	Generated at 150 to 400 kVp
Rad	4	The unit of absorbed dose. It is 100 erg/s.
Radiation protection officer	4	One who has the knowledge and responsibility to apply appropriate radiation protection regulations.
r.b.e.	4	Relative biological efficiency $= \dfrac{\text{biological efficiency of radiation under investigation}}{\text{biological efficiency of x-radiation of energy 250 kV}}$ $= \dfrac{\text{dose in rads to produce effect with 250 kV x-rays}}{\text{dose in rads to produce same effect with radiation under investigation}}$
Sarcoma	3	A malignant tumour of connective tissue or its derivative.
Stopping power	1	The energy loss suffered by the particle per unit length of its path in the material. It is usually expressed as MeV per cm and although often treated as a constant, it does in fact, vary along the path of the material.
Stopping power (mass)	4	The mass stopping power (s/p) of a material for charged particles is the quotient of dE_s by the product of dl and p, where dE_s is the average energy lost by a charged particle of specified

Word	*Source of definition*	*Definition*
		energy in traversing a path of length dl, and p is the density of the medium. $$\frac{S}{P}=\frac{l}{P}\frac{dE}{dl}$$ It may be measured in erg-cm²/g.
Tenth-value layer (TVL)	2	The thickness of a specified absorbing material which, when introduced into the path of a beam of radiation, reduces the dose rate to one-tenth of its original value.
Use factor	2	The fraction of the work-load during which the useful beam is pointed in the direction under consideration.
Work-load	2	A measure in suitable units of the amount of use of radiation equipment.

List of Manufacturers

1 Electronic Instruments Ltd., Lower Mortlake Road, Richmond, Surrey.
2 Baldwin Instrument Co. Ltd., Dartford, Kent.
3 Victoreen Instrument Co., Grove House, London Road, Isleworth, Middx.
4 Mullard Ltd., X-ray Division, New Road, Mitcham Junction, Surrey.
4*a* Mullard Equipment Ltd., Manor Royal, Crawley, Sussex.
5 20th Century Electronics Ltd., Centronics Works, King Henry's Drive, New Addington, Surrey.
6 Pye Unicam Ltd., York Street, Cambridge.
7 Bush Beach and Segner Bayley Ltd., 175, Tottenham Court Road, London, W.1.
8 D. Penneltier and Co. Ltd., 28, Hatton Garden, London, E.C.1.
9 Cuthbert Andrews, 5, High Street, Bushey Village, Herts.
10 Kodak Ltd., Kodak House, Kingsway, London, W.C.2.
10*a* Kodak Ltd., Medical Sales Division, Kodak House, Kingsway, London, W.C.2.
11 Ilford Ltd., Ilford, Essex.
11*a* Ilford Ltd., Dept. of Radiography and Medical Photography, Tavistock House North, Tavistock Square, London, W.C.2.
12 Industrial X-rays Ltd., 54, Walsall Street, Willenhall, Staffs.
13 Electronic and X-ray Applications Ltd., 13–17, Pennart Mews, London, W.1.
14 Gevaert Ltd., Gt. West Road, Brentford, Middx.
15 Sierex Ltd., 15–18, Clipstone Street, London, W.1.
16 Evans Electroselenium Ltd., 110, Potter Street, Harlow, Essex.
17 Joyce, Loebl and Co., Gateshead-on-Tyne.
18 Loxford Equipment Co. Ltd., 77/85, Tanner St., Barking, Essex.
19 Watson and Sons (Electro-medical) Limited, East Lane, North Wembley, Middx.
20 General Radiological Ltd., County Buildings, Honeypot Lane, Stanmore, Middx.
21 Philips Electrical Ltd., Medical Apparatus Division, 45, Nightingale Lane, Balham, London, S.W.12.
21*a* Philips Electrical Ltd., Medical X-ray Division, 45, Nightingale Lane, Balham, London, S.W.12.
22 Marconi-Elliott Ltd., St. Albans, Herts.
23 Thorn Electrical Industries Ltd., Gt. Cambridge Road, Enfield, Middlesex.
24 A. E. Dean and Company Ltd., Progress Way, Waddon, Surrey.
25 Exal, Houndmills Estate, Basingstoke, Hants.
26 Bengué and Co. Ltd., Mount Pleasant, Wembley, Middx.
27 Damancy and Co. Ltd., Ware, Herts.

28 Glaxo Laboratories Limited, Greenford, Middx.
29 Lewis Laboratories Ltd., Leeds 9.
30 May and Baker Ltd., Dagenham, Essex.
31 Pharmaceuticals (London) Ltd., Victoria Way, Burgess Hill, Sussex.
32 General Electric Co. Ltd., Trafford Square, Manchester.
33 Vickers Armstrong Ltd., South Marston Works, Swindon, Wilts.
34 Akashi Electron Microscopes; U.K. Agents A. Gallenkamp and Co. Ltd., Technico House, Christopher St., E.12.
35 Matchlett Laboratories Inc., Springdale, Connecticut, U.S.A. (*see* 19).
36 Rank Precision Industries Ltd., St. Pancras Way, London, N.W.1.
37 High Voltage Engineering Corp., Burlington, Mass., U.S.A.
38 Radiological Protection Service, Clifton Avenue, Belmont, Sutton, Surrey.
39 R. A. Stephen and Co. Ltd., 126, Lavender Avenue, Mitcham, Surrey.
40 Nuclear Enterprises Ltd., Bath Road, Beenham, Reading, Berks.
41 D. A. Pitman Ltd., 91, Heath Road, Weybridge, Surrey.
42 Unibrix Radiation Shieldings Ltd., 190, Piccadilly, London, W.1.
43 X-ray Engineering (London) Ltd., 19, Broad Court, Drury Lane, London, W.C.2.
44 Premise Engineering Co. Ltd., The Crescent, Abbots Langley, Herts.
45 Research and Control Instruments Ltd., (*see* 6).
46 Inspection Equipment Ltd., 19, Broad Court, Drury Lane, London, W.C.1.
47 G.E.C. Component Sales Department, Osram Division, East Lane, North Wembley, Middx.
48 Chance-Pilkington Optical Works, Glascoed Road, St. Asaph, Flintshire.
49 Radiatron, 7, Sheen Park, Richmond, Surrey.
50 Mini-Monitors, 36, Southampton Street, London, W.C.2.
51 W. S. Rothband and Co. Ltd., Bent Street, Manchester, 8.
52 Continental Distributors Ltd., 81, Cromwell Road, London, S.W.7.
53 Jencons Ltd., Mark Road, Hemel Hempstead, Herts.

Index

ABDOMEN, 263
Abnormalities, developmental, 249, 252
Absorption coefficient, 31, 36, 47, 207
  atomic, 32
  electronic, 34
  linear, 4, 32, 35, 47, 100, 126
  pair production, 47
  partial, 47, 48, 78
  photoelectric, 45
  mass, 34, 35, 208, 275
Absorption edge, 44, 160, 207, 208
Absorption edge filters, 205
Absorption limit, 44
Absorption, preferential, 208
Absorption, self-, 207
Accelerator, linear, 168, 212, 240, 242, 277, 278
  see also Electron, acceleration of
Action level, definition of, 251
Acts of Parliament, 256
Adrian Committee, 248, 281
Adults, 260, 279
Age, 164, 251, 252, 260, 261, 264
Air dose, 103
Alignment, 176, 263, 279
  see also Setting-up
Alloys, 154, 220, 228, 229, 276
American Society for Testing Materials, 209
Analysis, x-ray chemical, 202 ff
  absorption methods, 202, 207
  diffraction methods, 202, 208
  fluorescence methods, 202 ff, 206
  non-dispersive methods, 205
  quantitative, 195, 205 ff, 209, 210
  spectroscopic methods, 202 ff, 205
Ångstrom unit, 2, 10, 11
Animals, 213, 251, 252, 254, 279
Ankylosing spondylitis, 164, 249, 288
Annihilation radiation, 46
Anode, 221 ff, 227, 231, 235
Anode circuit, 228, 232
Anode current stability, 224
Anode temperature, 224, 225, 226, 235
Anoxia and radio-resistance, 165, 166
Antenatal radiography, 253
Antrum, 174
Anus, 169
Apparatus, care of, 224 ff, 239
  see also Failure
Applicator, 93, 94, 134, 169, 171
Aqueous media, 212
Art, applications of x-rays in, 160
Artefacts, 113, 115, 159
Asterism, 196
Astronomy, x-ray, 13
Atom arrangement in crystals, 190, 193, 199, 208
Atomic bombs, 252
Atomic cross-section, 32
Atomic Energy Authority Act, 1954, 256
Atomic number, 205, 223, 265, 275, 277, 278
  effective, 19, 50
Atomic size, 200
Atomic structure, 23 ff
Attenuation, 31, 47, 265, 269, 270, 271, 272, 273, 274, 275, 276
Attenuation processes, 35
Auger effect, 29
Automation, 171, 262
Avogadro's constant, 10, 34

BABIES, 279
Background elimination, 146, 148
Back pointer, 176
Backscatter factor (BSF), 99, 101
Badges, film, 122, 262
  see also Monitoring personal
Barriers, 264, 265 ff, 275, 276, 277, 278, 279, 282
  primary, 265 ff, 277
  see also Protection
Barium meal, 248
Barium salts, 150, 151, 275
Barium spray, 151
Barkla, 24

Barn (unit), 32
Bashandy, 275
Beam, 32, 265, 275
  contamination by electrons, 92
  direction, 159, 279
    see also Alignment
  divergence, 224
  limitation, 93, 94, 95, 134, 152, 169, 223
  scanning by, 206
  traps for, 275
  useful, 275, 276, 278 ff, 280, 281
Beryllium, 153, 219
Beryllium window, 169, 216, 219
Betatron, 240, 244, 245
Biological effects, 164, 166, 167, 249, 250, 251 ff, 258, 259, 261, 262
  see also Damage, Genetic effects, Somatic effects
Bladder cancer, 151
Blocking material, 134, 159
Blood, 249, 253, 260, 261, 281
Blood-forming organs, 253
  see also Bone marrow
Blood diseases, 248
Blood supply and radio-resistance, 165
Body contour, 166, 170, 171
Body inhomogeneities, 104, 171, 173
Body thickness, equivalent, 105
Bohr model of atom, 24
Bolus, 171
Bonding, chemical, 199, 200, 216, 217
Bone, 105, 151, 168
  permissible dose, 105
Bone marrow, 248, 253, 260, 262, 264
Bragg, 191, 203
Bragg angles, 195
Bragg planes, 11, 192, 193, 195, 196, 197, 198, 209
Bragg reflections, 11, 192, 193, 195, 196, 197, 198, 209
Bragg–Gray principle, 66
Bragg's equation, 157
Bragg's law, 8, 191, 192, 194, 195, 196, 201, 202
Brain, 252
Bravais lattice, 189, 190, 192, 194, 195, 196, 197, 198, 205
Bravais points, 189, 190, 196
Bravais planes, 197, 198, 205
Breast, 151, 166, 169
Buccal cavity, 169
Buildings, 275, 276, 277, 278, 282
  see also Barriers, Doors, Floors, Installations, Walls, Windows
Burns, skin, 248

CABLE, 228, 234, 235, 238, 239
  terminations, 228, 239
Cade, 253
Cadmium sulphide sheet, 139
Calcite crystal, 185, 190
Callier coefficient, 116
Camera, 194, 239, 274
  electron diffraction, 227
  powder, 194, 195, 208, 209
  Weissenberg, 198
Cancer, 151, 164, 165, 253
Capacitor, 219, 236, 238, 263
Carbon dioxide, 151, 165, 242
Carcinogenesis, 164, 253
Carcinoma, 169, 216, 253
Cardioangiography, 150
Cassette, 137, 138, 153, 159
Castaigne, 205
Casting metals, 153, 154, 155
Castings—
  defects in, 155 ff
  radiography of, 156
Castration, 252
Cataract, radiation-induced, 251
Cathode, 219, 222, 223, 227
Cathode-ray tube, 206
Cell—
  count, white, 249
  cycle, 166
  growth rate, 174
  population kinetics, 168
  volume, 166
Cells, germ, 254
Central axis depth dose, 102, 103, 166, 171
Ceramics, 200
Cervix uteri, 176
Chamber, ionization, *see* Ionization chamber
Characteristic curve of films and plates, 117, 127, 131
Characteristic radiation, *see* Radiation
Characteristic spectrum, 23
Characteristic wavelength, 205, 206, 208
Chemical analysis
  see Analysis, x-ray chemical
Chemical bonds, 199, 200, 216, 217
Chemical reactions, 212, 215–17
Chemical yield of radio-chemical reaction, 212
Chemotherapy, 175, 176
Chess board, 174
Chest, radiography of, 149, 280
Chest wall, 169
Childbearing age, 164, 260, 264
Childhood malignancies, 249, 252, 253
Children, 248, 253, 263, 264
Chromium, 223
Chromosome, 213, 249, 254
Cine-radiography, 149, 233, 278
Circuit, 233, 234, 235, 236, 237, 240, 241, 242
  anode, 228, 232
  constant voltage, 225

Circuit, *contd.*
full-wave, 233, 235
half-wave, 225, 233, 234, 242
safety, 226
transformer, 231
Classified worker, 260, 261
Clothing, protective, *see* Protective clothing
Cobalt-60, 212
Cockroft, 242
Codes of practice, 251, 255, 257, 258, 259, 260, 262
Cold emission, 219, 227
Collimator, 93, 95, 202
Colon, 151
Coma in x-ray reflection, 11
Compensating filter, 177
Compensator, metal, 105
Compression of tissue, 142, 170
Compton absorption, 151, 265
Compton process, 36, 127, 153, 275
Compton scatter, 40
Compton's experiments, 40
Computers, 193, 205, 209
Concrete, 273, 274, 275
Condenser chamber, 64
Conduction detectors, 139
Cone, beam-limiting, 93, 152
Cone of rays, 194, 196
Connective tissue, 151
Conservation of parity, 28
Contrast, 130, 135, 140, 151, 159
Contrast medium, 127, 143, 148, 150, 151
Contrast, photographic, 108, 110, 117, 119, 123
Contrast, subject, 126, 127, 129
Control in installations, 239, 277, 278
Controlled areas, 261, 282
Controlling authority, 258, 259
Conversion efficiency, energy, 224
Conversion factors (röntgens to rads), 98
Coolidge, 220, 221, 231
Cooling, 219, 225, 226, 234
Copper, 153, 154, 157, 158, 190, 219, 226
Copper K_α radiation, 208
Copper sulphate, 191, 195
Corona discharge, 141, 229
Correction factors, 122, 123
Corroded pots, 160
Cosmic rays, 248
Cosslett, 221
Counters, Geiger, *see* Geiger counter
Counters, proportional, 74 ff, 203, 205, 206
Counters, rectifying junction, 72
Counting methods, 208
Court Brown, 253
Cracks in castings and welds, 128, 133, 156, 157, 159
Crystallites, 193, 195
Crystallography, 119, 120, 185 ff, 190, 191, 194, 196, 223
Crystallography, *contd.*
beam limitation in, 95
oscillation and rotation methods, 196 ff
stationary crystal, 201
Crystals, 185, 190, 199, 200, 201, 209
binding forces in, 199
chemistry of, 199
curved, 203, 207
diffraction by, 195, 202
growth of, in metals, 190
habit of, 188
lattice, *see* Bravais lattice
mixtures of, analysis, 208
orientation of, 195, 196, 197, 209
planes in, 197, 202, 203
size of, 195
stationary, 201
structure of, 190, 192, 193, 194, 196, 199, 200, 201, 203, 207, 208
symmetry of, 195, 196
systems of, 189
Current, 225, 227, 229, 233, 235, 236
filament, 228, 233
tube, 221, 223
Current cycle, 234, 235, 236
Cut-outs, lead, 134
Cystography, contrast, 151

DAMAGE by radiation, 166, 224, 232, 248, 249, 251, 253
see also Biological effects
Darkroom, 114, 115
Dead time in counters, 76
Dealler, 278
de Broglie equation, 27
Debye-Scherrer rings, 195
Deep therapy, 168, 169
Defects in castings and welds, 156, 157, 159
Definition in radiographs, 127, 128, 130, 132, 134
Degassing tubes, 220
Dendrites, 155
Densitometer, 117, 120
Density of barrier materials, 265, 275
Density, photographic, 116, 117, 119, 157, 159
energy dependence of, 119
Dental use, code of practice, 257, 258, 260
Dental work, 147, 152, 279
Dentist, 279
Depilation, 164
Dermatitis, 110, 248
Designated persons, 260, 262
Detectors, conduction, 139
Developer, photographic, 110, 111, 112
Development of films, 109, 110, 111
Developmental abnormalities, 249, 252
Deviation, angle of, 192, 199, 197, 198

Diagnostic radiology, 248, 252, 253, 278, 280, 281
Diamond, 193, 200, 208
Diaphragm in linear accelerator, 243, 244
Diaphragm, Potter-Bucky, 136
Diaphragm, 93, 94, 95
Diffractometer, 96, 205, 207, 208, 209, 210
Diffracted beams, 195, 196
Diffracting crystal, 202
Diffraction, 9, 10, 190, 192, 195, 199, 203, 208, 210, 220, 227, 233, 235, 251, 275, 279
  electron, 201
Diffraction grating, 9, 191
Diffraction mottling, 157, 160
Diffraction patterns, 195, 202, 208, 209, 210
  card index of, 209, 210
Diffusable substance, 174
Diffusion current, 60
Discharge, electrical, 234
  corona, 229
  silent, 229
Disease resistance, 213
Doctor, appointed, 259
Donut, 246
Doors, 277, 278, 282
Dose, 52, 56, 66, 88, 98, 100, 166, 212, 213, 214, 215, 244, 248, 254, 260, 262, 263, 264, 278, 280, 281
  absorbed, 52, 55, 56, 100
  accumulated, 260, 261
  background, 248, 251, 261, 263, 264
  cataract-inducing, 251
  distribution of, 102, 172
  fractionation of, 167, 168, 252
  genetic, 249, 254
  integral, 165, 166
  large, 212
  lethal, 166, 213, 250, 251
  limits, 213, 251
  maximum permissible, *see* MPD
  modification by bone, fat and lung, 104, 105
  patient, 150, 166, 167, 278
  primary, 100
  prescription of, 171
  rate of, 216, 217, 244, 245, 254, 260, 262, 263
  scatter, 100
  single, 167
  small, 250, 254
  specification of, 177
  sterility-causing, 214
  threshold, 252, 253, 254
  tumour-treatment, 168, 171
  tumour-lethal, 166
  see also Air dose, Population dose, Skin dose
Dose equivalent, 53
Dosemeter, 61, 64, 68, 71 ff, 88, 232, 263
Dosemeter, *contd.*
  film, 120, 128, 262
  integrating, 278
  personal, 120, 262, 263
  x-ray, 64, 177, 233, 264, 278
  see also Monitor
Dosimetry, moving field, 103
Drift tube, 243
D–T reaction, 242
Duncomb, 206
Dushman, 231

EARTHING, 237, 239
Education—
  courses in health and safety, 259
  ionizing radiations in schools and colleges, 257
Education authority, local, 257
Efficiency of energy conversion, 21, 132, 219
Efficiency, quantum counting, 79
Efficiency, thermal, 234
Eggs, hens', 250
Ehrenburg, 221
Electric field, 205, 219, 243
Electro-luminescent layer, 136
Electromagnet, 245
Electrometer, 263, 281
Electromotive force, 229
Electron, 24, 46, 55, 219, 244, 245
  acceleration, 29, 219, 240, 243, 244, 245, 246
  diffraction, 20, 201
  emission, 219, 221
  range, 56
  streaming, 239
  transitions, 28, 202
  valency, 24
Electron beam, 205
Electron contamination of x-ray beams, 92
Electron density, 193
Electron equilibrium in walled chambers, 61
Electron miscroscope, 13, 219, 222
Electron probe microanalyser, 205, 207
Electron traps, 109
Electronic scattering coefficient, 41
Electroscope, 263
Electrostatic charge, 242
Electrostatic field, 227
Electrostatic force, 199, 231
Elements, 206, 207, 208
Embryo, 249, 263
Emulsion, photographic, 108 ff, 119, 121, 280
Energy—
  absorbed, total, 281
  absorption, 48, 50, 65
  analysis, 205
  conversion efficiency, 224

Energy-dependence of film sensitivity, 119, 120
Energy fluence, 4, 31
Energy flux density, 4, 31
Energy quantum, 5
Energy spectrum, 102
Entry ports, 172, 176
Environmental radiation, 253, 260, 263
Equipment, x-ray, 219
Ewald, 191
Exhibitions, code of practice, 257
Exhibition radiographs, 137
Exit beam, 176
Exciting radiation, 207
Exposure, 55, 117, 208, 225, 227, 233, 248, 253, 259, 260, 261, 279, 281
  abnormal, 261
  accidental, 251, 261, 278, 279, 282
  casual, 250
  categories for, 259
  integrated, 56
  measurement of, 87, 281
    see also Dosemeter, Monitor
  medical, 263, 275, 278
  occupational, 139, 140, 248, 249, 258, 260, 261, 262, 282
  of public, 251, 260, 263, 264, 278
  overexposure, 258, 260
  rate of, 84, 87 ff, 100, 169, 223, 228. 232, 242, 249, 251, 263, 264, 275, 278, 279, 280, 281
  timing of, 131
  transit, 104
  veterinary, 153
  voluntary, 261
Extrapolation chamber, 65
Eye, 281
  spectral sensitivity of, 114, 140

FACTORIES ACTS, 256, 258, 259, 260, 261
Failla, 252
Failure of apparatus, 221, 225, 228, 229, 235
Faraday, 10
Fat, 104, 151, 168
Fat patients, 167
Fermat's principle, 5
Fertility, *see* Reproductive capacity
F-factor, 120
Field—
  coulomb, 46
  periphery of, 171
  single, 171
  size of, 92, 95, 166
    limiters for, 278, 279
  uniformity of, 92
  visual indication of, 95
Field emission, 224, 227
Filament, 219, 220, 221, 222, 225, 231, 239
  failure of, 220, 221, 239
Filament current, 228, 233
Filament temperature, 221, 222, 224, 231, 234
Filament voltage, 228, 233
Film, 108, 109, 118, 119, 194, 208
  blackening, 116, 119, 262
  characteristic curve, 117, 127, 131
  x-ray, 108 ff
  density of, 116, 208
  double-coated, 121
  drying, 113
  grain of, 130
  for monitoring, 121, 282
  negative, 116
  processing, 111–13, 115
  sensitivity of, 108, 109, 117, 118, 119, 127
  speed of, 117, 119
  storage of, 115
Film badge, 262
Film badge holder, 122
Film base 108 ff
Film holder, lead-backed, 134
Film pack, 119
Filter, 134, 177, 280
  compensating, 177
  for film badge, 121
  light, 114
  wedge, 117, 174
Filtration, 92, 134, 195
Finch, 227
Fine-focus tube, 220, 221, 222, 226
Finger-tip monitoring, 263
Five-year survival, 166
Fixing films, 112 ff
Flaw sensitivity, 131
Flaws in castings and welds, 156, 157, 159
Floors, 275, 276, 278
Fluorescence, 45, 206
Fluorescent screen, 133, 140, 149, 159, 248, 279
Fluorescent x-radiation, 206, 207, 225
Fluorods, 72
Fluoroscope, shoe-fitting, 249, 257
Fluoroscopy, 140, 279, 280, 281
Focal distance, 85 ff, 245, 280
Focal-skin distance, 280
Focal spot, 84 ff, 127, 220, 221, 224, 225, 227
Focusing, 220 ff, 227, 242
Foetus, 249, 252, 253, 260, 263
Fogging of film, 109, 115, 117, 262
Food, preservation of, 214, 215
Foods, irradiated, 214
Forces, binding, in crystals, 199, 200
Forces, electrostatic, 199, 231
Fourier methods, 130, 193
Fresnel zone plate, 12, 13
Fricke dosemeter, 73
Friedrich, 191

G-VALUE, 73, 212
Gamma rays, 2, 275
Gamma value, 118
Gases, removal from tube, 220
Gas amplification factor, 75
Gas counter, 205
Gas gangrene, 151
Gas ionization, 227
Gassy tube, 220, 231, 234
Gastric cancer, 151
Gastro-intestinal tract, 150
Geiger counter, 13, 74, 76, 77, 78, 203, 263, 279
Geiger tubes, high current, 78
Gene, 1, 212, 213, 254
Gene mutation, 212, 213, 254
Generator, Van de Graaff *see* Van de Graaff generator
Genetic dose, 249, 263, 264
Genetic effects, 249, 251, 253, 254, 259
  see also Biological effects,
Genetic exposure, 248
Genetic hazards, 164, 248
Germ cells, 254
Germanium, 200
Germany, regulations, 214
Getters, 220
Glance-field technique, 169
Glancing angle, 8, 9, 10, 191, 195, 202
Glass, 199, 240, 278, 281
  borosilicate, 220
  lead, 140, 277, 278, 279, 281
Glass components for tubes, 169, 220
Gonads, 152, 248, 252, 254, 259, 260, 281
Goniometer, Weissenberg, 199
Graphite, 193, 208
Grating, 11, 191
Grätz full-wave circuit, 235
Greene, 275, 277
Greening, 91
Greinacher circuit, 238
Grenz rays, 65, 169
Grid, 134, 135, 136
Grid therapy, 174, 175
Groth, 191

HAEMANGIOMA, 169
Haematology, 258
Hagenow, 15
Half-value depth, 169
Half-value layer, HVL, 55, 89 ff, 99, 169, 172, 233, 263, 281
Half-value thickness, HVT, 89 ff
Hallwachs, 42
Homburg, 248
Hanawalt, 209, 210
Harbours, code of practice, 257
Harness, 279
Haüy, 185, 188
Haversian canals, 105
Hay, 139
Haybittle, 253
Hazards from radiation, 120, 122, 127, 239, 242, 248, 249, 255, 257, 259, 281
  genetic, 164, 248
  somatic, 248, 249
  to patients, 248
Health, Ministry of, 277
Health and safety courses, 259
Health physicist, 258
Health register, 261
Health surveillance, 260
Heat dissipation, 219, 224, 225, 226, 229
Heat-labile materials, sterilization, 212
Heating effect of x-rays, 212, 228, 232
Heel effect, 92
Henson, 264
Hereditary effects, 254
Hereditable characteristic, 212
Hero of Alexandria, 5
Hertz, 2, 42
Hewlett, 16
High frequency, 243, 245
High-voltage equipment, 220
High-voltage transformers, 229, 230, 231
High-voltage x-ray tube, 240
Hiroshima, 252
Homogeneity coefficient, 91
Hood, 221, 222, 223, 224
Hospitals, 258
Hoyle, 13
Hunt, 22
HVL *see* Half-value layer

ICRP, 250, 253, 255, 259, 260, 261, 263, 264, 266, 267, 268, 269, 270, 271, 272, 273, 274, 275, 277, 278, 280
Illumination for radiograph viewing, 126
Image—
  latent, 109, 136, 137
  photographic, 109 ff, 137
  pseudo-, 128
  radiographic, 125 ff, 159, 280, 281, 282
  registration of, 136
  visible, 136, 206
Image amplifier, 137, 139, 149
Image convertor, 139
Image intensifier, 280
Impedance, 228, 231, 232, 235
Imperfections, 201
Impurities in specimens, 201, 207, 209
Inclusions in metals, 157, 159, 206
Incoherent photons, 40
Indexing crystal structure, 197, 198, 209, 210
Inductance, 228

Industrial radiography, 125, 136, 138, 142, 143, 146, 153, 221
safety precautions, 257
safety regulations, 255
Industrial radiology, 125, 153 ff
Infertility, 212, 253
Inherent filtration, 92
Inhomogeneities, body, 104, 171, 173
Injury, *see* Biological effects, Damage
Insects, 212, 213, 214
Installations, 264, 276, 277, 278, 282
monitoring and survey, 276, 282
see also Barriers, Buildings, Warning devices
Insulating materials (electrical), 63, 64, 225, 229, 238, 242
Insulation, 228, 229, 233, 236, 239
Insulation breakdown, 229, 235, 239
Intensification, 132
Intensifier, image, 280
Intensifying screen, 108, 132 ff, 133, 137, 151, 159, 280
Intensity, 4, 31, 192, 202, 205, 206, 207, 208, 209, 210
Intensity maxima, 199
Interference, 8, 9, 191, 201
International Commission on Radiological Protection, *see* ICRP
International Organization for Standardization, 263
Intracavity sources, 176
Intravenous pyelography, 150
Inverse square law, 4, 5
Ionization, gas, 227, 234
Ionization, multiple, 29
Ionization by collision, 59
Ionization chamber, 56 ff, 263, 280, 281
walled, 61 ff
Ionization current, 59 ff
Ionizing Radiations (Sealed Sources) Regulations and Orders, 256, 261, 262
Ions, 59, 215, 231, 243
Iron, 154, 190, 208, 223, 229, 275
Isochromat, 22
Isodose curve, 101 ff, 102, 103, 171, 174

Jennings, 168
Jolles, 169
Jones, 277

K_α-RADIATION, 208
kX unit, 11
Kemp comparator, 90
Kenotron, 231
Kerley, 248
kerma, 53
Kiessig, 9
Kikuchi, 201
Kilovoltage, choice of, 127, 134, 223, 232
Klystron, 243
Kossel, 26
Kossel lines, 201

L-RADIATION, 207
Lactation changes, 151
Lambert's law, 4, 31
Langmuir, 219, 231
Larynx, 174
Latent period in cancer, 166, 253
Lattice, Bravais (or space), *see* Bravais lattice
Lattice, reciprocal, 197, 198
Lattice, space (or Bravais), *see* Bravais lattice
Lattice parameters, 201
Lattice plane, 194
Laue, 1, 120, 190, 191, 195, 196
Laue method, 195
Lawrence, 242
Law, *see* Legal aspects, Legislation
Layer lines, 197, 198
Lead, 153, 160, 263, 265, 269, 270, 271, 272, 275, 276, 281
Lead cut-outs, 134, 169
Lead tetraethyl, determination in petrol, 208
Leakage current, 60
Leakage radiation, 275, 276, 277, 282
Legal aspects, 255, 256, 257, 262
Legislation, 255, 256
LET, *see* Linear energy transfer
Lethargy, radiation, 212
Leukaemia, 248, 251, 252, 253, 254, 261, 264
Light, speed of, 244
Light device for field indication, 95
Life span, 251, 252, 254, 255
Limitations in usage of x-rays, 249
Lindop, 252
Line spectra, 203
Lines, spectral, 207, 208
Linear accelerator, 168, 212, 240, 242, 243 244, 277, 278
Linear energy transfer (LET), 51, 52, 53, 165, 168
Lining up, *see* Alignment
Linnik, 9
Lipoma, 151
Liquids, 190, 193, 199
Lloyd's mirror effect, 9
Localization, 143 ff
Loeck, 201
Lonsdale, 201
Low-angle scattering, 199
Lung, 104, 166
Lymphangiography, 150

MacMahon, 253
Magnetic field, 229, 231, 244, 245, 246
Magnetron, 243
Main supply, stabilization, 233
Mains frequency, 226, 228, 232
malignancies, 165, 248, 249, 252, 253
Martyrs, x-ray, 248
Massey, 275
Mastectomy, 175
Materials science, 193, 200, 216
Maximum permissible dose, *see* MPD
  see also Maximum permissible level
Maximum permissible level, 249, 251, 264, 282
Maxwell, 2
McClement, 252
Mean free path of electron, 51
Measurement, 143
Medical authority, 261
Medical examination, 258, 260
Medical exposure, 261, 263, 275, 278
  see also Radiography, medical
Medical officer, 258, 259
Medical supervision, 260
Medical use, code of practice, 257, 258, 260
Megarad, 212
Megavoltage radiation, 104, 239, 240,242, 252
Mendeleev, 24
Menopause, artificial, 253
Menstrual cycle, 151, 251, 263
Metals, 153, 154, 159, 160, 190, 200, 223, 226
  crystal structure of, 190, 195, 209
Metallic forces 200,
Metallurgy, 154, 159, 199, 205, 206, 208, 209
Metastasis, 165
Meters, 232, 233, 239, 278
Meter readings, misleading, 232
Michelson, 1
Micro-analyser, electron probe, 205, 207
Microbiology, 213, 214, 215
Microdensitometer, 117, 210
Microscope, electron, 13, 219, 222
Microscope, field-ion, 227
Microscope, x-ray, 12
Microscopy, x-ray, 149
Microwave, 243
Miller index, 197
Milliameter, 228, 232, 234, 239
Millikan, 10
Mirrors, 140, 279
  x-ray, 11, 13
Mitchell, 253
Mitotic cycle, 165
Mixtures of materials, analysis, 195, 208, 209
MLD, 213, 289
Mobile equipment, 238, 239, 280
Molecules, 199, 212, 215
Monitor, 65, 262, 263, 280
Monitoring, 260, 276, 282
Monitoring, personal, 120, 261, 262, 263, 282
Monitoring film, 121, 262
  fogging, 262
Monument to x-ray martyrs, 248
Moseley, 23, 24, 26
Movement blur, 128, 130, 153
Moving field therapy, 95, 103, 104, 173
MPD, 250, 251, 252, 253, 254, 255, 259, 260, 261, 262, 263, 275, 277
  see also Maximum permissible level
Multi-field technique, 172, 173, 175
Muscle, 151
Mutagenic agents, 212, 213, 254
Mutation, 212, 213, 254

NAGASAKI, 251
Nanometre, 11
National Radiation Protection Board 263 282
Natural selection, 213
Natural sources of radiation, 13, 248
Neary, 252
Necrosis, 165
Negatron, 46, 55
Negligence, 256, 257
Neutron, 242, 251
Neutron therapy, 165, 168
Nicholas, 2
Nitrogen mustards, 175
Nobel prize, 193
Non-malignant conditions, 249
Nuclear research, 242, 246

OBSERVATION of patients, 279
  see also Window, observation
Obstetric radiography, 133
Occupancy factor, 264, 265
Oil as electrical insulation, 225, 229, 230, 231
Operating theatres, 110, 115
Orbit in electron accelerators, 244, 245, 246
Orientation of crystals, 194, 195, 196, 197, 209
Orthovoltage radiation, 101, 104, 167, 168, 173, 174, 176, 280
Oscillation method, 196 ff
Osteoporosis, 151
Outlining effect, 141
Output, 221, 224, 225, 233, 234, 235, 238, 244, 265, 266, 267, 268, 278
Ovaries, 253
Owers, 59
Oxygen tension and radio sensitivity, 165, 213, 215

PACKING anomalies in specimens, 209
Pair production, 46, 47, 153, 265
Palliative treatment, 164, 169, 176
Panoramic tomography, 152
Pantograph, 170
Paralysis time of Geiger counter, 205
Patients, 150, 167, 171, 176, 248, 253, 279, 280, 281
  setting up, 171, 176, 279
Pattern theory, 188, 189, 190
Patterson, 165
Pelvis, 263
Pendulum therapy, 173
Penetrameter, 130, 131
Penetrameter sensitivity, 131
Penrose, 254
Penumbra, 85, 93, 94, 103, 127 ff
Periapical film, 152
Perineal air insufflation, 151
Periodic Table, 24
Periodic time, 3
Peterson, 253
Petrol, analysis of, 208
Phantom, 104
Phase, 8, 9, 16, 192, 193
Photoconductors, 139, 140
Photo-electric effect, 42, 43, 44, 45, 46, 119, 127, 133, 151, 153, 265
Photo-fluorography, 149, 150
Photographic emulsion, 108 ff, 280
Photographic image, 109 ff, 137
Photographic processing units, 115
Photography, x-ray, 108 ff
Photon, 40, 265, 280
Physicist, 170, 258
Piezo-electric effect, 190
Pin and arc, 176
Pinhole, 84, 86, 96
Planck's constant, 5, 22, 43
Plane-polarized radiation, 13
Plants and plant breeding, 213, 215
Plaster cast, use in setting-up, 176
Platinum, 219, 223
Pneumoencephalography, 151
Point projection micro-radiography, 149
Point-wedge, 174
Polarization of x-rays, 13, 14, 15, 16
Polaroid system, 115
Polymerization, 216, 217
Polymers, cross-linking, 216
Population dose, 249, 254, 260, 263, 264, 278
Porcelain, 160
Ports, code of practice, 257
Portable sets, 239
Positive ion bombardment, 219
Positron, 46, 55
Potassium-40, 248
Potter-Bucky diaphragm (grid), 136, 152
Powder cloud xerography, 141
Powder method (crystallography), 193, 194, 195, 209
Powder pattern, 195
Precautions, *see* Safety
Preferred orientation of metal crystals, 195, 209
Pregnant women, 248, 249, 253, 260, 263
Pressure chamber, 165
Print, photographing, 119
Prone position, 171
Protection against radiation, 228, 248 ff, 255, 258, 261, 264, 265, 276, 277, 278, 282
  recommendations for, *see* Recommendations
  see also Barriers, Safety, Shielding, Shields
Protective clothing etc., 116, 140, 279, 281, 282
Protective agents, 165, 167
Protein structure, 193, 200
Proteins, 200
Proximal skin, 151
Pseudo-image, 128
Public, 251, 260, 263, 264
Public health, 214
Publications, 177, 255, 257, 259, 261, 265
Pulse height analyser, 205
Pulsed output, 244
Pulsing, multiple, 76
Pumping x-ray tubes, 220
Pyelography, intravenous, 150

Q-FACTOR (Callier coefficient), 116
Q-value, 241
QF (quality factor), 53
Qualified expert, 258
Quantum counting efficiency, 79
Quantum numbers, 27, 28
Quantum theory, 1, 5, 21
Quenching agent, 76, 77

RAD, 52, 66 ff, 98, 212, 213, 215, 244
Rad, measurement of, 66
Radiation, 2, 234, 264
  biological effects of, *see* Biological effects
  characteristic, 19, 23, 45, 205, 207, 208, 223
  continuous, 19, 22
  exciting, 206
  intra-cavity sources of, 176
  leakage of, *see* Leakage radiation
  man-made sources of, 13, 248
  monochromatic, 194, 196, 201
  natural sources of, 13, 248, 249
  quality of, 55, 262

Radiation, *contd.*
scattered, 134, 223
spurious, 224
stray, 279
white, 120, 195, 207, 208, 223
see also under Megavoltage, Orthovoltage, Supervoltage, X-radiation
Radiation damage, *see* Damage
Radiation exposure, *see* Exposure
Radiation hazards, *see* Hazards
Radiation injury, *see* Damage
Radiation lethargy, 212
Radiation protection, *see* Protection
Radiation surveys, 276, 282
see also X-radiation, X-rays
Radicals, 73, 200, 215
Radioactive nuclide, 212
Radioactive Substances Act, 1948, 256
Radiobiological equivalent, 53
Radio-chemical reaction, 212
Radio-curability, 166
Radio-frequency, 246
Radiograph, 125 ff, 137, 232
for exhibition, 137
quality, 125, 127, 128, 130, 153
viewing conditions, 126, 159
Radiographic image, 125 ff, 159, 280, 281 282
Radiographic techniques, 142 ff
Radiography—
cine-, 149, 233, 278
of castings, 156, 157
dental, 143, 152
dynamic, 149, 150
flash, 140, 149, 153, 219, 225
industrial, *see* Industrial radiography
mass dental, 152
mass miniature, 149, 150, 248, 249
medical, 136, 142, 148, 149, 221, 249, 252, 278, 280
of metals, 153, 156, 157, 159, 160
micro-, 125, 149, 157
multiple (multi-section), 133
obstetric, 133
of soft tissue, 151
solid state, 139
subtraction, 148
veterinary, 129, 133
of welds, 159, 160
Radiological health and safety courses, 259
Radiological Protection Service, 123, 263, 282
Radiologist, 149, 170, 248, 252
Radiology, 223
diagnostic, 248, 252, 281
industrial, 125, 153 ff
Radio-protective drugs, 167
Radio-resistance, 165, 166
Radiosensitivity, 165, 166, 213, 215, 252
Radiosensitizer, 165, 175
Radiotherapist, 170
Radiotherapy, 62, 102, 164 ff, 169, 175, 176, 221, 232, 233, 244, 245, 249, 252, 253, 277, 278
Radio-thermoluminescence, 71
Radio-transparency, spurious, 128
Radium, 249
Ratemeter, 206
Rating, tube, 225
Rayleigh scattering, 39, 40, 41, 192
Recessive mutations, 254
Reciprocal lattice, 197, 198
Reciprocal point, 198
Reciprocal space, 198
Reciprocity rule, 118
Recommendations, radiation protection, 255, 257, 260, 282
Records, 201, 209, 262
Rectifier, 219, 227, 231, 232, 234, 235, 238
Rectifying junction counters, 72
Rectilinear propagation, 5
Recurrent growths, 165, 174
Reflected radiation, 196, 207
Reflection, 6, 7, 9, 188, 194, 210
Reflection, Bragg, *see* Bragg reflections
Reflections, diffuse, 210
Refraction, 6
Refractive index, 6, 7, 11, 12, 278
Registration, image, 136
Regulations, safety, 214, 256, 257, 258
Reisma, 65
Rem, 53, 252, 260, 264
Reproductive capacity, 212, 251, 252, 253, 254, 263
Reproductive organs, 254
see also Gonads
Research, code of practice, 257, 258, 259, 260
Respiratory stimulation, 165
Resolution, 119, 130, 152, 157, 226
Resolving power, 12
Resolving time, 77
Resonant circuit, 241
Resonant frequency, 241
Responsibility, 257, 258
Restraint, of patients, 153, 279
Restraint of workers, 279
Reticulation of emulsion, 112
Retina, 252
Rings, Debye-Scherrer, 195
Rings, diffuse, 208
Risks, *see* Hazards
Rodent ulcer, 169
Röntgen, 1, 6, 7, 53, 55, 56, 58, 62, 68, 98, 231
Röntgen, Frau, 125
Röntgen (unit), measurement of, 56
Rotation method, 196
Rotblat, 252
Rowland circle, 203

Rubber, 190, 193, 215, 216, 219, 228
Rubber, depolymerized, 176
Rutherford model of atom, 26
Rydberg constant, 25

SACRO-ILIAC area, 165
Sadler, 24
Safe system of working, 256
Safe-lights, 114
Safety, 255 ff
  codes of practice, 257
    see also Codes of Practice
  instruction courses in, 276
  see also Protection, Warning devices
Safety circuits, 226
Safety devices, 277, 278, 279, 280, 281, 282
Safety officers, 258, 259, 276
Sagnac, 1
Sarcoma, 253
Satellites, 29
Saturation current, 59
Saturation voltage, 59
Scan, beam, 206
Scatter, 99, 101, 134, 265, 276, 278
Scatter factor, 99
Scattered radiation, 134, 223
  barriers for, 276, 277, 278
Scatterer, 13, 14, 276, 277, 280
Scattering—
  coherent, 40, 202
  low-angle, 198, 199
  optical, 191
  Rayleigh, 39, 40, 41, 192, 202
  x-ray, 192, 199, 202, 276, 277
Scattering coefficient, electronic, 41
Schools, 257, 261, 263
Scintillation counters, 69, 70, 79, 203
Screen, fluorescent, *see* Fluorescent screen
Screen, intensifying, 108, 118, 132, 133, 151
Screen, long-persistance, 206
Screening constant, 25
Sealed sources, 256
Seeds, 213
Selection rules, electron transition, 28
Selenium, 140, 141
Self-absorption, 207
Self-quenching counter, 76
Self-rectifying, 225, 227, 240
Semiconductor, 201
Semiconductor detectors, 72
Sensitive age, 251
Sensitive volume, 56, 58
Sensitivity speck, 109, 118
Setting-up, 171, 176, 279
Shadow formation, 127
Shielding, 228, 249, 264, 265, 276, 278, 279
  see also Protection
Shields, body, 134, 152, 169, 281
Shock-proof tubes, 220, 225, 228
Shoe-fitting fluoroscopes, 249, 257
Short-wave cut-off, 21, 22
Siegbahn, 11
Sieve radiotherapy, 169, 174, 175
Sim, 255
Single crystal indexing, 209
Single field, 171
Sintering, 221
Skin, 169, 253, 280
Skin burns, 248
Skin conditions, 249
Skin dose, 141, 151, 166, 280, 281
Sloan, 242
Soft tissues, 151, 168
Solarization of film, 118
Solid state dosemeter, 68, 263
Solid state radiography, 139
Somatic effects, 248, 249, 251, 259, 264
  see also Biological effects
Sommerfeld, 21, 190, 191
Space charge voltage, 231
Space lattice, *see* Bravais lattice
Spear, 221, 222
Specimen, 279
  alignment, 209
  metallurgical, 157, 205, 208
  orientation, 157, 209
  preparation, 205, 210
Specimen holder, 209
Spectra, 44, 203, 205, 207, 208
Spectrometer, x-ray, 22, 202 ff, 205, 206, 207, 238
Spectroscopic analysis, x-ray, 202 ff
Spectrum characteristic, 19, 23, 44, 203, 205
Spine, 165, 248, 252
Sprouting, of vegetables, 214, 215
Stabilization, mains and output, 233
Stabilizing circuit, 278
Standards, 209, 210, 263, 281
Stenström, 8
Stephenson, 277
Stereo-fluoroscopy, 221
Stereo-radiography, 142, 143, 152
Stereoscope, 143
Sterility dose, 214
Sterilization of equipment and materials, 212, 215
Stevenson, 254
Stokes, 1, 10
Stopping power, 51, 67
Strain phenomena in crystals, 195
Stray electric fields, 205
Subject contrast, 126, 127, 129
Superficial lesions, treatment, 168, 169
Supervoltage radiation, 168, 169, 172, 173
Supine position, 171
Surge, 229, 235, 239
Surveys of installations, 276, 282
Survival, five year, 166
Synchrotron, 13, 240, 245, 246

TARGET, 84, 205, 207, 219, 223, 240, 245
contamination of, 207
evaporation of, 239
foil, 222
inherent beam filtration by, 92
pitting of, 224, 239
reflection, 92
transmission, 92, 239
tungsten, 207, 240
Target angle, 87, 221
Target material, 223, 240
Teaching, code of practice, 257, 258, 259, 260
Technique selection, 171
Techniques, radiographic, 142
Techniques, therapy, *see* Therapy
Teeth, 151, 152
Television, 139, 249, 263, 275, 278, 279
Tenth-value layer, 277
Testing, non-destructive, 202
Tetraethyl, lead, determination, 208
Textile fibres, 200
Therapy check list, 177
Therapy techniques—
deep, 168, 169
glance-field, 169
grid, 174, 175
moving-field, 103, 173
multi-field, 172, 173, 175
pendulum, 173
sieve, 174, 175
single-field, 171, 172
Thin patients, 167
Thompson, J. J., 13, 15, 42
Three-dimensional effect, 142
Thyratron, 233
Timers, 232, 233, 278
Tissue, soft, 168
Tissue–air ratio, 103
Tissue-equivalent chambers, 67
Tissue-simulating wax, 176
TLD, 71
Tolerance, 166
Tomography, 146, 147, 148, 152
Townsend avalanche, 75, 76
Training in radiation protection, 259
Transformer, 219, 221, 229, 230, 231, 232, 233, 234, 235, 240, 241
high-voltage, 229, 230, 231
resonance, 240
Transformer oil, 229, 230
Transit exposure, 104
Transmission through barriers, 264, 265, 276
Traps, beam, 275
Traps, electron, 71, 109
Treatment couch, 176
Treatment method, 168
Treatment planning, 141, 166, 168, 170, 171, 176
see also Therapy techniques
Treatment room, 277, 278
Treatment termination, 177
Triboluminescence, 71
Triplet production, 47
Tritium, 249
Trunk, 169, 252
Trunk bridge, 170
Tube—
cathode ray, 206
continuously evacuated, 220
cooling, 225
cracks in, 224, 225
demountable, 220, 221
drift, 243
field-emission, 227
fine-focus, 220, 221, 222, 226
gas-filled, 219, 227
glass, 220
operating conditions for, 225, 226
rod-anode, 222, 223
rotating-anode, 221, 225, 226
sealed-off, 220
shock-proof, 225
toroidal, 244
x-ray, 219 ff, 235, 240
see also X-ray tube
Tube casing, 228, 239
Tube characteristics, 233
Tube current, 221, 223, 232, 234, 237, 238 239, 241
Tube design, 222
Tube envelope, 220
Tube filtration, 195
Tube foci, 220, 221, 242
Tube head, 239
Tube housing, 277
Tube impedance, 235
Tube rating, 220, 225, 239
Tube voltage measurement, 238
Tube window, 169, 219
Tube-shift method, 143, 145, 146
Tumour dose, 167, 168
Tumour cell growth, 165, 166
Tumour-killing dose, 166, 167
Tumour recurrence, 165, 174
Tumours, 151, 166, 167
Tungsten, 219, 220, 221, 223, 224, 225, 276
Tungsten, oxide, 221
Tungsten deposit, 224
Tungsten evaporation, 224
Tungsten spectral lines, 207
TVL, 277

ULTRA-VIOLET radiation, 1, 2, 213, 215
Umbra, 127, 129
Undercutting, 128, 141
Underrunning, 239
Unit cells, 188, 189, 190, 192, 197

United Kingdom, 244, 248, 255, 261
United Kingdom Atomic Energy Authority, 255, 258
Unsharpness, 127, 130, 134, 140
Uranium, 153, 276
Use factor, 264, 265
U.S.S.R., regulations, 214

VACUUM, 219, 220, 229
Valve, rectifying, 231, 234
Valve, x-ray emission by, 231
Van de Graaff generator, 212, 219, 240, 241, 242
Van der Waals forces, 199
Vasodilation, 165
Veale Committee, 259
Vegetable sprouting 214, 215
Veterinary radiography, 129, 133, 152, 278
Viewing conditions, 126, 159
Villard circuit, 238
Vitamins, 214, 215
Voges, 201
Voltage, 225, 229, 232, 235, 238, 241
  filament, 228, 233
  see also High voltage, Kilovoltage, Megavoltage, Orthovoltage, Supervoltage
Voltage-doubling circuits, 237, 238, 242
Voltmeter, 232, 238

W-VALUE, 53, 56
Walls, ionization chamber, 61 ff
Walls of buildings, 275, 276, 278
Walton, 240, 242
Warning devices, 261, 263, 277, 279, 281, 282
Watches, luminous, 249, 263
Wave velocity in linear accelerator, 243
Wave-form, 219, 225, 233, 235, 236, 241
Wavelength, 3, 191, 192, 195, 202
  characteristic, 205, 206, 208
Wave-mechanical model, 26
Waves, 2, 3, 5, 191, 192, 193
  electromagnetic, 2, 3, 6, 13, 19, 20, 243
Wax, tissue-simulating, 176
Wehnelt, 231
Weissenberg camera, 198
Welding, 157, 158
Welds, 125, 157, 158, 159
  defects in, 157, 159
Welds, radiography of, 128, 133, 142, 159
  precautions in, 257
Wentzel, 29
Westman, 277
White cell count, 249
Whyte, 37, 39, 48, 98
Window, observation, 140, 277, 278
Window, x-ray tube, 169, 219
Window, counter, 207
Windows in buildings, 278
Women, 251, 260, 263
Work function, 43
Workers, 257, 259, 260, 261, 262, 264, 281
Workers, classified, 260

XEROGRAPHIC plate, 132, 141
Xerography, 140, 141
X-radiation, fluorescence, 206, 207
X-radiation, reflected, 207
X-radiation, supplementary, 176
X-ray absorption analysis, *see* under Analysis
X-ray astronomy, 13
X-ray chemical analysis, *see* under Analysis
X-ray diffraction, 10, 190 ff, 210, 227, 251
  see also Crystals, Crystallography
X-ray diffraction analysis, *see* under Analysis
X-ray equipment, 219 ff
X-ray film, 108 ff
X-ray fluorescence analysis, *see* under Analysis
X-ray machines, 256
  output, *see* Output
  see also Tubes, X-ray equipment, X-ray tubes
X-ray photography, *see* Photography, x-ray
X-ray production, 219
X-ray spectrometer, *see* Spectrometer
X-ray spectroscopic analysis, *see* under Analysis
X-ray tubes, 219 ff
  high-voltage, 240
  output, *see* Output
X-rays—
  calibration, 84
  discovery, 1, 248
  effective energy, 89
  emission by valves, 231
  nature of, 1
  polarization of, 13
  short-wave cut-off, 21, 22
X-unit, 11

ZONE axis, 196
Zone plate, Fresnel, 12, 13